Audio Bandwidth Extension

Audio Bandwidth Extension

Application of Psychoacoustics, Signal Processing and Loudspeaker Design

Erik Larsen
MIT, Speech and Hearing Bioscience and Technology, USA

Ronald M. Aarts
Philips Research Laboratories, The Netherlands

John Wiley & Sons, Ltd

Telephone (+44) 1243 779777

Email (for orders and customer service enquiries): cs-books@wiley.co.uk
Visit our Home Page on www.wileyeurope.com or www.wiley.com

Other Wiley Editorial Offices

John Wiley & Sons Inc., 111 River Street, Hoboken, NJ 07030, USA

Jossey-Bass, 989 Market Street, San Francisco, CA 94103-1741, USA

Wiley-VCH Verlag GmbH, Boschstr. 12, D-69469 Weinheim, Germany

John Wiley & Sons Australia Ltd, 33 Park Road, Milton, Queensland 4064, Australia

John Wiley & Sons (Asia) Pte Ltd, 2 Clementi Loop #02-01, Jin Xing Distripark, Singapore 129809

John Wiley & Sons Canada Ltd, 22 Worcester Road, Etobicoke, Ontario, Canada M9W 1L1

Wiley also publishes its books in a variety of electronic formats. Some content that appears in print may not be available in electronic books.

British Library Cataloguing in Publication Data

A catalogue record for this book is available from the British Library

ISBN 0-470-85864-8

Typeset in 10/12pt Times by Laserwords Private Limited, Chennai, India

Contents

Preface

Bandwidth extension (BWE) refers to methods that increase the frequency spectrum, or bandwidth, of electronic signals. Such frequency extension is desirable if at some point the frequency content of the signal has been reduced, as can happen, for example, during recording, transmission (including storage), or reproduction, mostly because of economical constraints. In this text, we limit the discussion to audio applications, which include music and speech. Most of the BWE methods heavily use signal processing – in fact, it is almost a premise that BWE is a signal-processing tool for achieving what is otherwise physically not possible. As Chapter 4 shows, a combination of mechanical engineering and signal processing can lead to interesting results as well.

BWE is a field that has seen increasing attention in recent years. Although some work was done in the early years of the twentieth century, a much more systematic and large-scale approach did not occur until recently. BWE for speech is the most mature area in this field, as the primary application (telephony) has existed for a long time. It is the objective of this book to gather most of the recent work into a single volume and present a coherent framework to the reader. It is the first time an entire book has been devoted to BWE theory, applications, and algorithms. It is intended as a broad introduction to BWE topics, but also discusses in detail various applications, thereby including material from scientific and patent literature, and also presents some previously unpublished work. The latter can be found in Sec. 2.4 (BWE using frequency tracking), most of Chapter 3 (low-frequency physical BWE systems), and Sec. 5.6 (high-frequency BWE using instantaneous compression).

Bandwidth reduction implies a decrease in perceptual quality, and therefore BWE algorithms are employed as tools to enhance the perceived quality of reproduced sound. In most cases, BWE methods are therefore post-processing algorithms, occurring just before sound reproduction, and the processing aims to compensate for the limited bandwidth that is available in a prior part of the chain. Sometimes, however, bandwidth reduction is actually desirable, for example, to enhance the coding efficiency of perceptual audio codecs. With a little additional complexity and data storage, required bit rates can be drastically reduced while maintaining subjective audio quality.

The main application areas discussed are bass enhancement for sound reproduction (Chapters 2, 3, and 4), high-frequency enhancement for general audio (Chapter 5) and speech (Chapter 6) applications. We include a short discussion on how BWE can be used as a very effective noise-abatement technology (Chapter 7), and present an overview of BWE patents (Chapter 8). Chapter 6, on BWE for speech, is contributed entirely by Peter Jax, who did most of the presented work as part of his doctoral dissertation [128].

In Appendix A, we present a brief overview of univariate and multidimensional scaling, which can be very useful techniques for analyzing the outcome of subjective experiments; an application example can be found in Sec. 2.5.

Although the BWE algorithms for the various application areas have many similarities on a conceptual level, there are interesting differences. For example, BWE to enhance bass reproduction on small loudspeakers, making use of the psychoacoustic effect of the 'missing fundamental', is largely focused on devising and analysing non-linear functions that are used to generate harmonics of very low-frequency signals. The goal of this technology is to allow a small loudspeaker to be used, giving a good percept of the entire audible bass range (down to 20 Hz), while the loudspeaker may only be able to physically radiate sound down to 100 Hz. A completely different approach is taken in Chapter 4, where inherently inefficient small loudspeakers are mechanically transformed into units that are highly efficient at only one frequency. Bass sounds from an entire bass frequency band are then mapped to the particular frequency at which the loudspeaker can radiate a considerable amount of energy. Thus, the problems of small loudspeakers at low frequencies can be ameliorated with different BWE methods. As another example, high-frequency enhancement of band-limited signals can be handled in two fundamentally different ways. In the first case, nothing is known about the missing high-frequency band, and one must resort to general procedures for recreating a 'reasonable' signal. The emphasis is again on devising proper non-linear functions, the output signals of which are then filtered to have a spectral envelope that is 'reasonable'. On the other hand, if some information is available about the missing high frequencies, the emphasis shifts to modeling the high-frequency spectral envelope as accurately as possible, as it is known that this is the dominating factor in achieving a high-quality signal. Both these approaches are discussed in Chapter 5. Chapter 6 deals with high-frequency enhancement as well, but for a very particular application, namely, telephonic speech. This application also demands an algorithm that works without information about the missing high frequencies, but because the signal is restricted to speech (and the band limitation is very well defined), a very specialized algorithm can be developed, which works well on speech (but not on other signals). It appears that the available low-frequency band contains information on the spectral envelope of the high-frequency band. Note that a considerable portion of what is presented in Chapter 2 (low-frequency psychoacoustic BWE systems) also applies to material of Chapter 3 (low-frequency physical BWE systems), and to a lesser degree, also to material of Chapter 5 (high-frequency BWE for audio). Therefore, Chapter 2 is considerably larger in size than the latter two chapters and much cross-referencing will be used to avoid repetition.

In most of the work that we present, we have tried to justify the approaches by considering psychoacoustical models of auditory perception. Because this may not be familiar to all readers, some psychoacoustics is reviewed in Chapter 1. This chapter is titled 'From physics to psychophysics' as we have presented a little background in all of the required disciplines: signal processing, statistics of audio signals, loudspeakers, and psychoacoustics. Of course, these background materials cover only the basic concepts that would be helpful for understanding the BWE topics in this book and are definitely not sufficient to cover all that may be of interest. Therefore, references are given, as has been done throughout the book, so that much of the relevant literature (both for background as well as for more specific BWE material) can easily be found.

As always, the work that was done and presented in this book, as well as the writing itself, is the product of thc influence and support of many people. A special acknowledgement goes to Dr Janssen for valuable mathematical contributions that had considerable impact on Secs. 1.3.3.2, 2.4.2, 2.6, and 5.6.2. Of the others who contributed, we can only mention a few names here: Arie Kaizer for introducing us to electro-acoustics; Erik Druyvesteyn for always being enthusiastic; Paul Boers, Erik van der Tol, Okke Ouweltjes, and Daniël Schobben for valuable support; Cathy Polisset, Stefan Willems, and Gerrit DePoortere for development work; John Vanderkooy for the cooperation on high-Bl loudspeakers and many interesting discussions; Nathan Thomas and Michael Danessis from Salford University for their help; Ronald van Rijn for proofreading the manuscript; and Jeannie and Doortje for being patient, supportive, and for just being there.

E. Larsen and R. M. Aarts
Champaign (IL, USA) & Geldrop (The Netherlands)

Introduction

I.1 BANDWIDTH DEFINED

The word 'bandwidth' can apply to different situations. The IEEE Standard Dictionary of Electrical and Electronics Terms [134] gives for the most relevant cases:

Definition 1 Bandwidth of a continuous frequency band: *The difference between the limiting frequencies.*

Definition 2 Bandwidth of a waveform: *The least frequency interval outside of which the power spectrum of a time-varying quantity is everywhere less than some specified fraction of its value at a reference frequency.*

Definition 3 Bandwidth of a signal transmission system: *The range of frequencies within which performance, with respect to some characteristic, falls within specific limits (usually 3 dB less than the reference or maximum value).*

Two of the above definitions show that what exactly the bandwidth of a signal or transmission system is will depend on a more or less arbitrary choice. For example, in Def. 3 the standard notion of '3 dB below the reference' is indicated. This problem arises because the power spectrum of a signal does not terminate abruptly, at least not for physically realizable signals. An extensive study of bandwidth and the relations between time limiting and frequency limiting was conducted by Slepian, Landau, and Pollack, and published in a series of landmark papers [250, 154, 155]; a short overview of this work is presented by Slepian [249].

'Bandwidth extension' (BWE) indicates the process of increasing the bandwidth of a signal. In the context of this book, BWE is usually achieved by a signal-processing algorithm. Sometimes, explicit use is made of the properties of the auditory system, and the signal processing is done in such a way as to let the actual BWE take place in the auditory system itself. The use of BWE implies that at some point bandwidth reduction has taken place, which is the opposite of bandwidth extension. Examples of where bandwidth reduction occurs are telephony, perceptual audio coding (at low bit rates), and sound reproduction with non-ideal transducers; these examples will be further explored in the various chapters of this book, and solutions in terms of BWE algorithms are also presented.

I.2 HISTORIC OVERVIEW

BWE methods are required because the systems they operate with are somehow sub-optimal, and usually so by design. For example, loudspeakers can be built such that they properly reproduce the entire audible frequency spectrum, down to 20 Hz; but such systems would be very expensive and also very bulky. As another example, digital storage and transmission of audio can be done without loss of information, but at a rather high bit rate. To achieve higher storage (coding) efficiency such that more audio can be stored with the same amount of bits, information has to be discarded. Telephony is another example where economic constraints led to the design of a transmission system that had the smallest bandwidth that could be used, while ensuring good speech intelligibility at the price of a markedly reduced quality.

The process of being ever more economical is still going on, but at the same time people demand the highest possible sound quality. Here, we briefly look at this from a historical perspective.

I.2.1 ELECTROACOUSTIC TRANSDUCERS

Loudspeakers have been around for a long time, but the practicality of loudspeakers is still limited. In 1954, Hunt [114] pointed out prosaically

> '*Electroacoustics is as old and as familiar as thunder and lightning, but the knowledge that is the power to control such modes of energy conversion is still a fresh conquest of science not yet fully consolidated.*'

This is still true today, and in fact one of the reasons for the need for BWE is the limited bandwidth of transducers, in particular, electroacoustic transducers (devices to convert electric energy into acoustic energy, or vice versa). Especially, low frequencies are difficult to reproduce efficiently. We can classify electroacoustic transducers in the following five categories:

- *Electrodynamic*: Movement is produced because of a current flow in a wire located in a fixed magnetic field. Most drivers in audio and TV sets are of this type, and will be discussed in greater detail in Sec. 1.3.2.
- *Electrostatic*: Movement is produced because of a force between two or more electrodes with a (high) voltage difference. Condenser microphones are of this type; for loudspeakers, they are mainly for Hi-Fi use.
- *Magnetic*: Movement is produced by attraction of metal due to an electromagnet. This is very common for doorbells and hearing aids, but it is not very much in use for loudspeakers.
- *Magnetostriction*: Movement is due to the magnetostriction effect – an effect arising in a variety of ferromagnetic materials whereby magnetic polarization gives rise to elastic strain, and vice versa.
- *Piezoelectric*: Movement is produced because of the direct and converse piezoelectric effect – an effect arising in a variety of non-conducting crystals whereby dielectric polarization gives rise to elastic strain, and vice versa. One usually sees this type of

loudspeakers for high-frequency units (tweeters) only, since with a low voltage only small movements can be achieved.

Each of these classes has its benefits and specific applications areas, but none can yield an overall desirable performance. Then there are some other, more exotic methods of sound transduction (which are not much in use), like:

- Laser loudspeakers, using the photoacoustic effect (Westervelt and Larson [296]).
- 'Audio spotlight', using interfering ultrasonic sound rays (Yoneyama and Fujimoto [300], Pompei [211]) to make a narrow beam of audible sound from a small acoustic source.
- 'Singing display', which is based on electrostatic forces between the plates of an LCD display (description in Chapter 8).
- Flame loudspeaker (Gander [84]) and Ionophone (Russell [230]), using pyroacoustic transduction.

After the discovery of electromagnetism in 1802 and Reiss' telephone in 1860, it was Bell, on 10 March 1876, who uttered the famous words 'Mr. Watson, come here, I want to see you!', in the first successful electromagnetic transmission of speech. Not long after Bell's invention of the telephone, Charles Cuttris and Jerome Redding [55] (see also Hunt [114]) filed a US patent application describing what appears as the first moving-coil electroacoustic transducer. Various principles were explored, but it took until 1925 before the loudspeaker came to its full growth, due to the work of Rice and Kellog [223]. This year is generally considered as the birth of the modern loudspeaker. An intimidating array of books, research papers and patents has been devoted to the science and technology of transducers since 1925, a few of which are Beranek [28], Borwick [36], Gander [84], Geddes and Lee [85], Hunt [114], McLachlan [172], and Olson [192]. Although a tremendous amount of energy has been devoted to increasing the performance of transducers, Chapter 4 presents, as a special case of a BWE system, an unusual loudspeaker design that has the curious property that it has a very high efficiency at one (low) frequency only. This frequency is then used to reproduce most of the low bass of the audio signal, together with appropriate signal processing.

I.2.2 SOUND QUALITY

There is an ever-continuing desire to increase sound quality (which often competes with economic constraints). From 1925 to 1926, the Edison Company sponsored 'tone tests', recitals in which phonographic 're-creations' of musicians, as reproduced by the Edison 'Diamond Disc Phonograph', were compared directly to live performances by those same musicians (Thompson [269]). In auditoriums and concert halls across the US, curious crowds gathered to engage in a very public kind of critical listening. Today, we can hardly believe that these re-creations were indistinguishable from the original. But two things can be observed: (1) Those gatherings can be considered as the start of the 'A/B' listening test, and (2) that most people increase their demands (or, perhaps, change their expectations) for quality as soon as they get used to a certain quality level. After the introduction of loudspeakers in the early 1920s, there was a demand for more bass and more volume (Read and Welch [221, p239]), and this demand has never gone away.

As electrical engineering advanced over the years, and especially with the advent of digital technology, sound enhancement became possible through electronic means; nowadays, audio engineering relies heavily on signal-processing techniques. BWE is one of the methods that can be used to enhance the quality of sound, which is especially attractive in areas such as consumer electronics. In this market, sound quality is often sub-optimal because of economic constraints on the size and cost of components. Most manufacturers want to produce as cheaply as possible, yet retain a high subjective quality. Nonetheless, quality does suffer, and in many cases a bandwidth reduction results. Electronic means (such as BWE systems) are comparatively cheap and flexible, and play an ever-larger role in determining the sound quality of audio systems. Chapter 2 presents several signal-processing methods that allow a small loudspeaker to be used for reproducing a wide low-frequency bandwidth, and is thus a prime example of how signal processing can be used to circumvent physical/acoustical difficulties. Bandwidth reduction due to audio compression can be (partially) negated by BWE algorithms discussed in Chapter 5, and in Chapter 6, we show how speech quality can be improved using the existing narrow-band telephone network.

I.3 BANDWIDTH EXTENSION FRAMEWORK

I.3.1 INTRODUCTION

Here we introduce BWE from a general point of view, adapted from Aarts *et al.* [10]. We focus on requirements that BWE algorithms need to comply with, both perceptual as well as implementational (in a broad sense; more specific treatments are given in the appropriate chapters). This overview is also useful because various BWE algorithms are very similar from a conceptual point of view, even though in detail they can be quite different. Keeping the general picture of BWE in mind makes it easier to find connections between the topics discussed in the various chapters.

An obvious way to categorize various BWE methods is based on the frequency range of interest. Some methods extend the low end of the audio spectrum, other methods extend the high end of the spectrum. The classifications 'low' and 'high' in this sense are relative to the remaining audio spectrum, and should not be considered in absolute sense. The second categorization is to realize 'where' the signal bandwidth is actually extended: in the auditory system or in the reproduced waveform. In other words, psychoacoustic or physical. These four categories are indicated in Table I.1 (and in Fig. I.1), together with references to the chapters where the various BWE categories are discussed. Finally, we

Table I.1 The four categories of BWE as function of frequency band and type, with reference to the chapters that cover that kind of BWE application. Chapter 1 covers background material, and Chapter 8 presents an overview of BWE patents

	Low band	High band
Psychoacoustic	Chapters 2, 7	N/A
Physical	Chapters 3, 4	Chapters 5, 6

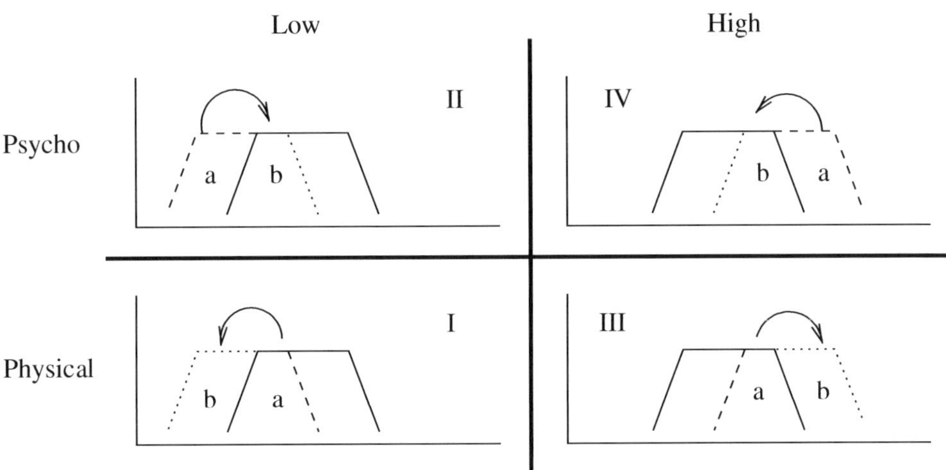

Figure I.1 The four categories of bandwidth extension. Low-frequency psychoacoustic BWE in the upper left panel, high-frequency psychoacoustic BWE in the upper right panel (which as yet has no practical implementation), low-frequency physical BWE in the lower left panel, and high-frequency BWE in the lower right panel. Energy from the dashed frequency range 'a' is shifted to the dotted frequency range 'b'. Adapted from Aarts *et al.* [10]

can identify BWE algorithms that use a priori information about the desired frequency components that need to be resynthesized, and those that use no such information. The latter class is termed 'blind', the other 'non-blind'. Generally, psychoacoustic BWE algorithms are never blind, because it is not the signal that is bandlimited, but the transducer. For the low-frequency psychoacoustic case, the application area is small loudspeakers, which cannot radiate very low frequency components. These low frequencies are present in the received signal, but have to be modified in such a way that the loudspeaker can reproduce a signal that has the same pitch percept, yet does not physically contain very low frequencies. High-frequency psychoacoustic BWE does not exist; it would require an algorithm that can yield a 'bright' (in terms of timbre) sound percept without reproducing the required high-frequency components. No known psychoacoustic effect can do this, and therefore this technology does not exist[1]. Physical BWE methods can be either blind or non-blind, because in these cases it is the signal that is bandlimited, not the transducer. The BWE algorithm has to resynthesize the missing frequency components from the narrow-band input signal. This can be done quite well with a priori information about the high frequencies, but it is also possible without such information (although usually at a lower quality). In this book, we almost exclusively deal with blind BWE systems, the exception being a kind of high-frequency BWE for audio, discussed in Sec. 5.5. It appears that the four classes of blind algorithms (of which three have practical applications), all

[1] Therefore, we simply use the term high-frequency BWE for what is in terms of the categorization of Table I.1 properly called high-frequency physical BWE.

have similar requirements, and can be implemented in a broadly similar fashion (although varying greatly on a more detailed level).

Blind algorithms can only use statistical information about the expected signals. This calls for a more general approach than what non-blind algorithms would require, and in the following text, we show how this generality can be exploited to cast the various BWE categories into a generalized signal-processing framework. A signal-processing framework is developed, which can be used to design BWE algorithms for many applications.

I.3.2 THE FRAMEWORK

I.3.2.1 Bandwidth Extension Categories

The four categories of BWE were already presented, and have been arranged in matrix form in Fig. I.1, where the columns indicate either low- or high-frequency extension, and the rows indicate psychoacoustic or physical BWE type. Each of the four graphs indicates a stylized power spectrum of an audio signal. The arrow indicates the action of the BWE algorithm: energy from the dashed frequency range 'a' is shifted to the dotted frequency range 'b'. Such 'shifting' of energy from one frequency range to the other obviously needs to be done in a special way; this will be elaborately discussed in the remainder of the book. The four indicated categories of BWE have the following characteristics:

1. *Low-frequency physical BWE category*: The lowest frequency components of the signal are used to extend the lower end of the signal's spectrum. Such an algorithm can be used if the low-frequency bandwidth of the signal has been reduced in storage or transmission; alternatively, the algorithm can be used for audio enhancement purposes, even if no prior bandwidth reduction had taken place. The loudspeaker will need to have an extended low-frequency response to reproduce the synthesized low frequencies.
2. *Low-frequency psychoacoustic BWE category*: The lowest frequency components of the signal cannot be reproduced by the loudspeaker, and are shifted to above the loudspeaker's low cut-off frequency. This must be done in such a way as to preserve the correct pitch and loudness of the low frequencies (and timbre as well, but this is not entirely possible).
3. *High-frequency BWE category*: The highest frequency components of the signal are used to extend the higher end of the signal's spectrum. Such an algorithm can be used if the high-frequency bandwidth of the signal has been reduced in storage or transmission; alternatively, the algorithm can be used for audio enhancement, even if no prior bandwidth reduction had taken place. The loudspeaker will need to have an extended high-frequency response to reproduce the synthesized high frequencies (which is usually not a problem).
4. *High-frequency (psychoacoustic) BWE category*: the highest frequency components of the signal cannot be reproduced by the loudspeaker, and are shifted to below the loudspeaker's high cut-off frequency. This must be done in such a way as to preserve the correct pitch, timbre, and loudness of the high frequencies. However, there is no known psychoacoustic effect that evokes a bright timbre percept when only lower-frequency components are present. Therefore, this category of BWE has no known implementation.

I.3.2.2 Perceptual Considerations

All of the BWE algorithms derive from one part of a signal's spectrum, a second signal in a different frequency range, which is then added to the input signal. The sum of these two signals should blend together to form an enhanced version of the original. The analysis by the auditory system should therefore group these two signals into the same stream, yielding a single percept. Bregman [38] gives some clues as to what signal characteristics are important in this grouping decision: pitch, timbre, and temporal modulation. If any one of these parameters differs 'too much' between the two signals, the signals will be segregated and be heard as two separate streams. This would constitute a failure of the BWE algorithm. Therefore, we must ensure that all the said signal characteristics of the synthetic signal remain as similar as possible to those of the original signal. On the other hand, a slight dissimilarity between, say, the pitches of two signals, can be 'overcome' by strong similarity in temporal modulation. Indicating the synthetic signal (output of BWE algorithm) by $y(t)$, and the input signal by $x(t)$, we have the following considerations:

Pitch: $x(t)$ and $y(t)$ should have a similar tonal structure, that is, a common fundamental frequency f_0. If the signals are atonal (noise), then $x(t)$ and $y(t)$ should have similar moments (at least up to second order). We shall see that we can design efficient algorithms such that the pitch of $y(t)$ matches that of $x(t)$.

Timbre: Timbre is usually associated with the spectral envelope, although temporal envelope and spectral phase also have an influence. In the BWE algorithms, we can control timbre to some extent by the correct design of filters.

Loudness: Similar temporal modulations for $y(t)$ and $x(t)$ are required for covarying loudness of both signals, and can be achieved by ensuring that the amplitudes of $y(t)$ and $x(t)$ are (nearly) proportional. The BWE algorithms described later on will usually be linear in amplitude, so this is automatically taken care of.

Because there is little objective data available on how 'close' or 'similar' the mentioned psychoacoustic parameters must be for $x(t)$ and $y(t)$ to be grouped, especially for realistic audio signals, the tolerance of BWE algorithms for these grouping and segregation effects is, to some degree, a matter of trial and error.

I.3.2.3 Implementational Considerations

Besides perceptual constraints, there are some constraints on the implementation of the algorithms. These are not necessarily exclusive to BWE methods, but to most signal-processing algorithms for use in consumer electronic applications. These constraints are:

1. Low computational complexity and low memory requirements.
2. Independence of signal format.
3. Applicable to music, speech and, preferably, both.

The first constraint is important for the algorithm to be a feasible solution for consumer devices, which typically have very limited resources. Although the use of digital signal processors is becoming more widespread in consumer electronics, many signal-processing features are usually implemented together, and each one may only use a fraction of the total computing power.

Independence of signal format means that the algorithm is applied to a PCM-like signal. Dependence on a certain coding or decoding scheme would limit the scope of the algorithm. Of course, in some cases, this can lead to higher-quality output signals, as for the SBR technology in high-frequency BWE application, discussed in Chapter 5.

The third constraint determines if we can use a detailed signal model or not. If the application is limited to speech only, a speech model that allows a more accurate BWE (Chapter 6) can be used. Signals not well described by the speech model, such as music, would not give good results with this algorithm. Thus, if the nature of the processed signals is unclear, a general algorithm must be used. Specialized BWE algorithms for music would be more difficult to devise than those for speech, because the statistics of musical signals depend heavily on the instrument being used; also, typically many instruments are active simultaneously, whereas in speech applications the signal can be assumed to derive from one sound source only. If we can decide for a given signal whether it is music or speech, which can be done with a speech-music discriminator (Aarts and Toonen Dekkers [6]), we can use BWE for music with strategies as discussed in Chapter 5, and BWE for speech with strategies as discussed in Chapter 6.

I.3.2.4 Processing Framework

Figure I.2 presents the signal-processing framework that we propose for all categories of BWE described here, and covers most of the BWE methods described elsewhere. The general algorithm consists of two branches: one in which the input signal is merely delayed, and one in which the bandwidth extension takes place. This bandwidth extension is done by bandpass filtering the input signal (FIL1) to select a portion of the audio signal (indicated by the letter 'a' in Fig. I.1). This portion is then passed to a non-linear device (NLD), which 'shifts' the frequencies to a higher or lower region by a suitable non-linear operation, according to the particular application. Subsequently, the signal is bandpass filtered (FIL2), to obtain a suitable spectrum and timbre (the signal now has frequencies in the range 'b', as shown in Fig. I.1). The resulting signal is amplified or attenuated as desired and added back to the (delayed) input signal to form the output.

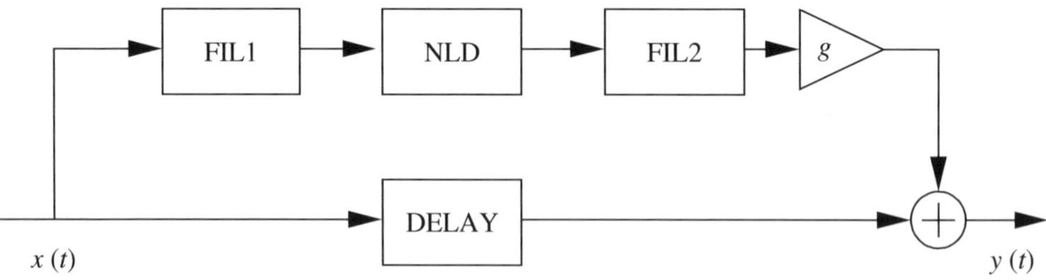

Figure I.2 General BWE framework, also shown as high-frequency BWE structure in Fig. 5.2. Similar structures are shown for low-frequency psychoacoustic BWE as Fig. 2.4 and for low-frequency physical BWE as Fig. 3.1

A somewhat different approach is taken in Secs. 2.4 and 3.3.2.4, where a frequency tracker is used and the harmonics signal is generated explicitly (not through non-linear processing of the filtered input). For speech BWE methods, discussed in Chapter 6, the same structure is used, but some of the processing steps are much more involved, in particular, the filtering by FIL2. This filter is derived adaptively from the input signal. For the non-blind BWE approach taken in Sec. 5.5, FIL2 is also adaptively determined, in this case by control information embedded in a coded audio bitstream. In Sec. 5.6, an extremely simplified algorithm (instantaneous compression) that only uses an NLD, without any filters, is used; this method is only suitable in particular circumstances and for particular signals.

Specific implementations for the NLD, which 'shifts' frequency components from low to high values (or vice versa), will be given in later chapters covering the specific BWE categories. They are based on generating harmonics or subharmonics of the signal passed by FIL1. Nearly all of the NLDs discussed in the following chapters implicitly determine the frequency of the incoming signal by its zero crossings (except the frequency tracker of Secs. 2.4 and 3.3.2.4). For a pure tone of frequency f_0, the situation is unambiguous, and there are $2f_0$ zero crossings per second. But because FIL1 is a bandpass filter of finite bandwidth, the signal going into the NLD will possibly contain more than one frequency component, which will disturb the zero crossing rate γ (number of zero crossing per second) of the signal. Fortunately, zero crossings appear to be quite robust in reflecting the dominant frequency of a signal, which is known as the 'dominant frequency principle'. This principle can be made explicit by the 'zero crossing spectral representation' (Kedem [141]) as

$$\cos(\pi\gamma) = \frac{\displaystyle\int_0^{\pi} \cos\omega \, \mathrm{d}F(\omega)}{\displaystyle\int_0^{\pi} \mathrm{d}F(\omega)}, \tag{I.1}$$

which holds for weakly stationary time series, with spectral distribution $F(\omega)$. For a pure tone of frequency $f_0 = \omega_0/2\pi$, we set $F(\omega) = \delta(\omega - \omega_0)$ in Eqn. I.1, and we find that $\cos(\pi\gamma) = \cos\omega_0$, which gives us the expected $\gamma = 2f_0$. If the signal passed by FIL1 has multiple frequency components, one of which is dominant, the NLD will 'detect' this frequency by the zero crossings of the signal, and construct a harmonics signal on the basis of this dominant frequency.

1

From Physics to Psychophysics

This first chapter does not present new material, but rather presents some backgrounds for the various BWE topics. The selection of the material to be included in this chapter is mainly motivated by what is considered to be a useful reference if unfamiliar concepts are encountered in later chapters. To keep this backgrounds section concise, it will not always contain all desired information, but references are provided for further reading (as is done throughout the remainder of the book).

The topics covered in this chapter are basics in signal processing in Sec. 1.1, statistics of speech and music in Sec. 1.2, acoustics (mainly concerning loudspeakers) in Sec. 1.3, and auditory perception in Sec. 1.4.

1.1 SIGNAL THEORY

This section reviews some preliminaries for digital signal processing, most notably the concepts of linearity versus non-linearity, and commonly used digital filter structures and some of their properties. An understanding of these concepts is essential to appreciate the algorithms presented in later chapters, but knowledge of more advanced signal processing concepts would be very useful. If necessary, reviews of digital signal processing theory and applications can be found in, for example, Rabiner and Schafer [217], Rabiner and Gold [218], or Oppenheim and Schafer [194]. Van den Enden and Verhoeckx [281] also present a good introduction into digital signal processing. Golub and Van Loan [93] is a good reference for matrix computations, which are extensively used in Chapter 6.

1.1.1 LINEAR AND NON-LINEAR SYSTEMS

Consider a system that transforms an input signal $x(t)$ into an output signal $y(t)$. Assume that this transformation can be described by a function g such that

$$y(t) = g(x(t)). \tag{1.1}$$

Time invariance means that we must have

$$y(t-\tau) = g(x(t-\tau)), \quad \tau \in \mathbb{R}, \tag{1.2}$$

Audio Bandwidth Extension E. Larsen and R. M. Aarts
© 2004 John Wiley & Sons, Ltd ISBN 0-470-85864-8

that is, the output of a time-shifted input is the time-shifted output. The system is linear iff

$$\begin{aligned} y_1(t) &= g(x_1(t)), \\ y_2(t) &= g(x_2(t)), \\ y_1(t) + y_2(t) &= g(x_1(t) + x_2(t)), \quad (1.3) \\ a y_1(t) &= g(a x_1(t)), \quad a \in \mathbb{R}. \quad (1.4) \end{aligned}$$

Equation 1.3 is called the superposition property, that is, the output of the sum of two signals equals the sum of the outputs. Equation 1.4 is called the homogeneity property, that is, the output of a scaled input equals the scaled output. If any of these two properties (or both) do not hold, the system is non-linear. Vaidyanathan [278] introduces the terminology *homogeneous time-invariant* for a time-invariant system, where Eqn. 1.3 is not true, while Eqn. 1.4 is true; such systems are an important class for BWE algorithms, as we shall see in Chapter 2 and following chapters. Note that these comments and equations are valid for both continuous as well as discrete-time (sampled) systems.

Many mathematical techniques exist for analyzing properties of linear time-invariant (LTI) systems, and some basic ones will be discussed shortly. For non-linear or time-variant systems, such analysis often becomes very complicated or even impossible, which is why one traditionally avoids dealing with such systems (or makes linear approximations of non-linear systems). Nonetheless, non-linear systems can have useful properties that LTI systems do not have. For BWE purposes, an important example is that LTI systems cannot introduce new frequency components into a signal; only the amplitude and/or phase of existing components can be altered. So, if the frequency bandwidth of a signal needs to be extended, the use of a non-linear system is inevitable (and thus desirable). Note that non-linearities in audio applications are often considered as generating undesirable distortion, but controlled use of non-linearities, such as in BWE algorithms, can be beneficial.

1.1.2 CONTINUOUS-TIME LTI (LTC) SYSTEMS

A continuous-time LTI system will be abbreviated as LTC. Beside the input–output function g (Eqn. 1.1), an LTC system can be fully described by its impulse response $h(t)$, which is the output to a delta function $\delta(t)$ input[1] ($h(t) = g(\delta(t))$). Because the system is linear, an arbitrary input signal $x(t)$ can be written as an infinite series of delta functions by

$$x(t) = \int_{-\infty}^{\infty} x(\tau)\delta(t - \tau)\,\mathrm{d}\tau. \quad (1.5)$$

Using this principle, we can calculate the output signal $y(t)$ as

$$y(t) = \int_{-\infty}^{\infty} x(\tau)h(t - \tau)\,\mathrm{d}\tau, \quad (1.6)$$

[1] A delta function is the mathematical concept of a function that is zero everywhere except at $x = 0$, and which has an area of 1, that is, $\int_{-\infty}^{\infty} \delta(x)\,\mathrm{d}x = 1$.

which is called the convolution integral, compactly written as

$$y(t) = x(t) * h(t). \tag{1.7}$$

An LTC is called stable iff

$$\int_{-\infty}^{\infty} |y(t)|\,\mathrm{d}t < \infty, \tag{1.8}$$

and causal iff

$$h(t) = 0 \qquad \text{for } t < 0. \tag{1.9}$$

The Fourier transform of the impulse response $h(t)$ is called the frequency response $H(\omega)$ – like $h(t)$, $H(\omega)$ gives a full description of the LTC system. Here, ω is the *angular* frequency, related to frequency f as $\omega = 2\pi f$. It is convenient to calculate the frequency response (also called frequency spectrum, or simply spectrum) of the output of the system as

$$Y(\omega) = X(\omega)H(\omega). \tag{1.10}$$

We see that convolution in the time domain equals multiplication in the frequency domain. The reverse is also true, and therefore the Fourier transform has the property, which is sometimes called "convolution in one domain is equal to multiplication in the other domain". A slightly more general representation is through the Laplace transform, yielding $X(s)$ as

$$X(s) = \int_{-\infty}^{\infty} x(t)\mathrm{e}^{-st}\,\mathrm{d}t, \tag{1.11}$$

where s is the Laplace variable. For continuous-time, physical frequencies ω lie on the y-axis, thus we can write $s = i\omega$ (note that in practice the i is often dropped when writing $H(i\omega)$). We can also write the LTC system function as

$$H(s) = c\frac{\prod_{i=1}^{N}(s - z_i)}{\prod_{j=1}^{M}(s - p_j)}, \tag{1.12}$$

where c is a constant. The z_i are the N zeros of the system, and p_j are the M poles, either of which can be real or complex. The system will be stable if all poles lie in the left hemifield, that is, $\Re\{p_j\} < 0$.

As an example, a simple AC coupling filter may be considered: a capacitance of C F connecting input and output, and a resistance of R Ω from output to the common terminal. This system transfer function can be written as

$$H(s) = \frac{RCs}{1 + RCs}, \tag{1.13}$$

where the frequency response follows by substituting $s = i\omega$. Comparing Eqn. 1.12 with Eqn. 1.13 shows that $c = RC$, $N = 1$, $z_1 = 0$, $M = 1$, and $p_1 = -\frac{1}{RC}$.

The *magnitude* and *phase* of an LTC system function $H(s)$ are defined as

$$|H(s)| = \left[\Re\{H(s)\}^2 + \Im\{H(s)\}^2\right]^{1/2}, \tag{1.14}$$

$$\angle H(s) = \tan^{-1}\frac{\Im\{H(s)\}}{\Re\{H(s)\}}, \tag{1.15}$$

respectively. From the phase response, we can define the group delay $\tau_d(\omega)$, which can be interpreted as the time delay of a frequency ω between input and output and is given by

$$\tau_d(\omega) = -\frac{d}{d\omega}\angle H(i\omega). \tag{1.16}$$

1.1.3 DISCRETE-TIME LTI (LTD) SYSTEMS

In DSP systems, the signals are known only at certain time instants kT_s, where $k \in \mathbb{Z}$ and $T_s = 1/f_s$ is the sampling interval, with f_s being the sample frequency or sample rate. Thus, these systems are known as discrete-time LTI systems, abbreviated as LTD. According to the *sampling theorem*[2], we can perfectly reconstruct a continuous-time signal $x(t)$ from its sampled version $x(k)$ if f_s is at least twice the highest frequency occurring in $x(t)$. To convert $x(k)$ (consisting of appropriately scaled delta function at the sample times) to $x(t)$ (the continuous signal), a filter that has the following impulse response needs to be applied

$$h(t) = \frac{\sin(\pi t/T_s)}{\pi t/T_s}. \tag{1.17}$$

In the frequency domain, this corresponds to an ideal low-pass filter ('brick-wall' filter), which only passes frequencies $|f| < f_s/2$. System functions in the discrete-time domain are usually described in the z domain (z being a complex number) as $H(z)$, rather than the s domain. Likewise, the corresponding input and output signals are denoted as $X(z)$ and $Y(z)$ (Jury [138]). $X(z)$ can be obtained from $x(k)$ through the Z-transform

$$X(z) = \sum_{k=-\infty}^{\infty} x(k)z^{-k}. \tag{1.18}$$

Normalized physical frequencies Ω in discrete time lie on the unit circle, thus $z = e^{-i\Omega}$; therefore $|\Omega| \leq \pi$.

There are various ways to convert a known continuous-time system function $H(s)$ to its discrete-time counterpart $H(z)$, all of which have various advantages and disadvantages. The most widely used method is the bilinear transformation, which relates s and z as

$$s = \frac{2}{T_s}\frac{1 - z^{-1}}{1 + z^{-1}}. \tag{1.19}$$

[2] The sampling theorem is frequently contributed to Shannon's work in the 1940s, but, at the same time, Kotelnikov worked out similar ideas, and, a few decades before that, Whitaker. Therefore, some texts use the term WKS-sampling theorem (Jerri [135]).

Applying this to the example of Eqn. 1.13, we find

$$H(z) = \frac{k(1 - z^{-1})}{1 + k + (1 - k)z^{-1}}, \tag{1.20}$$

with $k = 2RC/T_s$, a dimensionless quantity. Like Eqn. 1.12, we can write $H(z)$ in a similar manner, with a zero at $z_1 = 1$ and a pole at $p_1 = \frac{k-1}{k+1}$. Substituting $z = e^{-i\Omega}$ in Eqn. 1.20 gives the frequency response $H(e^{-i\Omega})$. Stability for LTD systems requires that all poles lie within the unit circle, that is, $|p_j| < 1$. Magnitude, phase, and group delay of an LTD system are defined analogously as for LTC systems (Eqns. 1.14–1.15).

1.1.4 OTHER PROPERTIES OF LTI SYSTEMS

There are a few other properties of LTI systems that are of interest. We will discuss these using LTD systems, but analogous equations hold for LTC systems.

We have already found that stability requires that all poles must lie within the unit circle, that is, $|p_j| < 1$. Where the locations of zeros z_j are concerned, the system is called minimum phase if all $|z_j| < 1$. This is of interest because a minimum-phase system has a stable inverse. To see this, note that

$$H^{-1}(z) = \left[c \frac{\prod_{i=1}^{N}(z - z_i)}{\prod_{j=1}^{M}(z - p_j)} \right]^{-1} = c^{-1} \frac{\prod_{j=1}^{M}(z - p_j)}{\prod_{i=1}^{N}(z - z_i)}, \tag{1.21}$$

that is, the poles become zeros, and vice versa. We can conclude that a stable minimum-phase system has an inverse, which is also stable and minimum phase. A non-minimum-phase system cannot be inverted[3]. A 'partial inversion' is possible by splitting the system function into a minimum-phase and a non-minimum-phase part and then inverting the minimum-phase part (Neely and Allen [184]).

A linear-phase system $H(z)$ is one for which $\angle H(z) = -a\omega$ $(a > 0)$, that is, the phase is a linear function of frequency. It implies that the group delay is a constant $\tau_d(\omega) = a$, which is often considered to be beneficial for audio applications. For $H(z)$ to be linear phase, the impulse response $h(k)$ must be either symmetric or anti-symmetric.

1.1.5 DIGITAL FILTERS

The simplest digital filter is $H(z) = z^{-1}$, being a delay of one sample. If one cascades $N + 1$ such delays and sums all the scaled delayed signals, one gets a filter as depicted in Fig. 1.1. Such a filter is called a finite impulse response (FIR) filter, since its impulse response is zero after N time samples. Its system function is written as

$$H(z) = \frac{Y(z)}{X(z)} = \sum_{i=0}^{N} b_i z^{-1}, \tag{1.22}$$

[3] A practical example of a non-minimum-phase system is a room impulse response between a sound source and a receiver (unless they are very close together).

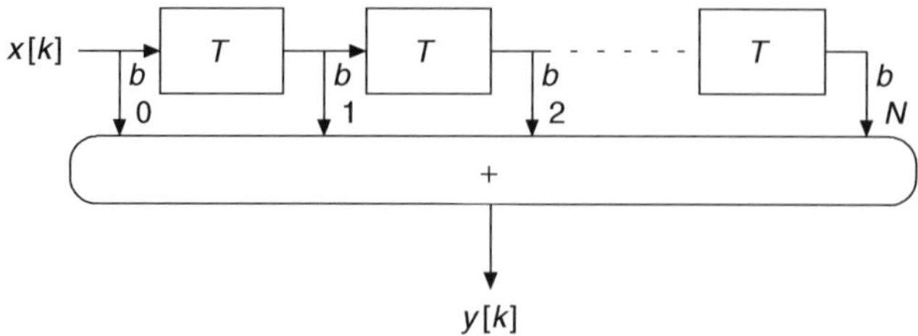

Figure 1.1 A finite impulse response (FIR) filter. The input signal $x(k)$ is filtered, yielding the output signal $y(k)$. The boxes labeled 'T' are one sample (unit) delay elements. The signal at each tap is multiplied with the corresponding coefficients b

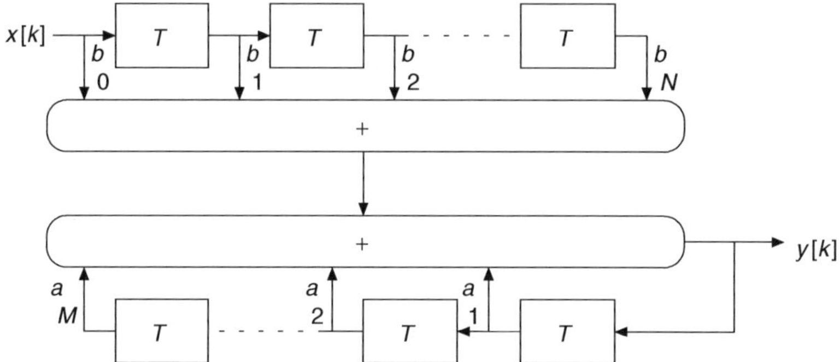

Figure 1.2 An infinite impulse response (IIR) filter in direct form-I structure. The input signal $x(k)$ is filtered by the filter, yielding the output signal $y(k)$. The boxes labeled 'T' are one sample (unit) delay elements. The signal at each forward tap is multiplied with the corresponding coefficients b, while the signal in the recursive (feedback) part is multiplied with the corresponding coefficients a, respectively

and the frequency response can be found by substituting $z = \mathrm{e}^{-i\Omega}$. It is obvious that an FIR filter has only zeros and no poles. Therefore, it is guaranteed to be stable. Note that an FIR filter need not be minimum phase, and therefore a stable inverse is not guaranteed to exist.

By applying feedback to a filter, as shown in Fig. 1.2, a filter with an infinite impulse response (IIR) can be made. Its system function is written as

$$H(z) = \frac{Y(z)}{X(z)} = \frac{\sum_{i=0}^{N} b_i z^{-1}}{1 - \sum_{j=1}^{M} a_i z^{-1}}. \tag{1.23}$$

This shows that, in general, an IIR filter has both zeros and poles; therefore, an IIR filter would be unstable if any of the poles lie outside the unit circle. An IIR filter can be both minimum- and non-minimum phase. While the structure of Fig. 1.2 can implement any IIR filter, it is, for practical reasons (like finite word-length calculations), more customary to partition the filter by cascading 'second-order sections' (IIR filters with two delays in both forward and feedback paths), also known as biquads.

FIR and IIR filters have a number of differences, as a result of which each of them has specific applications areas, although overlap exists, of course. The most important differences (for BWE purposes) are:

- *Phase characteristic*: Only FIR filters can be designed to have linear phase; IIR filters can only approximate linear phase at a greatly increased complexity. In Sec. 2.3.3.3, it is discussed why a linear-phase characteristic is beneficial for a particular BWE application.
- *Computational complexity*: For both FIR and IIR filters, the computational complexity is proportional to the number of coefficients used. However, for an FIR filter, the number of coefficients directly represents the length of the impulse response, or, equivalently, the frequency selectivity. An IIR filter has the advantage that very good frequency selectivity is possible using only a small number of coefficients. Therefore, IIR filters are generally much more efficient than FIR filters. For low-frequency BWE applications, filters with narrow passbands are often required, centered on very low frequencies; in such cases, IIR filters are orders of magnitude more efficient than FIR filters.

It will be apparent that the choice of FIR or IIR depends on what features are important for a particular application. Sometimes, as in some BWE applications, both linear phase *and* high frequency selectivity are very desirable. In such cases, IIR filters can be used, but in a special way and at a somewhat increased computational complexity and memory requirement (see e.g. Powell and Chau [213]).

1.2 STATISTICS OF AUDIO SIGNALS

For BWE methods, it is important to know the spectrum of the audio signal. Because these signals are generally not stationary, the spectrum varies from moment to moment, and the spectrogram is useful to visualize this spectro-temporal behaviour of speech and music. First we will consider speech, and then music. For both cases, we will make it plausible that certain bandwidth limitations can be overcome by suitable processing. Some of the material in this chapter is taken from Aarts *et al.* [10].

1.2.1 SPEECH

Speech communication is one of the basic and most essential capabilities of human beings. The speech wave itself conveys linguistic information, the speaker's tone, and the speaker's emotion. Information exchange by speech clearly plays a very significant role in our lives. Therefore, it is important to keep speech communication as transparent as possible, both to be intelligible as well as to be natural. Unfortunately, owing to bandwidth limitation, both aspects can suffer. But, because a lot is known about the statistics of speech, specialized BWE algorithms can be used to restore, to a large extent, missing frequency components, if the available bandwidth is sufficiently large. This is explored in detail in Chapter 6.

Speech can be voiced or unvoiced (tonal or noise-like), see, for example, Furui [80], Olive *et al.* [191], Rabiner and Schafer [217], and Fig. 6.4. A voiced sound can be modelled as a pulse source, which is repeated at every fundamental period $1/f_p$ (where f_p

is the pitch), while an unvoiced sound can be modelled as a white noise generator. The loudness of speech is proportional to the peak amplitudes of the waveform. Articulation can be modelled as a cascade of filters – these filters simulate resonant effects R_i (formants) in the vocal tract, which extends from the vocal cords to the lips, including the nasal cavity. Consequently, any harmonic of the series of pulses with frequency kf_p that happens to lie close to one of the R_i is enhanced. To make the various vowel sounds, a speaker or singer must change these vocal tract resonances by altering the configuration of tongue, jaw, and lips. The distinction between different vowel sounds in Western languages is determined almost entirely by R_1 and R_2, the two lowest resonances, that is, vowels are created by the first few broad peaks on the spectral envelope imposed on the overtone spectrum, by vocal tract resonances. For the vowel sound in 'hood', pronounced by a male speaker, $R_1 \approx 400\,\mathrm{Hz}$ and $R_2 \approx 1000\,\mathrm{Hz}$. In contrast, to produce the vowel in 'had', R_1 and R_2 must be raised to about 600 and 1400 Hz respectively, by opening the mouth wider and pulling the tongue back. For women, the characteristic resonance frequencies are roughly 10% higher. But for both sexes, the pitch frequency f_p in speech and singing is generally well below R_1 for any ordinary vowel sound – except when sopranos are singing very high notes, in which case they raise R_1 towards f_p (Goss Levi [160], Joliveau *et al.* [137]). Finally, radiation of speech sounds can be modelled as arising from a piston sound source attached to an infinite baffle, like a loudspeaker model discussed in Sec. 1.3.2. The range of frequencies for speech is roughly between 100 and 8 kHz (whereas the ordinary telephone channel is limited between 300 and 3400 Hz).

An important parameter for speech is the fundamental frequency or pitch. Furui [80] presents a statistical analysis of temporal variations in the pitch of conversational speech for individual talkers, which indicates that the mean and standard deviation for a female voice are roughly twice those of a male voice. This is shown in Fig. 1.3. The pitch distributed over talkers on a logarithmic frequency scale (not the linear scale of Fig. 1.3) can be approximated by two normal distributions that correspond to the male and female voice, respectively. The mean and standard deviation for a male voice are 125 and 20.5 Hz, respectively, whereas those for a female voice are twice as large. Conversational speech includes discourse as well as pause, and the proportion of actual speech periods relative to the total period is called 'speech ratio'. In conversational speech, the speech ratio for individual talkers is about 1/3 (Furui [80]), and can be used as a feature for speech detection in a speech–music discriminator (Aarts and Toonen Dekkers [6]).

In order to gain insight in the long-term average of speech spectra, six speech fragments of utterances of various speakers of both sexes were measured. Figure 1.4 shows the power spectra of tracks 49–54 of the SQAM disk [255]. To parameterize the spectra in the plot, we derived the following heuristic formula

$$|H(f)| \approx \frac{\left(\dfrac{f}{f_\mathrm{p}}\right)^6}{1+\left(\dfrac{f}{f_\mathrm{p}}\right)^6} \frac{1}{1+\dfrac{f}{1000}}, \tag{1.24}$$

were f is the frequency and f_p the pitch of the voice. Byrne *et al.* [42] have shown that there is not much difference in the long-term speech spectra of different languages. The first factor in the product of Eqn. 1.24 denotes the high-pass behavior, and the second

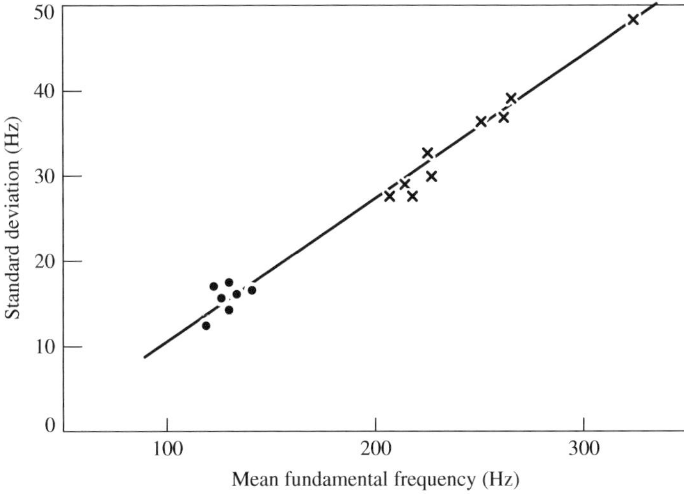

Figure 1.3 Male (•)/female pitch (×), adapted from Furui [80]. The mean and standard deviation for male voice pitch is 125 ± 20.5 Hz and are twice these values for female voice pitch

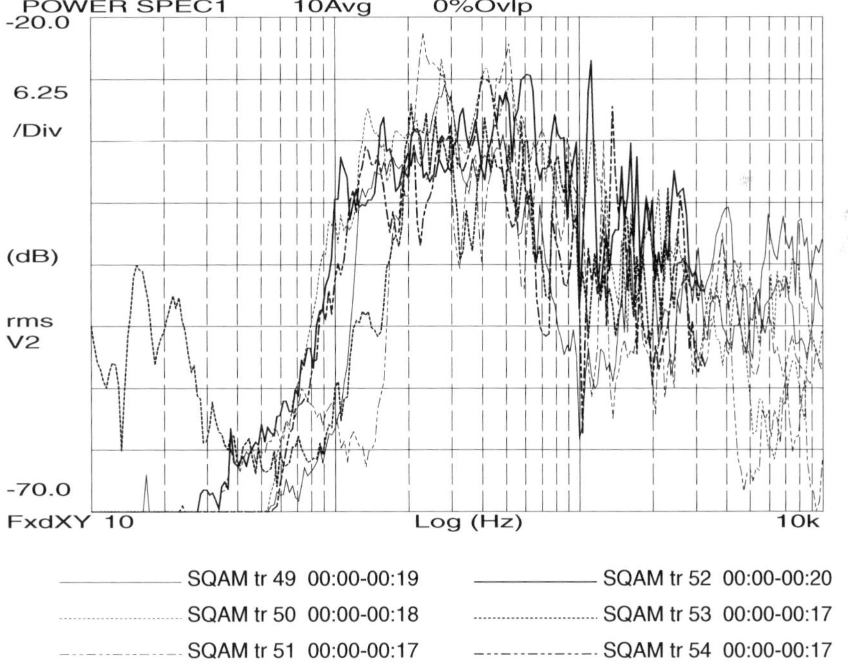

Figure 1.4 Long-term spectrum of speech, for a total of six talkers, both sexes (SQAM disc [255], tracks 49–54)

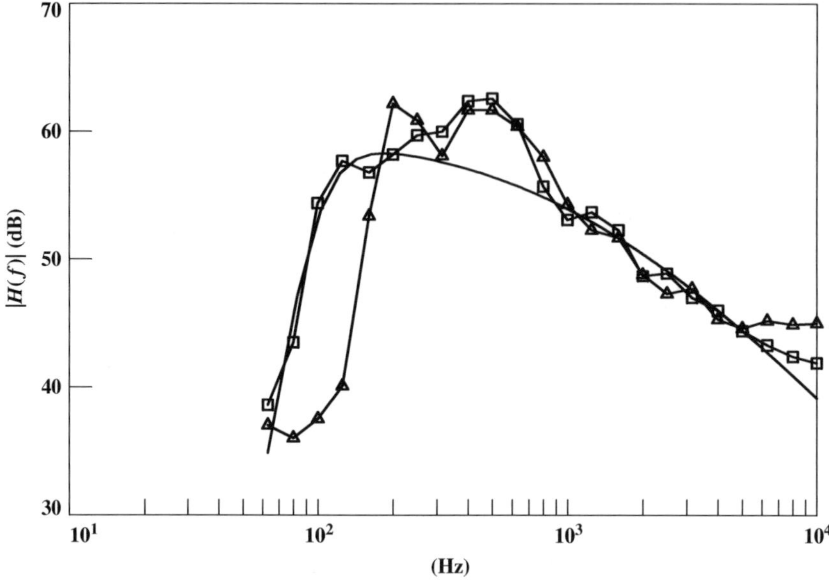

Figure 1.5 Equation 1.24 is plotted (curve without marker) for a male voice (f_p = 120 Hz, offset by 60 dB), together with the data from Byrne *et al.* [42] for males (squares) and females (triangles). Note that the parameterization of Eqn. 1.24 fits the empirical data for the male voice quite well (the female voice would have to be modelled with a higher pitch value)

one the low-pass behaviour. It shows that there is a steep slope at low frequencies, and a rather modest slope at high frequencies, as is depicted in Fig. 1.5. The lowest pitch of the male voice is about 120 Hz (see Fig. 1.3), which corresponds very well with the high-pass frequency in Eqn. 1.24, and Fig. 1.4. The telephone range is 300–3400 Hz, which clearly is not sufficient to pass the high-frequency range, nor the lowest male fundamental frequencies (and also not most female fundamental frequencies). A BWE algorithm that can recreate the fundamental (and possibly the lower harmonics) and the high-frequency band of speech signals should be able to create a more natural sounding speech for telephony.

Because the speech spectrum changes over time, it is instructive to compute spectra at frequent intervals and display the changing spectra. A spectrogram is shown for a particular speech sample in Fig. 1.6 (b); the pitch of the voice is time-varying, as can be seen in (c). A BWE algorithm must be able to follow these pitch changes and change its output frequencies accordingly. Fig. 1.7 shows a waveform and spectrogram (a and b) of the same speech utterance, and an 8-kHz low-pass filtered version thereof, which could occur in perceptual audio coders at very high compression rates; the telephone channel would low-pass filter the signal even more severely at 3.4 kHz. A high-frequency BWE algorithm can resynthesize an octave of high-frequency components, as shown in Fig. 1.7(c); note the similarities and differences with respect to the original spectrogram. High-frequency BWE algorithms are discussed in Chapters 5 (audio) and 6 (speech).

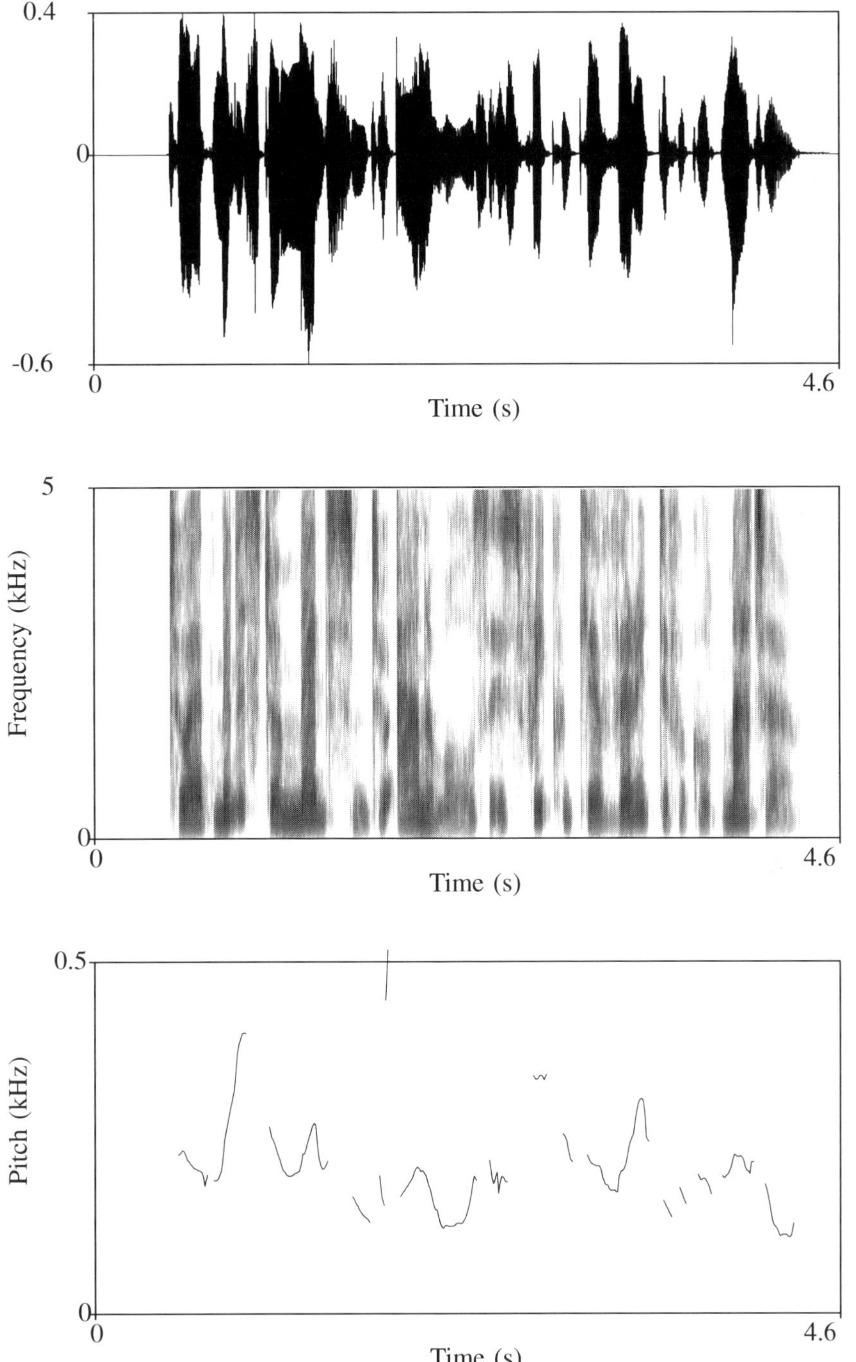

Figure 1.6 For a given speech waveform (a), the spectrogram is displayed in (b) (white–black indicating low–high energy, dB scale). (c) shows the pitch of the voice as determined by a pitch tracker ('Praat', Boersma and Weenink [35])

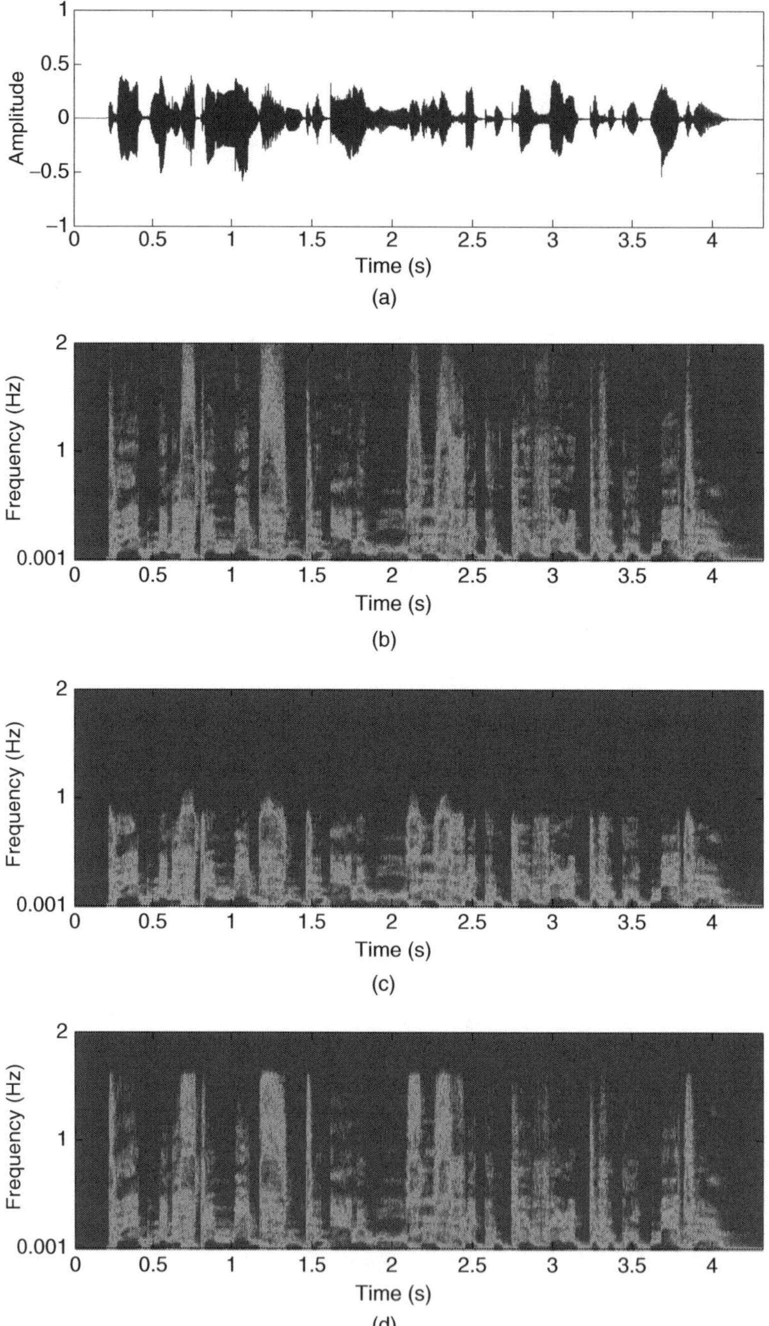

Figure 1.7 High-frequency BWE for a female voice (waveform in a, spectrogram in b) that has been low-pass filtered at 8 kHz (spectrogram in c). The processed signal (extended to 16 kHz) is displayed as a spectrogram in d

1.2.2 MUSIC

More than 70 years ago, Sivian *et al.* [248] performed a pioneering study of musical spectra using live musicians and – for that time – innovative electronic measurement equipment. Shortly after the introduction of the CD, this study was repeated by Greiner and Eggers [99] by using modern digital equipment and modern source material, at that time CDs. The result of both studies was a series of graphs showing for each instrument or ensemble the spectral amplitude distribution of the performed musical passage. The findings were that, in general, musical spectra have a bandpass characteristic, the exact shape of which is determined by the music and the instrument. As in speech, the fundamental frequency (pitch) is time varying. A complicating factor is that various instruments may be playing together, creating a superposition of several complex tones.

An example is shown in Fig. 1.8, where the variable time–frequency characteristic of a 10-s excerpt of music is shown ('One', by Metallica). The waveform is shown in (a), and (b) shows a spectrogram (frequencies 0–140 Hz) of the original signal. The energy extends down to about 40 Hz. By using a low-frequency physical BWE algorithm, we can extend this lower limit to about 20 Hz (c), which requires a subwoofer of excellent quality for correct reproduction. Because the resulting synthetic frequencies have similar spectro-temporal characteristics as the original low frequencies, they will be perceived as an integral part of the signal (lowering the pitch of the bass tones to 20 Hz). Because of the very low frequencies that are now being radiated, it will also add 'feeling' to the music. Low-frequency physical BWE algorithms are discussed in Chapter 3.

Another study (Fielder and Benjamin [70]) was conducted to establish design criteria for the performance of subwoofers to be used for the reproduction of music in homes. The focus on subwoofers was motivated by the fact that low frequencies play an important role in the musical experience. A first conclusion of that study was that recordings with audible bass below 30 Hz are relatively rare. Second, these very low frequencies were generated by pipe organs, synthesizers, or special effects and environmental noise. Other instruments, such as bass guitar, bass viol, tympani, or bass drum, produce relatively little output below 40 Hz, although they may have very high levels at or above that frequency. Fielder and Benjamin [70] gave an example that for an average listening room of 68 m^3, the required acoustic power for reproduction is 0.0316 W (which yields a sound pressure level of 97 dB), which requires a volume displacement of 0.685 l at 20 Hz. This requires an excursion of 13.5 mm for a 10 in. (0.25 m) woofer. These are extraordinary requirements, and very hard to fulfil in practice. An alternative is to use low-frequency psychoacoustic BWE methods, where frequencies that are too low to reproduce are shifted to higher frequencies, in such a way that the pitch percept remains the same. These methods are discussed in Chapter 2. If we consider Fig. 1.8 (c) as the original signal, we could think of such BWE as shifting the frequency band 20–40 Hz to above 40 Hz. The spectrogram of the resulting signal would resemble that of Fig. 1.8(b).

1.3 LOUDSPEAKERS

1.3.1 INTRODUCTION TO ACOUSTICS

BWE methods are closely related to acoustics, particularly acoustics of loudspeakers, so here we will review some basic concepts in this area. Extensive and more general

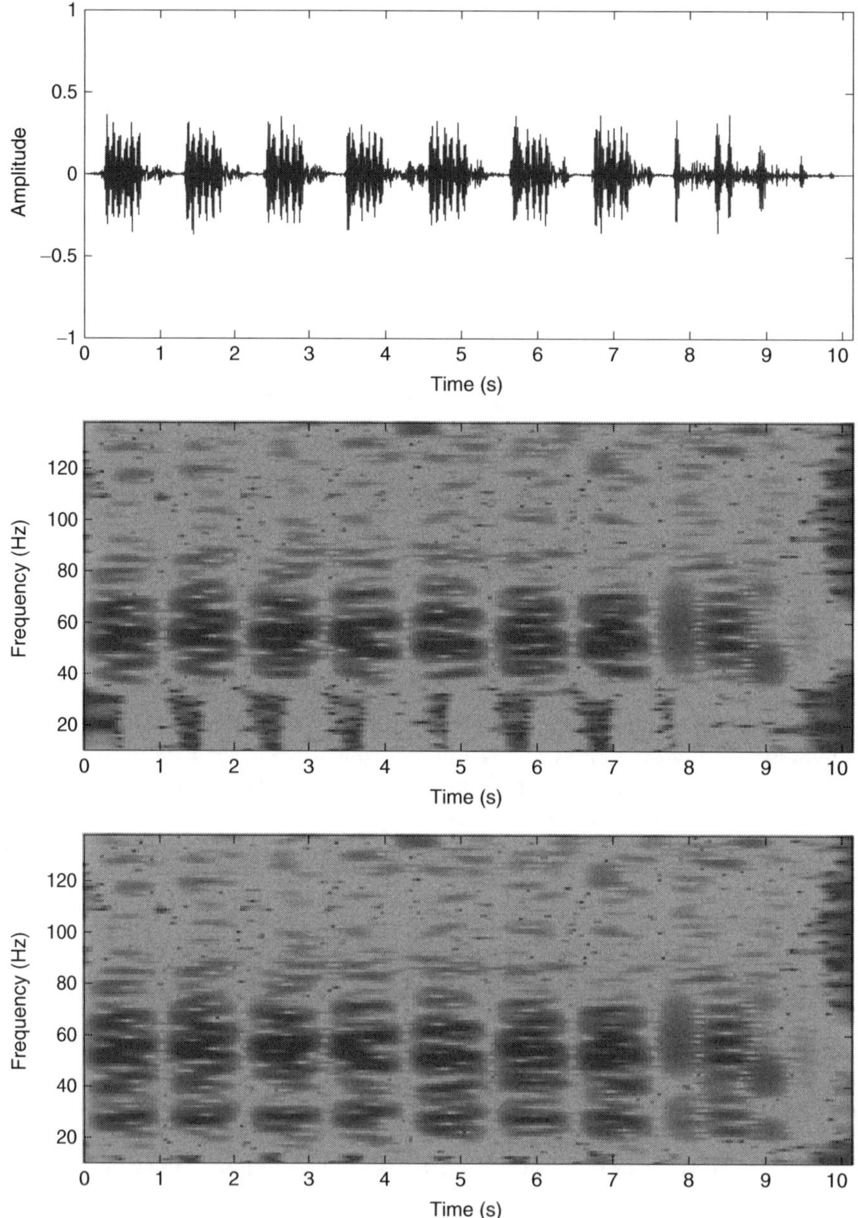

Figure 1.8 A 10-s excerpt of music (a), and its spectrogram for frequencies 0–140 Hz (b), which shows that the lowest frequencies contained in the signal are around 40 Hz. A low-frequency physical BWE algorithm can extend this low-frequency spectrum down to about 20 Hz, as shown in (c). Because the additional frequency components have a proper harmonic relation with the original frequency components, and have a common temporal modulation, they will be perceived as part of the original sound. In this case, the pitch of the bass notes will be lowered to 20 Hz

treatments of acoustics can be found in textbooks such as Kinsler *et al.* [142], Pierce [208], Beranek [28], Morse and Ingard [180]. Acoustics can be defined as the generation, transmission, and reception of energy in the form of vibrational waves in matter. As the atoms or molecules of a fluid or solid are displaced from their normal configurations, an internal elastic restoring force arises. Examples include the tensile force arising when a spring is stretched, the increase in pressure in a compressed fluid, and the transverse restoring force of a stretched wire that is displaced in a direction normal to its length. It is this elastic restoring force, together with the inertia of the system, that enables matter to exhibit oscillatory vibrations, and thereby generate and transmit acoustic waves. Those waves that produce the sensation of sound are of a variety of pressure disturbances that propagate through a compressible fluid.

1.3.1.1 The Wave Equation

The wave equation gives the relation between the spatial ($\mathbf{r}$) and temporal (t) derivates of pressure $p(\mathbf{r}, t)$ as

$$\nabla^2 p(\mathbf{r}, t) = \frac{1}{c^2}\frac{\partial^2 p(\mathbf{r}, t)}{\partial t^2} \tag{1.25}$$

where c is the speed of sound, which for air at 293 K is 343 m/s. Equation 1.25 is the linearized, loss-less wave equation for the propagation of sounds in linear inviscid fluids. As a special case of the wave equation, we can consider the one-dimensional case, where the acoustic variables are only a function of one spatial coordinate, say along the x direction. Equation 1.25 then reduces to

$$\frac{\partial^2 p(x, t)}{\partial x^2} = \frac{1}{c^2}\frac{\partial^2 p(x, t)}{\partial t^2}. \tag{1.26}$$

The solution of this equation yields two wave fields propagating in $\pm x$ directions, which are called plane (progressive) waves. Sound waves radiated by a loudspeaker are considered to be plane waves in the 'far field'.

1.3.1.2 Acoustic Impedance

The ratio of acoustic pressure in a medium to the associated particle velocity is called the specific acoustic impedance[4] $z(\mathbf{r})$

$$z(\mathbf{r}) = \frac{p(\mathbf{r})}{u(\mathbf{r})}. \tag{1.27}$$

For plane progressive waves (Eqn. 1.26), this becomes

$$z = \rho_0\, c, \tag{1.28}$$

independent of x, where ρ is the density of the fluid, being 1.21 kg/m^3 for air at 293 K.

[4] Similar to Ohm's law for electrical circuits.

Table 1.1 Typical sounds and their corresponding SPL values (dB)

Threshold of hearing	0
Whispering	20
Background noise at home	40
Normal talking	60
Noise pollution level	90
Pneumatic drill at 5 m	100
1 m from a loudspeaker at a disco	120
Threshold of pain	140

1.3.1.3 Decibel Scales

Because of the large range of acoustical quantities, it is customary to express values in a logarithmic way. For sound pressure, we define the sound pressure level (SPL) L_p in terms of decibel (dB), as

$$L_\mathrm{p} = 20\log(p/p_0), \tag{1.29}$$

where p_0 is a reference level (the log is base 10, as will be used throughout the book), for air $p_0 = 20$ μPa is used. This level is chosen such that it corresponds to the just-noticeable sound pressure level of a 2-kHz sinusoid for an 18-year-old person with normal hearing, see Fig. 1.18 and ISO 226-1987(E) [117]. Table 1.1 lists some typical sounds and their corresponding SPL values. It is convenient to memorize some dB values for the ratio's $\sqrt{2}/2$, 2, 10, and 30 as approximately 3, 6, 20, and 30 dB.

1.3.2 LOUDSPEAKERS

1.3.2.1 Electrodynamic Loudspeakers

Electroacoustic loudspeakers have been around for quite some time. While the first patent for a moving-coil loudspeaker was filed in 1877, by Cuttriss and Redding [55], shortly after Bell's [27] telephone invention, the real impetus to a commercial success was given by Rice and Kellog [223] through their famous paper, so that we can state that the classical electrodynamic loudspeaker as we know now (depicted in Fig. 1.10), is over 80 years old. All practical electroacoustical transducers are limited in their capabilities, owing to their size and excursion possibilities. Among those limitations, there is one in the frequency response, which will be the main topic in the following sections. To study these limitations, we will scrutinize the behaviour of transducers for various parameters. It will appear later that the 'force factor' (Bl) of a loudspeaker plays an important role. To have some qualitative impression regarding the band limitation, various curves are shown in Fig. 1.9. We clearly see that there is a band-pass behaviour of the acoustical power P_a (fifth curve in Fig. 1.9), and a high-pass response for the on-axis pressure p (third curve in Fig. 1.9).

First, we will discuss the efficiency of electrodynamic loudspeakers in general, which will be used in a discussion about a special driver with a very low Bl value in Sec. 4.3. This driver can be made very cost efficient, low weight, flat, and with high power

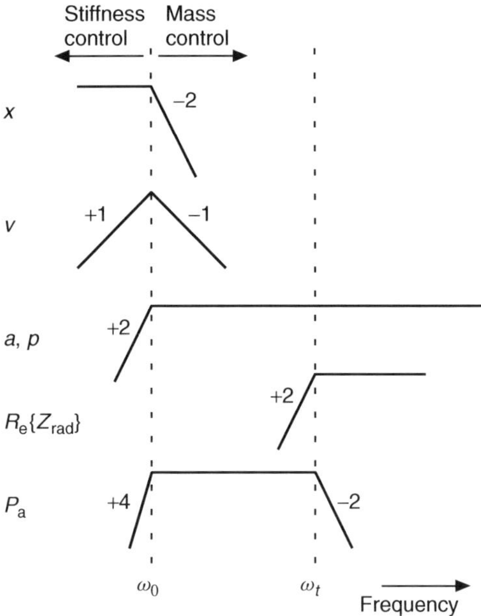

Figure 1.9 The displacement x, velocity v, acceleration a, together with the on-axis pressure p, the real part of the radiation impedance $\Re\{Z_{\mathrm{rad}}\}$, and the acoustical power P_{a} of a rigid-plane piston in an infinite baffle, driven by a constant force. The numbers denote the slopes of the curves; multiplied by 6, these yield the slope in dB/octave

efficiency. But first (Sec. 1.3.2.3), we show that sound reproduction at low frequencies with small transducers, and at a reasonable efficiency, is very difficult. The reasons for this are that the efficiency is inversely proportional to the moving mass and proportional to the square of the product of cone area and force factor Bl.

1.3.2.2 Construction

An electrodynamic loudspeaker, of the kind depicted in Fig. 1.10, consists of a conical diaphragm, the cone, usually made of paper, being suspended by an outer suspension device, or rim, and an inner suspension device, or spider. The suspension limits the maximum excursion of the cone so that the voice coil remains inside the air gap of the permanent magnet. This limitation can lead to non-linear distortion; see for example, Tannaka *et al.* [264], Olson [193], Klippel [143, 144], Kaizer [140]. The voice coil is attached to the voice coil cylinder, generally made of paper or metal, which is glued to the inner edge of the cone. In most cases, the spider is also attached to this edge. The voice coil is placed in the radial magnetic field of a permanent magnet and is fed with the signal current of the amplifier. For low frequencies, the driver can be modelled in a relatively simple way, as it behaves as a rigid piston. In the next section the electronic behavior of the driver will be described on the basis of a lumped-element model in which the mechanical and acoustical elements can be interpreted in terms of the well-known properties of

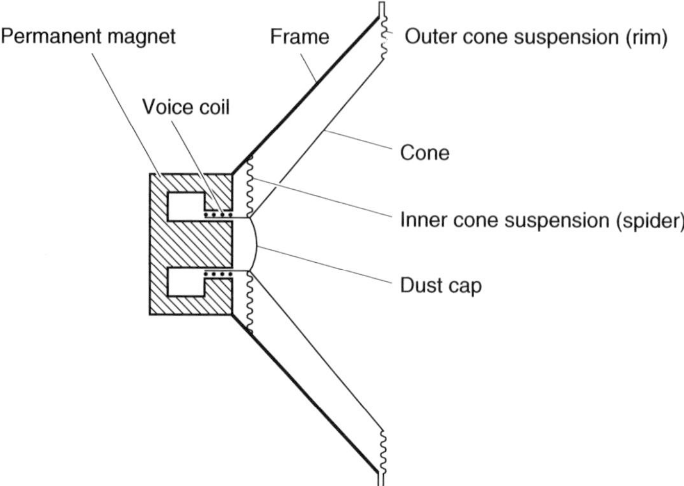

Figure 1.10 Cross section of an electrodynamic cone loudspeaker

their analogous electronic-network counterparts. At higher frequencies (above the cone break-up frequency), deviations from this model occur, as the driver's diaphragm is then no longer rigid. Both transverse and longitudinal waves then appear in the conical shell. These waves are coupled and together they determine the vibration pattern, which has a considerable effect on the sound radiation. Although this is an important issue, it will not be considered here (see e.g. Kaizer [140], Frankort [75], van der Pauw [282]).

An alternative construction method is to have the voice coil stationary, and a moving magnet; this will be discussed in Sec. 4.3.1.

1.3.2.3 Lumped-element Model

For low frequencies, a loudspeaker can be modelled with the aid of some simple elements, allowing the formulation of some approximate analytical expressions for the loudspeaker sound radiation due to an electrical input current, or voltage, which proves to be quite satisfactory for frequencies below the cone break-up frequency. The extreme accelerations experienced by a typical paper cone above about 2 kHz, cause it to flex in a complex pattern. The cone no longer acts as a rigid piston but rather as a collection of vibrating elements.

The forthcoming loudspeaker model will not be extensively derived here, as that has been done elsewhere; see for example, Olson [192], Beranek [28], Borwick [36], Merhaut [173], Thiele [268], Small [252], Clark [51]. We first reiterate briefly the theory for the sealed loudspeaker. In what follows, we use a driver model with a simple acoustic air load. Beranek [28] shows that for a baffled piston this air load is a mass of air equivalent to $0.85a$ in thickness on each side of a piston of radius a. In fact, the air load can exceed this value, since most drivers have a support basket, which obstructs the flow of air from the back of the cone, forcing it to move through smaller openings. This increases the acceleration of this air, augmenting the acoustic load.

Table 1.2 System parameters of the model of Fig. 1.11

R_e	Electrical resistance of the voice coil
L_e	Inductance of the voice coil
I	Voice coil current
U	Voltage induced in the voice coil
B	Flux density in the air gap
l	Effective length of the voice coil wire
Bl	Force factor
F	Lorentz force acting on the voice coil
V	Velocity of the voice coil
k_t	Total spring constant
m_t	Total moving mass, without air load mass
R_m	Mechanical damping
R_d	Electrical damping $= (Bl)^2/R_e$
R_t	Total damping $= R_r + R_m + R_d$
Z_{rad}	Mechanical radiation impedance $= R_r + jX_r$

The driver is characterized by a cone or piston of area

$$S = \pi a^2, \tag{1.30}$$

and various other parameters, which will be introduced successively, and are summarized in Table 1.2. The resonance frequency f_0 is given by

$$k_t = (2\pi f_0)^2 m, \tag{1.31}$$

where m is the total moving mass (which includes the air load), k_t is the total spring constant of the system, including the loudspeaker and possibly its enclosing cabinet of volume V_0. This cabinet exerts a restoring force on the piston with equivalent spring constant

$$k_B = \frac{\gamma P_0 S^2}{V_0} = \frac{\rho c^2 S^2}{V_0}, \tag{1.32}$$

where γ is the ratio of the specific heats (1.4 for air), P_0 is the atmospheric pressure, ρ, the density of the medium, and c, the speed of sound. The current $i(t)$ taken by the driver when driven with a voltage $v(t)$ will be given by equating that voltage to the voice coil resistance R_e and the induced voltage

$$v(t) = i(t)R_e + Bl\frac{dx}{dt} + L_e\frac{di}{dt} \tag{1.33}$$

where Bl is the force factor (which will be explained later on), x is the piston displacement, and L_e the self-inductance of the voice coil. The term in dx/dt is the voltage induced by the driver piston velocity of motion. Using the Laplace transform, Eqn. 1.33 can be written as

$$V(s) = I(s)R_{\mathrm{e}} + BlsX(s) + L_{\mathrm{e}}sI(s), \tag{1.34}$$

where capitals are used for the Laplace-transformed variables, and s is the Laplace variable, which can be replaced by $i\omega$ for stationary harmonic signals. The relation between the mechanical forces and the electrical driving force is given by

$$m\frac{\mathrm{d}^2x}{\mathrm{d}t^2} + R_{\mathrm{m}}\frac{\mathrm{d}x}{\mathrm{d}t} + k_{\mathrm{t}}x = Bli, \tag{1.35}$$

where at the left-hand side, we have the mechanical forces, which are the inertial reaction of the cone with mass m, the mechanical resistance R_{m}, and the total spring force with total spring constant k_{t}; at the right-hand side, we have the external electromagnetic Lorentz force $F = Bli$ acting on the voice coil, with B, the flux density in the air gap, i, the voice coil current, and l being the effective length of the voice coil wire. Combining Eqns. 1.34 and 1.35, we get

$$X(s)\left[s^2m + s(R_{\mathrm{m}} + \frac{(Bl)^2}{L_{\mathrm{e}}s + R_{\mathrm{e}}} + k_{\mathrm{t}}\right] = \frac{BlV(s)}{L_{\mathrm{e}}s + R_{\mathrm{e}}}. \tag{1.36}$$

We see that besides the mechanical damping R_{m}, we also get an electrical damping term $(Bl)^2/(L_{\mathrm{e}}s + R_{\mathrm{e}})$, and this term plays an important role. If we ignore the inductance of the loudspeaker, the effect of eddy currents[5] induced in the pole structure (Vanderkooy [283]), and the effect of creep[6], we can write Eqn. 1.36 as the transfer function $H_x(s)$ between voltage and excursion

$$H_x(s) = \frac{X(s)}{V(s)} = \frac{Bl/R_{\mathrm{e}}}{s^2m + s(R + (Bl)^2/R_{\mathrm{e}}) + k_{\mathrm{t}}}. \tag{1.37}$$

We use an infinite baffle to mount the piston, and in the compact-source regime $(a/r \ll c/(\omega a))$ the far-field acoustic pressure $p(t)$ a distance r away becomes

$$p(t) = \rho S(\mathrm{d}^2x/\mathrm{d}t^2)/(2\pi r), \tag{1.38}$$

proportional to the volume acceleration of the source (Morse and Ingard [180], Kinsler *et al.* [142]). In the Laplace domain, we have

$$P(s) = s^2\rho SX(s)/(2\pi r). \tag{1.39}$$

[5] Owing to the eddy current losses in the voice coil, the voice coil does not behave as an ideal coil, but it can be modelled very well by means of $L_{\mathrm{e}} = L_0(1 - j\alpha)$, where α is in the order of magnitude of 0.5.

[6] With a voltage or current step as the input, the displacement would be expected to reach its steady-state value in a fraction of a second, according to the traditional model. The displacement may, however, continue to increase. This phenomenon is called creep. Creep is due the viscoelastic effects (Knudsen and Jensen [145], Flügge [74]) of the spring (spider) and edge of the loudspeaker's suspension.

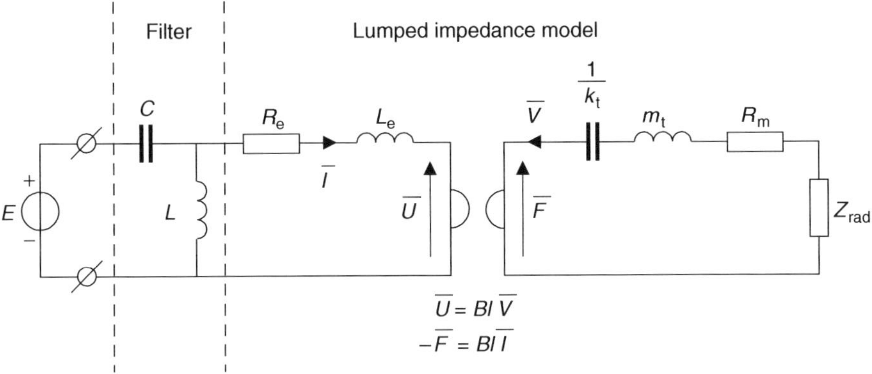

Figure 1.11 Lumped-element model of the impedance-type analogy of an electrodynamic loudspeaker preceded by an LC high-pass crossover filter, which is not part of the actual model. The coupling between the electrical and mechanical parts is represented by a gyrator. The system parameters are given in Table 1.2, from Aarts [4]

Using Eqn. 1.39 and neglecting the self-inductance L_e, we can write Eqn. 1.36 as the transfer function from excursion to pressure

$$H_p(s) = \frac{X(s)}{P(s)} = \frac{s^2 \rho S/(2\pi r) Bl/R_e}{s^2 m + s(R + (Bl)^2/R_e) + k_t}. \tag{1.40}$$

Using Eqns. 1.33 and 1.34, we can make the so-called lumped-element model as shown in Fig. 1.11, which behaves as a simple second-order mass-spring system. We have, for harmonic signals,

$$F = (R_m + i\omega m_t + \frac{k_t}{i\omega} + Z_{rad})V. \tag{1.41}$$

With the aid of Eqn. 1.41 and the properties of the gyrator as shown in Fig. 1.11, the electrical impedance of the loudspeaker (without X_r, which is the imaginary part of the mechanical radiation impedance[7]) can be calculated as follows

$$Z_{in} = R_e + i\omega L_e + \frac{(Bl)^2}{(R_m + R_r) + i\omega m_t + k_t/(i\omega)}. \tag{1.42}$$

Using the following relations

$$\begin{aligned} Q_m &= \sqrt{k_t m_t}/R_m, & Q_e &= R_e\sqrt{k_t m_t}/(Bl)^2, \\ Q_r &= \sqrt{k_t m_t}/R_r, & \omega_0 &= \sqrt{k_t/m_t}, \\ \nu &= \omega/\omega_0 - \omega_0/\omega, & \tau_e &= L_e/R_e, \\ Q_{mr} &= Q_m Q_r/(Q_m + Q_r), \end{aligned} \tag{1.43}$$

[7] For $\omega \ll \omega_t$ (defined in Eq. 1.49) X_r/ω can be taken into account in m_t.

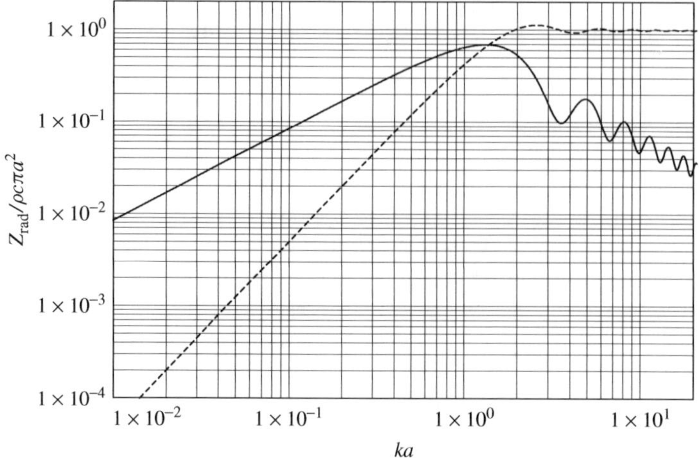

Figure 1.12 Real (dashed line) and imaginary (solid line) parts of the normalized radiation impedance of a rigid disk with a radius a in an infinite baffle

we can write Z_{in} as

$$Z_{\text{in}} = R_{\text{e}}\left[1 + i\omega\tau_{\text{e}} + \frac{Q_{\text{mr}}/Q_{\text{e}}}{1 + iQ_{\text{mr}}\nu}\right], \tag{1.44}$$

if we neglect L_{e}, we get at the resonance frequency ($\nu = 0$) the maximal input impedance

$$Z_{\text{in}}(\omega = \omega_0) = R_{\text{e}}(1 + Q_{\text{mr}}/Q_{\text{e}}) \approx R_{\text{e}} + (Bl)^2/R_{\text{m}}. \tag{1.45}$$

The time-averaged electrical power P_{e} delivered to the driver is then

$$P_{\text{e}} = 0.5|I|^2\Re\{Z_{\text{in}}\} = 0.5|I|^2 R_{\text{e}}\left[1 + \frac{Q_{\text{mr}}/Q_{\text{e}}}{1 + Q_{\text{mr}}^2\nu^2}\right]. \tag{1.46}$$

The radiation impedance of a plane-circular rigid piston[8] with a radius a in an infinite baffle can be derived as (Morse and Ingard [180, p. 384])

$$Z_{\text{rad}} = \pi a^2\rho c[1 - 2J_1(2ka)/(2ka) + i2\mathbf{H}_1(2ka)/(2ka)], \tag{1.47}$$

where $\mathbf{H}_1$ is a Struve function (Abramowitz and Stegun [12, 12.1.7]), J_1 is a Bessel function and k is the wave number ω/c. The real and imaginary parts of Z_{rad} are plotted in Fig. 1.12.

[8] The radiation impedance of rigid cones and that of rigid domes is studied in, for example, Suzuki and Tichy [259, 260]. They appeared to be significantly different with respect to rigid pistons for $ka > 1$, revealing that Z_{rad} for convex domes is generally lower than that for pistons and higher than that for concave domes.

Z_{rad} can be approximated as

$$Z_{\text{rad}} \approx \begin{cases} \pi a^2 \rho c[(ka)^2/2 + i8\,ka/(3\pi)], & \omega \ll \omega_{\text{t}} \\ \pi a^2 \rho c[1 + i2/(\pi ka)], & \omega \gg \omega_{\text{t}}, \end{cases} \tag{1.48}$$

where

$$\omega_{\text{t}} = 1.4\,c/a \tag{1.49}$$

is the transition frequency (−3 dB point). A full-range approximation of $\mathbf{H}_{\text{l}}$ is given in Sec. 1.3.3.2 and in Aarts and Janssen [9]. However, for low frequencies, we can either neglect the damping influence of Z_{rad}, or use

$$\Re\{Z_{\text{rad}}\} \approx a^4 f^2/4, \tag{1.50}$$

which follows immediately from Eqn. 1.48. The real part of Z_{rad} is qualitatively depicted in Fig. 1.9, but more precisely in Fig. 1.12. The time-averaged acoustically radiated power can then be calculated as follows

$$P_{\text{a}} = 0.5|V|^2 \Re\{Z_{\text{rad}}\}, \tag{1.51}$$

and with the aid of Eqns. 1.41–1.51 as follows

$$P_{\text{a}} = \frac{0.5(Bl/(R_{\text{m}} + R_{\text{r}}))^2 I^2 R_{\text{r}}}{1 + Q_{\text{mr}}^2 \nu^2}, \tag{1.52}$$

as depicted in Fig. 1.9, which clearly shows the bandwidth limitation similar to a bandpass filter. The acoustic pressure in the far field at distance r and azimuth angle θ is

$$p(r,t) = i\frac{f\rho_0 V \pi a^2}{r}\left[\frac{J_1(ka\sin\theta)}{ka\sin\theta}\right] e^{i\omega(t-r/c)}, \tag{1.53}$$

assuming an axis of symmetry at $\theta = 0$ rad, where V is the velocity of the piston (Beranek [28], Kinsler *et al.* [142]), and J_1 is a Bessel function, see Sec. 1.3.3.1. Assuming a velocity profile as depicted in Fig. 1.9, we can calculate the magnitude of the on-axis response ($\theta = 0$). This is also depicted in Fig. 1.9, which shows a 'flat' SPL for $\omega \gg \omega_0$. However, owing to the term in square brackets in Eq. 1.53, the off-axis pressure response ($\theta \neq 0$) decreases with increasing ka. This yields an upper frequency limit for the acoustic power, together with the mechanical lower frequency limit ω_0, and is the reason why a practical loudspeaker system needs more than one driver (a multi-way system) to handle the whole audible frequency range.

The power efficiency can be calculated as follows

$$\eta(\nu) = P_{\text{a}}/P_{\text{e}} = [Q_{\text{e}} Q_{\text{r}}(\nu^2 + 1/Q_{\text{mr}}^2) + Q_r/Q_{\text{mr}}]^{-1}. \tag{1.54}$$

In Fig. 4.13, some plots for $\eta(\nu)$ for various drivers are shown. For low frequencies[9], so that $Q_{\mathrm{mr}} \approx Q_{\mathrm{m}}$, the efficiency can be approximated as

$$\eta(\nu) = P_{\mathrm{a}}/P_{\mathrm{e}} \approx [Q_{\mathrm{r}}\{Q_{\mathrm{e}}(\nu^2 + 1/Q_{\mathrm{m}}^2) + 1/Q_{\mathrm{m}}\}]^{-1}. \tag{1.55}$$

A convenient way to relate the sound pressure level L_{p} to the power efficiency η is the following. For a plain wave, we have the relation between sound intensity I and sound pressure p

$$I = \frac{p^2}{\rho c}, \tag{1.56}$$

and the acoustical power is equal to

$$P_{\mathrm{a}} = 2\pi r^2 I \tag{1.57}$$

or

$$P_{\mathrm{a}} = \frac{2\pi r^2 p^2}{\rho c}. \tag{1.58}$$

Using the above relations, we get

$$L_{\mathrm{p}} = 20 \log \left(\sqrt{\frac{P_{\mathrm{a}} \rho c}{2\pi r^2}} / p_0 \right), \tag{1.59}$$

where we assume radiation into one hemifield (solid angle of 2π), that is, we only account for the pressure at one side of the cone, which is mounted in an infinite baffle. For $r = 1$ m, $\rho = 415$, $P_{\mathrm{a}} = 1$ W, and $p_0 = 20\ 10^{-6}$, we get

$$L_{\mathrm{p}} = 112 + \log \eta\,. \tag{1.60}$$

If $\eta = 1$ (in this case $P_{\mathrm{a}} = P_{\mathrm{e}} = 1$ W), we get the maximum attainable L_{p} of 112 dB. Equation 1.60 can also be used to calculate η if L_{p} is known, for example, by measurement.

1.3.3 BESSEL AND STRUVE FUNCTIONS

Bessel and Struve functions occur in many places in physics and quite prominently in acoustics for impedance calculations. The problem of the rigid-piston radiator mounted in an infinite baffle has been studied widely for tutorial as well as for practical reasons, see for example, Greenspan [98], Pierce [208], Kinsler *et al.* [142], Beranek [28], Morse and Ingard [180]. The resulting theory is commonly applied to model a loudspeaker in the audio-frequency range. For a baffled piston, the ratio of the force amplitude to the

[9] It should be noted that Q_{r} depends on ω, but using Eqns. 1.43 and 1.48 for $\omega \ll \omega_{\mathrm{t}}$, we can approximate $Q_{\mathrm{r}} \approx 2c\sqrt{k_{\mathrm{t}} m_{\mathrm{t}}}/(\pi a^4 \rho \omega^2)$.

normal velocity amplitude, which is called the piston mechanical radiation impedance, is given by

$$Z_{\mathrm{m}} = -i\frac{\omega\rho}{2\pi}\int\int\int\int R^{-1}\mathrm{e}^{ikR}\,\mathrm{d}x_{\mathrm{s}}\,\mathrm{d}y_{\mathrm{s}}\,\mathrm{d}x\,\mathrm{d}y. \tag{1.61}$$

Here $R = \sqrt{(x - x_{\mathrm{s}})^2 + (y - y_{\mathrm{s}})^2}$ is the distance between any two points, $(x_{\mathrm{s}}, y_{\mathrm{s}})$ and (x, y), on the surface of the piston. The integration limits are such that $(x_{\mathrm{s}}, y_{\mathrm{s}})$ and (x, y) are within the area of the piston. The four-fold integral in Eqn. 1.61, known as the Helmholtz integral, was solved by Rayleigh [219, 302] and further elaborated in Pierce [208], with the result

$$Z_{\mathrm{m}} = \rho c\pi a^2[R_1(2ka) - iX_1(2ka)], \tag{1.62}$$

where

$$R_1(2ka) = 1 - \frac{2J_1(2ka)}{2ka} \tag{1.63}$$

and

$$X_1(2ka) = \frac{2\mathbf{H}_1(2ka)}{2ka} \tag{1.64}$$

are the real and imaginary parts of the radiation impedance, respectively. In Eqns. 1.63 and 1.64, J_1 is the first-order Bessel function of the first kind (Abramowitz and Stegun [12, 9.1.21]), and $\mathbf{H}_1(z)$ is the Struve function of the first kind (Abramowitz and Stegun [12, 12.1.6]). Bessel functions are solutions to the homogeneous Bessel equation

$$z^2y'' + zy' + (z^2 - \nu^2)y = 0, \tag{1.65}$$

where a particular kind of solution J_n is discussed in Sec. 1.3.3.1. Struve functions are solutions to the inhomogeneous Bessel equation

$$z^2y'' + zy' + (z^2 - \nu^2)y = \frac{4(z/2)^{\nu+1}}{\sqrt{\pi}\,\Gamma(\nu + 1/2)}, \tag{1.66}$$

and are discussed in Sec. 1.3.3.2; Γ is the gamma function. Also, some useful formulas for Bessel and Struve functions are given and an effective and simple approximation of $\mathbf{H}_1(z)$, which is valid for all z (from Aarts and Janssen [9]).

1.3.3.1 Bessel Functions $J_n(z)$

The Bessel function of order n can be represented (Abramowitz and Stegun [12, 9.1.21]) by the integral

$$J_n(z) = \frac{1}{\pi}\int_0^{\pi}\cos(z\sin\theta - n\theta)\,\mathrm{d}\theta, \tag{1.67}$$

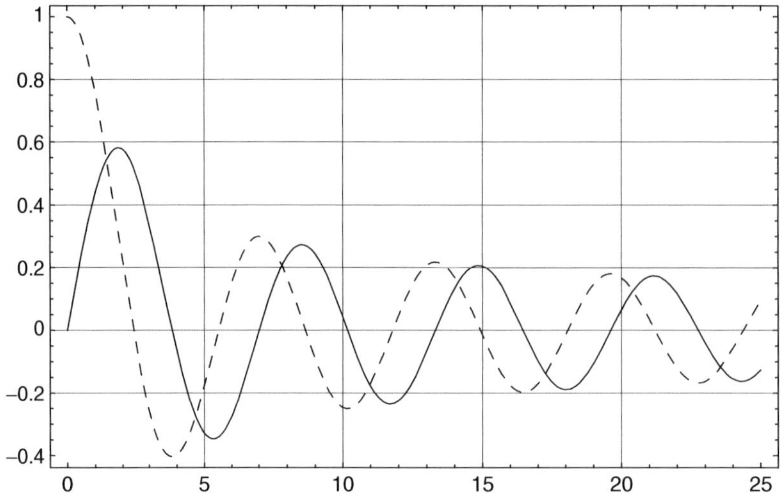

Figure 1.13 Plot of Bessel functions $J_0(x)$ (dashed), and $J_1(x)$ (solid)

and is plotted in Fig. 1.13 for $n = 0, 1$. There is the power series expansion (Abramowitz and Stegun [12, 9.1.10])

$$J_n(z) = \left(\frac{z}{2}\right)^n \sum_{k=0}^{\infty} \frac{\left(\frac{-z^2}{4}\right)^k}{k!\,\Gamma(n+k+1)}, \tag{1.68}$$

which yields

$$J_0(z) = 1 - \frac{\frac{1}{4}z^2}{(1!)^2} + \frac{\left(\frac{1}{4}z^2\right)^2}{(2!)^2} - \frac{\left(\frac{1}{4}z^2\right)^3}{(3!)^2} + \cdots, \tag{1.69}$$

and

$$J_1(z) = \frac{z}{2} - \frac{z^3}{16} + \frac{z^5}{384} - \frac{z^7}{18\,432} + \cdots. \tag{1.70}$$

For the purpose of numerical computation, these series are only useful for small values of z. For small values of z, Eqns. 1.63 and 1.70 yield

$$R_1(ka) \approx \frac{(ka)^2}{2}, \tag{1.71}$$

where we have substituted $ka = z$; this is in agreement with the small ka approximation as can be found in the references given earlier, see also Fig. 1.12. Furthermore, there is the asymptotic result (Abramowitz and Stegun [12, 9.2.1 with $\nu = 1$]), see Fig. 1.13,

$$J_1(z) = \sqrt{\frac{2}{\pi z}}\,(\cos(z - 3\pi/4) + O(1/z)), \quad z \to \infty, \tag{1.72}$$

but this is only useful for large values of z. Equation 1.63 and the first term of Equation 1.72 yield for large values of z (again substituting ka)

$$R_1(ka) \approx 1, \tag{1.73}$$

which is in agreement with the large ka approximation, as can be found in the given references as well. The function $J_n(x)$ is tabulated in many books (see e.g. Abramowitz and Stegun [12]), and many approximation formulas exist (see e.g. Abramowitz and Stegun [12, 9.4]). Another method to evaluate J_n is to use the following recurrent relation

$$J_{n-1}(x) + J_{n+1}(x) = \frac{2n}{x} J_n(x), \tag{1.74}$$

provided that $n < x$, otherwise severe accumulation of rounding errors will occur (Abramowitz and Stegun [12, 9.12]). However, $J_n(x)$ is always a decreasing function of n when $n > x$, so the recurrence can always be carried out in the direction of decreasing n. The iteration is started with an arbitrary value zero for J_n, and unity for J_{n-1}. We normalize the results by using the equation

$$J_0(x) + 2J_2(x) + 2J_4(x) + \cdots = 1. \tag{1.75}$$

A heuristic formula to determine the value of m to start the recurrence with $J_{\mathrm{m}} = 1$ and $J_{m-1} = 0$ is ($\lceil\cdot\rfloor$ indicates rounding to the nearest integer)

$$\begin{aligned} m &= \left\lceil \frac{6 + \max(n, p) + \frac{9p}{p+2}}{2} \right\rfloor \\ p &= \frac{3x}{2}. \end{aligned} \tag{1.76}$$

1.3.3.2 The Struve Function $\mathbf{H}_1(z)$

The first-order Struve function $\mathbf{H}_1(z)$ is defined as

$$\mathbf{H}_1(z) = \frac{2z}{\pi} \int_0^1 \sqrt{1 - t^2} \sin zt \, \mathrm{d}t \tag{1.77}$$

and is plotted in Fig. 1.14. There is the power series expansion (Abramowitz and Stegun [12, 12.1.5])

$$\mathbf{H}_1(z) = \frac{2}{\pi} \left[\frac{z^2}{1^2 3} - \frac{z^4}{1^2 3^2 5} + \frac{z^6}{1^2 3^2 5^2 7} - \cdots \right]. \tag{1.78}$$

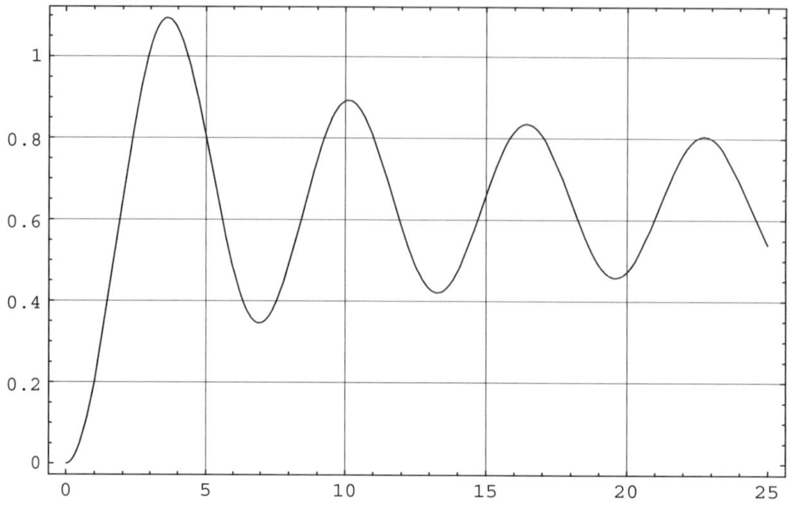

Figure 1.14 Plot of Struve function $\mathbf{H}_1(x)$

For the purpose of numerical computation, this series is only useful for small values of z. Eqns. 1.64 and 1.78 yield, for small values of z (substituting ka for z)

$$X_1(ka) \approx \frac{8\,ka}{3\pi}, \tag{1.79}$$

which is in agreement with the small ka approximation as can be found in the references given earlier, see also Fig. 1.12. Furthermore, there is the asymptotic result (Abramowitz and Stegun [12, 12.1.31, 9.2.2 with $\nu = 1$]).

$$\mathbf{H}_1(z) = \frac{2}{\pi} - \sqrt{\frac{2}{\pi z}}\,(\cos(z - \pi/4) + O(1/z))\,, \quad z \to \infty, \tag{1.80}$$

but this is only useful for large values of z. Eqn. 1.64 and the first term of Eqn. 1.80 yield for large values of ka

$$X_1(ka) \approx \frac{2}{\pi ka}, \tag{1.81}$$

which is in agreement with the large ka approximation, as can also be found in the earlier given references. An approximation for all values of ka was developed by Aarts and Janssen [9]. Here, only a limited number of elementary functions is involved:

$$\mathbf{H}_1(z) \approx \frac{2}{\pi} - J_0(z) + \left(\frac{16}{\pi} - 5\right)\frac{\sin z}{z} + \left(12 - \frac{36}{\pi}\right)\frac{1 - \cos z}{z^2}. \tag{1.82}$$

The approximation error is small and decently spread out over the whole z-range, vanishes for $z = 0$, and its maximum value is about 0.005. Replacing $\mathbf{H}_1(z)$ in Fig. 1.12 by the approximation in Eqn. 1.82 would result in no visible change. The maximum relative error appears to be less than 1%, equals 0.1% at $z = 0$, and decays to zero for $z \to \infty$.

1.3.3.3 Example

A prime example of the use of the radiation impedance is for the calculation of the radiated acoustic power of a circular piston in an infinite baffle. This is an accurate model for a loudspeaker with radius a mounted in a large cabinet (Beranek [28]). The radiated acoustic power is equal to

$$P_{\mathrm{a}} = 0.5|V|^2\Re\{Z_{\mathrm{m}}\}, \tag{1.83}$$

where V is the velocity of the loudspeaker's cone. The use of the just-obtained approximation for $\mathbf{H}_1$ is to calculate the loudspeaker's electrical input impedance Z_{in}, which is a function of Z_{m} (see Beranek [28]). Using Z_{in}, the time-averaged electrical power delivered to the loudspeaker is calculated as

$$P_{\mathrm{e}} = 0.5|I|^2\Re\{Z_{\mathrm{in}}\}, \tag{1.84}$$

where I is the current fed into the loudspeaker. Finally, the efficiency of a loudspeaker, defined as

$$\eta(ka) = P_{\mathrm{a}}/P_{\mathrm{e}}, \tag{1.85}$$

can be calculated. These techniques are used in Chapter 4 when analyzing the behavior of loudspeakers with special drivers.

1.4 AUDITORY PERCEPTION

This section reviews the basic concepts of the auditory system and auditory perception, insofar as they relate to BWE methods that will be discussed in later chapters. The treatment here is necessarily concise, but there are numerous references provided for further reading, if necessary or desired. Reviews of psychoacoustics can be found in, for example, Moore [177, 178], Yost *et al.* [302]; physiology of the peripheral hearing system is discussed in, for example, Geisler [86].

1.4.1 PHYSICAL CHARACTERISTICS OF THE PERIPHERAL HEARING SYSTEM

The peripheral hearing system consists of outer, middle, and inner ear, see Fig. 1.15. Sound first enters via the pinna, which has an irregular shape that filters impinging sound waves. This feature aids in sound localization, which is not further discussed here (see e.g. Batteau [26], Blauert [34]). Next, sound passes into the auditory canal and on to the eardrum, or tympanic membrane, which transmits vibrations in the air to the three middle ear bones, the ossicles (malleus, incus, and stapes). The stapes connects to the oval window, the entrance to the fluid-filled cochlea. The system from tympanic membrane to oval window serves as an impedance-matching device so that a large portion of sound energy at the frequencies of interest, in the air, is transmitted into the cochlea. Muscles connect the malleus and stapes to the bone of the skull, and contraction of these muscles

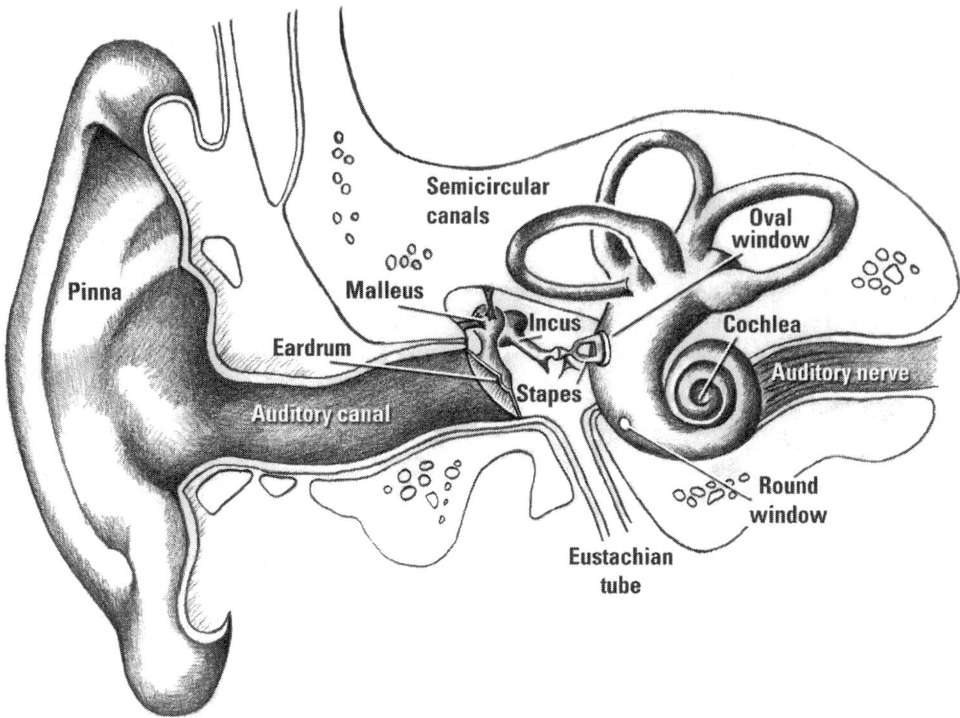

Figure 1.15 Sketch of the peripheral part of the hearing system, showing outer (pinna, auditory canal, eardrum), middle (malleus, incus, stapes), and inner ear (cochlea)

can be used to attenuate high-level sounds (primarily low frequency). Pressure equalization in the middle ear is achieved through the Eustachian tube, which connects the middle ear cavity to the throat.

The cochlea is a spiral-shaped cavity in the bone of the skull, filled with cerebro-spinal fluid; a cross section in shown in Fig. 1.16. The cochlea is wide at the oval window (the base) and narrows towards the other extreme (the apex). It is divided into three parts by the basilar membrane (BM) and Reisner's membrane; from top to bottom are the scala vestibuli, scala media, and scala tympani. The scala vestibuli and scala tympani are connected at the apex. Vibrations of the oval window are transmitted as a travelling wave through the cochlea, and also vibrate the BM. Locations on the BM have a very sharp frequency tuning because the variation in mechanical properties lead to different resonant frequencies at each location; the sharp frequency tuning is also achieved by active processes occurring within the cochlea. Between the BM and the tectorial membrane, in the organ of Corti, are rows of hair cells (outer hair cells and inner hair cells) with attached stereocilia. These oscillate along with the BM motion, which finally leads to a neural response from the inner hair cells. This response is propagated through the auditory nerve onto the cochlear nucleus and subsequent neural processing centres. There is also a descending pathway, from the brain to the outer hair cells, but this is not further discussed here.

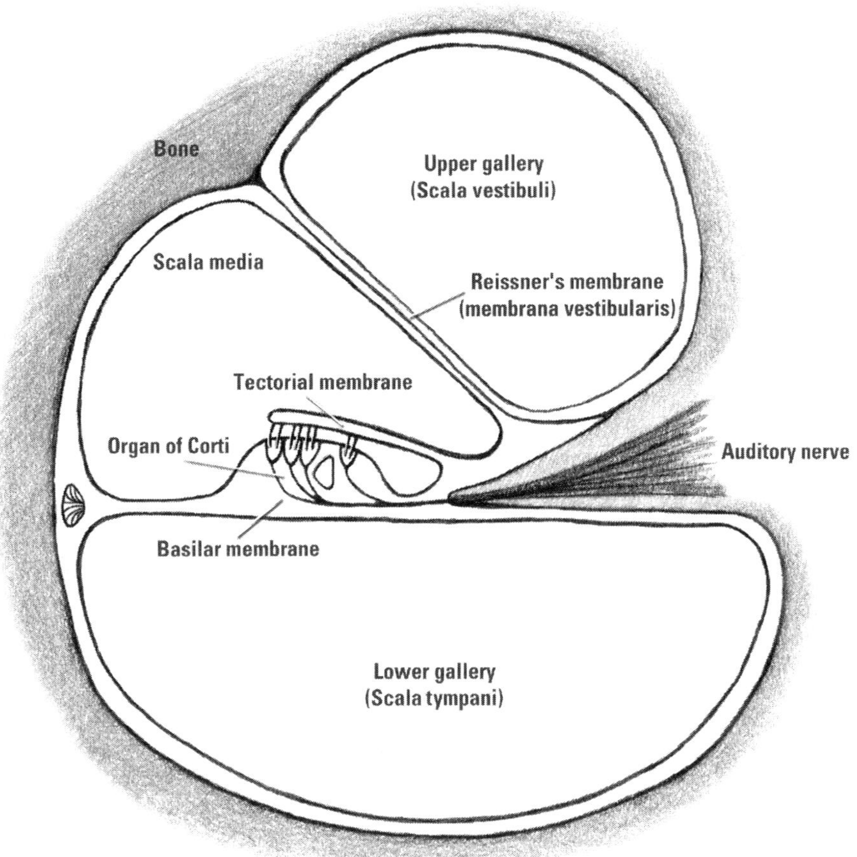

Figure 1.16 Cross section of the cochlea, showing the various membranes and cavities, and the organ of Corti. The hair cells are supported between the organ of Corti and the tectorial membrane, to which they are attached by stereocilia. The movement of the stereocilia, in response to sound entering the ear, causes a neural response, carried to the brain by the auditory nerve

The mechanical properties of the BM vary across the length of the cochlea: the BM is widest and has the lowest compliance at the apex, thereby causing each portion of the BM to respond maximally to a specific frequency (this is somewhat dependent on signal level). High-frequency sound vibrates the BM near the base, and low-frequency sound near the apex. These features were first elucidated by the investigations of the Nobel prize winner, Georg von Békésy [290]. The ordering of frequencies from low to high along the spatial extent of the BM is known as a tonotopic organization. The relation of the position y (distance in cm from the stapes, range approximately from 0–3.5 cm) of the maximum peak amplitude can be well approximated by

$$f = 2.5 \times 10^{4-0.72y}, \tag{1.86}$$

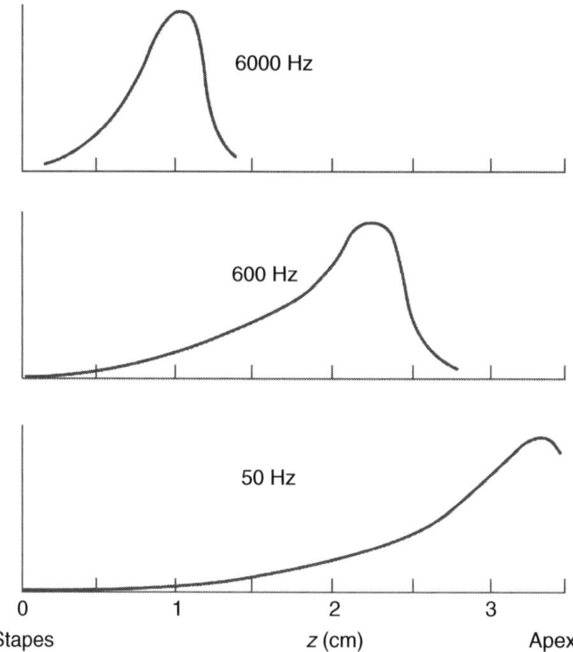

Figure 1.17 Peak displacement amplitude of the BM in response to pure tones of various frequencies

for frequencies f below 200 Hz. Below this frequency, a pattern is produced, which extends all along the basilar membrane, but reaches a maximum before the end of the membrane. Figure 1.17 illustrates for pure tones of various frequencies the peak displacement amplitude along the BM. The response of neurons along the BM generated by the BM motion is qualitatively similar; this neural response is called the excitation pattern.

1.4.2 NON-LINEARITY OF THE BASILAR MEMBRANE RESPONSE

An essential aspect of the cochlear response is that it is non-linear. Therefore, the shape of the graphs in Fig. 1.17 change somewhat, depending on the level of the stimulus. Also, the BM motion (BMM) is strongly compressed at moderate signal levels. This makes it possible for a normal ear to have a useable dynamic range of about 120 dB, while the variation in BMM is much smaller. It is thought that the outer hair cells (OHC) are largely responsible for this non-linear behavior.

An interesting consequence of this non-linearity is that if two tones are presented to the ear at sufficient sound pressure level, distortion products will be generated in the cochlea. These distortion products are also called combination tones (CT), see for example, Goldstein [91], Smoorenburg [253], Zwicker [307]. Assuming that the input frequencies are f_1 and f_2, CTs will appear at frequencies

$$f(n) = (n+1)f_1 - nf_2, \quad f_n > 0. \tag{1.87}$$

These combination tones are audible and, as we shall see in Chapter 2, may serve a useful purpose for certain kinds of BWE applications. The cubic CT ($n = 1$) is usually largest in amplitude, and is relatively independent of frequency, being about 15–20 dB below the level of each component of the two-tone complex (assumed identical). However, this is only true if $f_1/f_2 \approx 1.1$, that is, for closely spaced components. For a ratio $f_1/f_2 = 1.4$, the level of the cubic CT drops to about 40 dB below the level of the components in the two-tone stimulus.

Another distortion product is the difference tone (DT) $f_2 - f_1$ (Goldstein [91]). The DT level does not depend very much on the ratio f_1/f_2, but varies greatly with the level of the two-tone complex. The DT is barely audible at 50 dB and maintains a level of about 50 dB below the level of the components of the two-tone complex. If $f_1 = 2f_2$, the DT and the cubic CT will coincide.

1.4.3 FREQUENCY SELECTIVITY AND AUDITORY FILTERS

It was just shown how the BMM (and correspondingly, the excitation pattern) varies in position, depending on the frequency of a pure tone (Fig. 1.17). When considering a fixed position on the BM, we can use the excitation patterns of pure tones to determine the frequency response of the excitation pattern at that position. One then obtains a filter for each position on the BM; the resulting filters are called auditory filters. These describe how a position on the BM responds to pure tones of various frequencies. Important parameters of these filters are:

- *Characteristic frequency* (CF): This is the frequency (in Hz) of a pure tone, which yields the maximum response, and is also given (approximately) by Eqn. 1.86.
- *Equivalent rectangular bandwidth (*ERB*)*: This is the bandwidth (in Hz) of a rectangular filter having a passband amplitude equal to the maximum response of the auditory filter, that passes the same power of a white noise signal. A small ERB implies a narrow filter, and hence high frequency selectivity. Instead of ERB, sometimes a measure of tuning, Q, is used, to specify frequency selectivity. It is defined as $Q = \mathrm{CF}/\Delta f$, where Δf is some measure of bandwidth, usually the −3-dB bandwidth. Large Q values imply high frequency selectivity.

A relation describing ERB as a function of frequency f is given by (Glasberg and Moore [89])

$$\mathrm{ERB}(f) = 24.7(4.37 \times 10^{-3} f + 1). \tag{1.88}$$

Different experimenters have sometimes found different values for the bandwidth of auditory filters, and there is thus no universally agreed-upon value; the equation given here for auditory filter bandwidth is a widely used version. It is noted, however, that the original measurement of auditory filter bandwidth by Fletcher [72] yielded smaller bandwidths (Fletcher used the concept of 'critical band' analogously to the ERB as just described); recent experiments (Shera *et al.* [247]; Oxenham and Shera [195]) using novel techniques for measuring auditory filter bandwidth seem to agree better with Fletcher's original results than with Eqn. 1.88.

A commonly used model for auditory filter shape is by means of the gammatone filter (Patterson *et al.* [204], Hohman [110]) $g_t(t)$

$$g_t(t) = at^{n-1}e^{-2\pi b \cdot \mathrm{ERB}(f_c)t} \cos(2\pi f_c t + \phi). \tag{1.89}$$

Here a, b, n, f_c, and ϕ are parameters and $\mathrm{ERB}(f_c)$ is as given in Eqn. 1.88. The gammatone filters are often used in auditory models (e.g. AIM; see Sec. 1.4.8) to simulate the spectral analysis performed by the BM.

The frequency selectivity of the auditory filters is thought to have a large influence on many auditory tasks, such as understanding speech in noise or detecting small timbre differences between sounds. For BWE applications, another interesting aspect of auditory filters is their presumed influence on pitch perception, discussed in Sec. 1.4.5. For the moment, we mention that depending on the ERB of an auditory filter, it may pass one or more harmonics of a complex tone, also depending on the fundamental frequency of that tone. It turns out that roughly up to harmonic number 10, auditory filters pass only one harmonic, that is, these harmonics are *spectrally resolved*. At higher harmonic numbers, the ERB of the auditory filters become wider than the harmonic frequency spacing, and therefore the auditory filters pass two or more harmonics. These harmonics are thus *spectrally unresolved*, and the output of the auditory filter is a superposition of a number of these higher harmonics. Whether a harmonic is resolved or not will have a large influence on the subsequently generated neural response. In the following, we will alternately use the terms harmonic and partial, both referring to one component of a harmonically related complex tone.

1.4.4 LOUDNESS AND MASKING

1.4.4.1 Definitions

Loudness is related to the level, or amplitude, of a sound, but depends in a complicated manner on level and also frequency content. The following definitions are used by the ISO [117]

Definition 4 *Loudness: That attribute of auditory sensation in terms of which sounds may be ordered on a scale extending from soft to loud. Loudness is expressed in sone, where one sone is the loudness of a sound, whose loudness level is 40 phon.*

Definition 5 *Loudness level: Of a given sound, the sound pressure level of a reference sound, consisting of a sinusoidal plane progressive wave of frequency 1 kHz coming from directly in front of the listener, which is judged by otologically normal persons to be equally loud to the given sound. Loudness level is expressed in phon.*

Definition 6 *Critical bandwidth: The widest frequency band within which the loudness of a band of continuously distributed random noise of constant band sound pressure level is independent of its bandwidth.*

Note that the critical *bandwidth* so defined is intimately related to the ERB of the auditory filters (Eqn. 1.88) and also Fletcher's critical *band*.

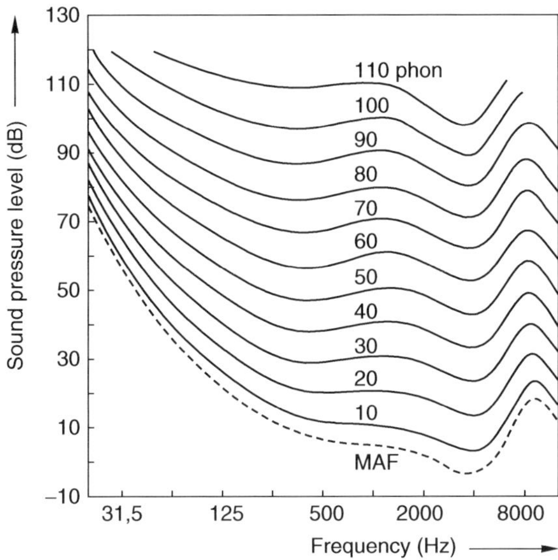

Figure 1.18 Normal equal-loudness level contours for pure tones (binaural free-field listening, frontal incidence. Data from ISO 226-1987(E) [117, Fig. 1]. MAF indicates minimum audible field, the just-noticeable sound pressure level

The threshold of audibility is the minimum perceptible free-field listening intensity level of a tone that can be detected at each frequency over the entire range of the ear. The average threshold of audibility for the normal ear is shown as the curve labeled MAF (minimum audible field) in Fig. 1.18; the other curves show the equal-loudness contours at various phon levels. The frequency of maximum sensitivity is near 4 kHz. Below this frequency, the threshold rises to about 70 dB. For high frequencies, the threshold rises rapidly, which also strongly depends on age, as is shown in Fig. 1.19. This has some implication for high-frequency BWE methods, as will be discussed in Chapters 6 and 5. Elderly listeners might, on average, not benefit as much (or at all) from processing strategies that increase high-frequency content of audio signals. The curves are also level dependent, especially at low frequencies, where they are compressed; this is illustrated in Fig. 1.20, which shows the normalized difference between the 80-phon contour and the 20-, 40-, 60-, and 100-phon contours. The compression of the equal-loudness contours at low frequencies implies that small changes in SPL lead to large changes in loudness.

1.4.4.2 Scaling of Loudness

Several experimenters have made contributions to the scaling of loudness. The earliest published work seems to be that credited to Richardson and Ross [224], who required an observer to rate one of two tones of different intensities, which he heard as a certain multiple or fraction of the other. Since then, various methods of evaluating loudness of complex sounds from objective spectrum analysis have been proposed. The earliest attempt to use algebraic models in psychophysical measurement is probably that of Fletcher and

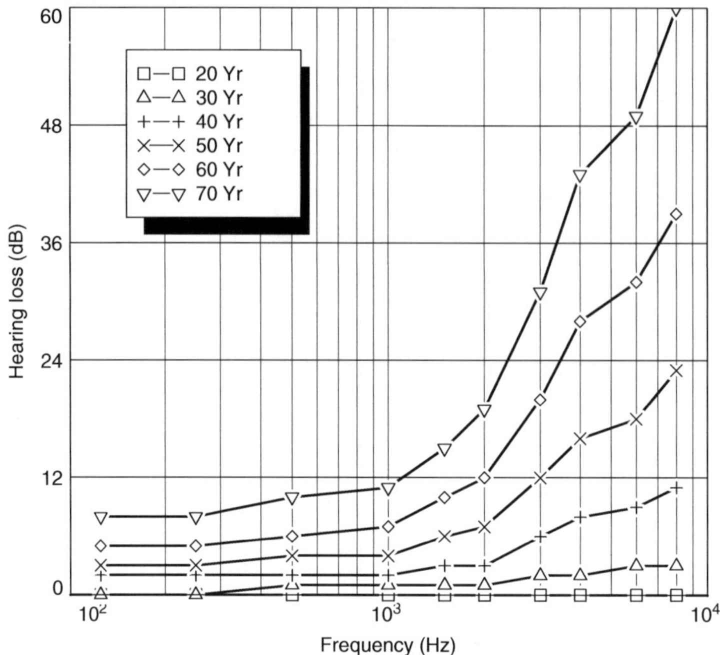

Figure 1.19 Hearing loss (with respect to threshold of hearing) for a group of normal males of various ages. For each age, the 50th percentile is shown. Data from ISO 7029-1984(E) [118]

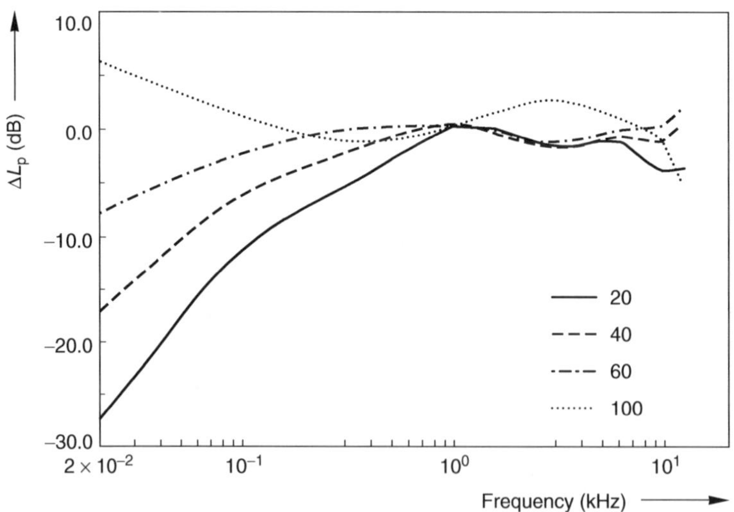

Figure 1.20 Differences between the 80-phon contour and the 20-, 40-, 60-, and 100-phon contours, respectively. The differences have been normalized to 0 dB at 1 kHz. From Aarts [4]

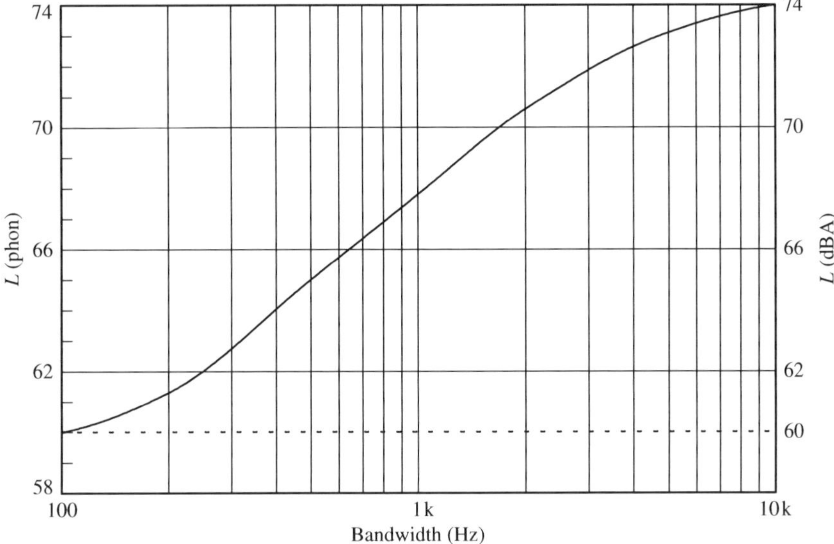

Figure 1.21 Increasing loudness (solid curve) versus bandwidth of white noise, while the dB(A) Level (dashed line) was kept constant. From Aarts [4]

Munson [73]. However, there is still much interest in this subject, and nowadays there are two standardized procedures for calculating loudness levels. The first is based on a method developed by Stevens [256], hereafter referred to as method 532A and the second one, 532B, by Zwicker [305, 310]. A-weighting is a widely used method, traditionally applied in sound level meters to measure the loudness of signals or to determine the annoyance of noise. It is based on an early 40-phon contour and is a rough approximation of a frequency weighting of the human auditory system. However, considerable differences are ascertained between subjective loudness ratings and the A-weighted measurements. For example, the loudness of noise increases when the dB(A) level is kept constant and the bandwidth of the noise is increased. As depicted in Fig. 1.21, with increasing bandwidth the loudness has increased from 60 to 74 phon (solid curve), while the dB(A) level (dashed line) was kept constant. The effect has been studied by Brittain [39], and is a striking example that for wideband signals the A-weighted method is generally too simple. As another example, consider the loudness of a tone of 200 Hz, 2000 Hz, or both combined, as in Fig. 1.22 (a, b, and c, respectively). Each part shows four dB values: the top value is that computed by a loudness model (ISO532B, to be discussed hereafter), the second and third by A- and B-weighting respectively, and the last value is the acoustic SPL. B-weighting is obtained by approximating the inverse of the 80-phon contour. Note that the A-weighting underestimates the perceived loudness for the 200-Hz tone and also for the combination. The B-weighting works better for the 200-Hz tone. Neither weighting procedure works for the combination, though.

The sone scale The loudness level is expressed in phon. However, loudness values expressed on this scale do not immediately suggest the actual magnitude of the sensation.

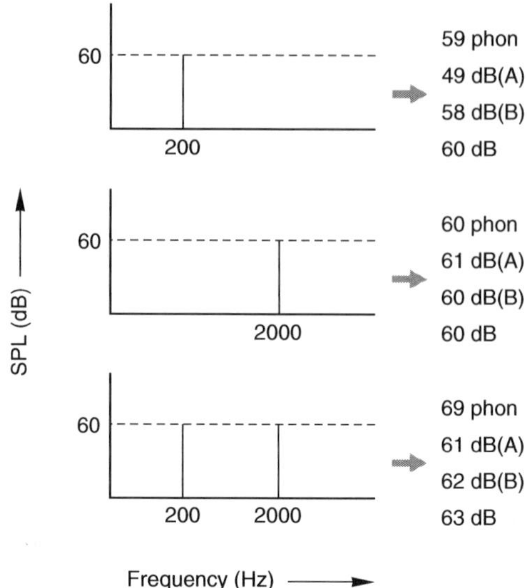

Figure 1.22 Levels of tones of 200 Hz (a), 2000 Hz (b), and the two tones simultaneously (c). The phon value is computed with a loudness model; dB(A) and dB(B) represent A- and B-weighted levels respectively. The lowest value in each part gives the acoustic SPL. From Aarts [4]

Therefore the sone scale, which is the numerical assignment of the strength of a sound, has been established. It has been obtained through subjective magnitude estimation using listeners with normal hearing. As a result of numerous experiments (Scharf [231]), the following expression has evolved to calculate the loudness S of a 1-kHz tone in sone:

$$S = 0.01 \times (p - p_0)^{0.6} \tag{1.90}$$

where $p_0 = 45$ μPa approximates the effective threshold of audibility and p is the sound pressure in μPa. For values $p \gg p_0$, Eqn. 1.90 can be approximated by the well-known expression

$$S = 2^{(P-40)/10} \tag{1.91}$$

or

$$P = 40 + 10 \log_2 S \tag{1.92}$$

where P is the loudness in phon.

ISO532A and 532B The ISO532A method is equal to the Mark VI version as described in Stevens [256]. However, Stevens refined the method, resulting in the Mark VII version

[257], which is not standardized. Here, the 532A method is discussed briefly. The SPL of each one-third octave band is converted into a loudness index using a table based on subjective measurements. The total loudness in sone S is then calculated by means of the equation

$$S = S_{\mathrm{m}} + F\left(\sum S_i - S_{\mathrm{m}}\right) \tag{1.93}$$

where S_{m} is the greatest of the loudness indices and $\sum S_i$ is the sum of the loudness indices of all the bands. For one-third octave bands the value of F is 0.15, for one-half octave bands it is 0.2, and for octave bands it is 0.3.

An early version of the ISO532B method is described in Zwicker [305] and later it has been refined, see for example, Zwicker and Feldtkeller [311], Paulus and Zwicker [205], and Zwicker [308]. The essential steps of the procedure are as follows. The sound spectrum measured in one-third octave bands is converted to bands with bandwidth roughly equal to critical bands. Each critical band is subdivided into bands of 0.1 Bark (which is an alternate measure of auditory filter bandwidth) wide. The SPL in each critical band is converted, by means of a table, into a loudness index for each of its sub-bands. In order to incorporate masking effects, contributions are also made to higher bands. The total loudness is finally calculated by integrating the loudness indices over all the sub-bands, resulting in the loudness in sone. The total loudness may be converted into loudness level in phon using Eqn. 1.92 (of course this can also be done for loudness as computed using method 532A).

Zwicker's method is elegant because of its compatibility with the accepted models of the human ear, whereas Stevens' method is based on a heuristic approach. Zwicker's procedure tends to give values systematically larger than Stevens'.

Time-varying loudness model The main drawback of both Stevens' and Zwicker's loudness models is that they are, in principle, only valid for stationary signals. This would seriously limit their applicability, but fortunately both models seem to correlate quite well with subjective judgements, even for realistic time-varying signals, see Sec. 1.4.4.4. Nonetheless, Glasberg and Moore [90] devised a method to predict loudness for time-varying sounds, building on the earlier models. The time-varying model was designed to predict known subjective data for stationary sounds, amplitude-modulated sounds, and short-duration (<100 ms) sounds. Broadly speaking, Glasberg and Moore's model is similar to Zwicker's, but loudness is temporally integrated to account for the time-varying nature of the signals. Specifically, the momentary excitation pattern generated by the sound at a specific time is used to compute the excitation pattern, and from this the 'instantaneous' loudness. This quantity is not consciously observable, but might correspond to total activity in the auditory nerve, for example. The instantaneous loudness is then 'smoothed' to obtain the short-term loudness, with a relatively fast attack and slower decay time. The short-term loudness is observable, for example, as would be perceivable for a 10-Hz amplitude-modulated signal. The short-term loudness is smoothed again, with larger time constants, to obtain the 'long-term' loudness. The long-term loudness corresponds to the overall loudness percept of the signal.

The model seems appropriate for use with audio signals, which are always time varying. The main limitation the authors mention is the fact that the relative phases of harmonics are not taken into account; the crest factor (peak-to-rms ratio) of waveforms on the BM can differ substantially for complex tones with identical power spectra but different phases, which may lead to loudness differences. This might be of some importance for BWE applications, where in some cases the harmonic structure of signals is modified. However, in practice, relative phases of harmonics are unpredictable, because of the randomizing effect of room reflections (unless the distance to the speaker is very small such that reflections are small compared with direct sound, or with headphone presentation).

1.4.4.3 Sensitivity to Changes in Intensity

Sensitivity to changes in intensity can be measured in several different ways, which mostly give similar trends. Results are usually expressed as the smallest detectable increment, or just-noticeable difference (JND), ΔI of a sound with intensity I, expressed as $\Delta L = 10\log([I + \Delta I]/I)$. For wideband noise, $\Delta L \approx 0.5$–$1.0\,$dB over most of the dynamic range of the auditory system. Thus, $\Delta I/I \approx 0.13$–0.25; this ratio is called the *Weber fraction*. A constant Weber fraction implies that sensitivity to changes in the stimulus is proportional to the magnitude of the stimulus; this property is known as Weber's law. Weber's law does not hold for pure tones, where it has been found that sensitivity increases with increasing intensity. Much work has been done to explain how intensity variations are coded by the auditory system, and how to explain the intensity versus loudness curves for various signals, but no definite theory exists as yet; Moore [178] presents a review. Allen and Neely [19] present a model that does account for the intensity JND of pure tones and wideband noise, on the basis of the assumption that the intensity JND is related to the variance of an internal loudness variable.

1.4.4.4 Loudness Issues for Listening Tests

Although the BWE algorithms to be discussed later can be analyzed in objective ways, the ultimate quality test is of course through subjective experiments. For this, not only is the quality of the algorithms important but, perhaps, also the quality of the loudspeaker. The perceived sound quality of a loudspeaker and its relation to its various physical properties have been a subject of discussion and research for a long time, see for example, Toole [272, 273, 274, 275], Gabrielsson and Lindström [82], Tannaka and Koshikawa [263]. In this regard, it is important that reproduction levels are chosen appropriately for the various signals tested, in particular, if different loudspeakers are used in the same test. Although normally one would prefer to use the same experimental hardware throughout one listening test, there are situations where this is not desirable. For example, to evaluate certain low-frequency enhancement algorithms (discussed in Chapter 2), one might want to subjectively compare a processed signal reproduced on a flat-panel loudspeaker with an unprocessed signal on a high-quality electrodynamic loudspeaker. Loudness matching across loudspeakers is especially important as it is well known that a higher reproduction level, or loudness level, of a loudspeaker can lead to a higher appreciation score than that of another one of the same quality, or even the same loudspeaker. The importance of equal-loudness levels of the sounds being compared is shown by a striking investigation of

Illényi and Korpássy [116]. They found that the rank order of the loudspeakers, according to the subjective quality judgements, was in good agreement with the rank order obtained by the corresponding calculated loudness.

In Aarts [4], it was found that the ISO 532B method was the most suitable of the two ISO methods to adjust interloudness levels of loudspeakers, while the simple B-weighting gave the most satisfactory results of all the tested methods (both ISO and A–D weightings). The widely used A-weighting gave poor results, though (see related comments in Sec. 1.4.4.2). It was also found that loudness levels were hardly influenced by the choice of the repertoire, more specifically that a varied repertoire, on average, sounds equally loud, as was computed for pink noise. This considerably facilitates the computation of appropriate loudness levels if multiple loudspeakers are used.

1.4.4.5 Masking

Masking is defined by the American Standards Association as [20]

Definition 7 *The process, and amount, by which the threshold of audibility for one sound is raised by the presence of another (masking) sound. The unit customarily used is the decibel.*

Masking and frequency selectivity are intimately related; it has been known for a long time that a sound is masked most easily by another sound that has similar frequency components (Wegel and Lane [295]). In fact, Fletcher [72] assumed, in his studies of the critical bandwidth, that masking is only possible if the masker and masked signal (maskee) fall within the same critical band, even though it was known that masking is possible at greater frequency separations.

For BWE applications, masking may be of interest to consider the audibility of distortion components, which are generated by some of the algorithms, which is a form of tone-on-tone masking. This kind of masking is known as energetic, simultaneous masking. Energetic masking refers to the fact that the detection threshold is determined by the power spectra of masker and maskee (power spectrum model of masking); masking that cannot be explained by the power spectrum model is informational masking. It is thought that this involves higher-level (attentional) processes. Simultaneous masking refers to the fact that masker and maskee occur at the same time; masking is also possible if the masker precedes the maskee (forward masking), or if the maskee precedes the masker (backward masking). Both informational and non-simultaneous masking do not seem very relevant for BWE applications.

Masking effects have not generally been studied in relation to BWE methods; it might have some use in connection with audibility of distortion components that some of the algorithms generate. The energy of these unwanted components can be analyzed and compared with respect to the energy of desired frequency components, and quantified as a kind of 'signal-to-noise' ratio. This is then used to assess the performance of various algorithms; see for example, Sec. 2.3.2.1, and following sections, for such analysis. However, this 'signal-to-noise' ratio is a purely physical description that does not factor in any perceptual effects, such as masking. This implies that conclusions thus reached are to be considered with some caution. A better modelling of BWE performance could be achieved if masking effects were also considered.

1.4.5 PITCH

1.4.5.1 Factors Influencing Pitch

According to the American Standards Association [20] pitch is defined as follows

Definition 8 *Pitch: that attribute of an auditory sensation in terms of which sounds may be ordered on a scale extending from low to high.*

According to Moore [177], there are various ways how the pitch of a pure tone depends on its frequency. One can obtain a pitch–frequency relation by various methods, the classical result being the mel scale. It has an arbitrary pitch reference of 1000 mel at a frequency of 1000 Hz. A tone that sounds, on average, twice as high receives a value of 2000 mel, whereas a tone that sounds only half as high has a pitch of 500 mel. Although the mel scale suggests that the pitch of a pure tone is simply determined by its frequency, the perceived pitch also depends on some other factors, one of them being the intensity. If one measures for a group of subjects how, on average, the pitch of a pure tone changes with the tone's intensity, one typically finds that (1) for tones below 1000 Hz the pitch decreases with increasing intensity (about 15%), (2) for tones between 1000 and 2000 Hz the pitch remains rather constant, and (3) for tones above 2000 Hz the pitch rises with increasing intensity (about 20%). This effect varies considerably between listeners and also depends on the duration of the tone. Hartmann [105] found that the pitch of short-duration tones ($\approx$100 ms) with decaying envelopes is higher than the frequency of the tone. The upward pitch shift seems to increase with decreasing frequency, being 2.6% at a frequency of 412 Hz (the lowest frequency used by Hartmann). The shift at even lower frequencies could be considerably higher, although there is no data to support this hypothesis.

For a complex tone, consisting of more than one frequency component, the situation is more complicated. Pitch should then be measured by psychophysical experiments. A pitch that is produced by a set of frequency components, rather than by a single sinusoid, is called a residue. Even if in a harmonic complex the fundamental frequency is missing, it will still be perceived as a residue pitch, which in this case is sometimes called virtual pitch, because the frequency corresponding to the pitch is absent. There is a vast literature on pitch perception and residue pitch. Some of the earlier systematic investigations are described in Bilsen and Ritsma [33], de Boer [58], Houtsma and Goldstein [113], and Schouten [239, 240]). The fact that low-order harmonics need not be physically present to evoke a pitch percept at the fundamental[10] is an attractive option to enhance low-pitched sounds reproduced by (small) loudspeakers; in Chapter 2, we shall see how this can be exploited. One factor that remains unclear is the strength of the residue pitch at very low frequencies ($<$100 or 200 Hz); most investigators have looked at higher frequencies. Ritsma [225, 226] has studied the existence region above 200 Hz.

Repetition pitch Some noise-like sounds do evoke pitch sensations. An example was described by Huygens [115] in the seventeenth century. He noticed that the noise of a water fountain, reflected by marble stairs, produced a distinct musical pitch equal to that of an organ pipe whose length matched the depth of the stairs. He essentially discovered

[10] In fact, the residue pitch can be heard even if there is a masking noise present in the frequency region of the fundamental such that it would normally be masked (Licklider [161]).

that when one or more consistently delayed images of a sound interfere with the original sound, one hears a pitch that corresponds to the inverse of the delay. The original sound can be noise, music, speech, or just about any other sound. Because the frequency response characteristic of such a time-delay system has a periodic comb-like structure, this process is often referred to as comb filtering (Boff *et al.* [232], Bilsen [32]).

Special cases However, the clearest pitch sensations are evoked by sounds that are periodic or, equivalently, sounds that have line spectra of harmonically related frequencies. Most string and wind instruments produce near-periodic sounds and are therefore very efficient in conveying pitch information. Other instruments, such as bells or chimes produce line spectra with inharmonically related frequencies that evoke the ambiguous sensations, characteristic of these instruments. Still other instruments, such as the snare drum or cymbals, produce sounds with continuous spectra that evoke a sensation of noise without any pitch. Accordingly, these are instruments used for rhythmic rather than for melodic or harmonic purposes. Other pitch phenomena are edge pitch (Small and Daniloff [251], Kohlrausch and Houtsma [146]) – which refers to a weak pitch sensation evoked by low-pass or high-pass filtering noise with a sufficiently sharp spectral edge; adaptation pitch (Zwicker [306]) is heard when one is exposed to wideband noise with a spectral notch of about half an octave. A weak tonal afterimage is heard when the tone is abruptly switched off.

1.4.5.2 Sensitivity to Changes in Frequency

Sensitivity to changes in stimulus frequency is remarkably high for pure tones of moderate frequency. Around 500–1000 Hz, the *difference limen for frequency* (DLF) is about 0.2–0.3% (Sek and Moore [244]); this means that two pure tones differing in frequency by the DLF will be correctly discriminated 75% of the time (or some other threshold). At low and high frequencies the DLF increases, and above 4–5 kHz the DLF exceeds 1%. Also, for short-duration tones (<100 ms) the DLF increases, being roughly 5 times larger than the values quoted previously for tones of 6.25 ms duration (Moore [176]).

Sensitivity to a modulation in frequency, the frequency modulation difference limen (FMDL), is less frequency dependent. The FMDL is about 0.5–1% over most of the audible frequency range. The FMDL seems the more suitable value to use when considering if frequency differences will be perceived in continuous musical or speech sounds.

One can also study the detectability of deviations from a perfect harmonic relationship between partials in a complex tone, see for example, Moore *et al.* [179]. In Le Goff *et al.* [158], experiments were performed to study the effect of mistuning only the fundamental component in a harmonic complex tone. Subjects had to distinguish the complex tone with its lowest harmonic at the fundamental frequency from a complex with the lowest harmonic shifted in frequency. Thresholds were determined for fundamental frequencies of 60 and 100 Hz. Complexes had either a flat spectrum, or components were generated with a spectral slope of −5 or −10 dB/octave. Conditions with the second or both the second and third harmonic omitted from the complex were also included. In additional conditions, the complex was slowly amplitude modulated and/or presented with a simultaneous distracting sound. The results showed a range of detectability from better than 0.5 to 7%, depending on the various conditions. Presenting the sound with a spectral slope strongly lowers thresholds. Adding a distractor or applying amplitude modulation, both lead to higher thresholds. The aim of this study was to investigate the perceptual consequences of a special frequency mapping technique, that is used to exploit the high efficiency of a loudspeaker design (trading high efficiency at the resonance frequency for decreased efficiency at higher frequencies), described in Sec. 4.3. The more complex stimuli used in the subjective experiment were thought to mimic signals that would typically be reproduced through this special driver.

1.4.5.3 Pitch Theories

A number of theories have been developed to explain the pitch of complex tones, which led to what are now known as the place and periodicity theories of pitch perception. Among these are the early theories of Ohm [190] and Seebeck [243]. Others, including Helmholtz [108], assumed that if the fundamental frequency of a harmonic stimulus was absent, non-linear distortion in the middle ear could recreate that fundamental as a difference tone.

Place theories of pitch perception Briefly, the place theory of pitch perception assumes that the cochlea performs a spectral analysis of the sound and maps different frequency components along the BM (tonotopic organization, see Sec. 1.4.1); then the various locations where spectrally resolved partials are detected are used to derive a pitch percept. This implies that tones without resolved partials have no definite pitch. Several models for such pitch extraction have been put forward, some of which are:

- *Closest matching subharmonic of lowest partial (Walliser [293])*: The first step is to determine the stimulus envelope repetition rate; this corresponds to the frequency difference of the partials. Next, a subharmonic of the lowest partial is determined that is closest to the frequency found in the first step.
- *Most frequently occurring subharmonic*: Terhardt [266] elaborated upon Walliser's model, and assumed that each partial elicits a number of subharmonics that are pitch 'candidates'. For a complex tone, there will be one subharmonic that occurs most frequently, which determines the perceived pitch. Ambiguous pitch percepts are possible if several candidate subharmonics have roughly equal number of occurrences or if subharmonics cluster around multiple values.
- *Optimum processor (Goldstein [92])*: The estimated frequency values of the partials are fed into a kind of pattern recognition device, which first estimates the corresponding harmonic numbers and then derives the best-matching subharmonic to fit the observed partials; that subharmonic will be the perceived pitch. This model explains ambiguous pitch percepts by assuming that errors can be made in the estimation of harmonic number.

It is noted that in all of these models the use of place information is not an absolute requirement. The frequencies of the partials could be derived through timing information.

These models predict pitch values not only for 'normal' complex tones, but also for complex tones where each partial is shifted in frequency by a fixed amount, creating an inharmonic complex that does not have a 'proper' fundamental frequency. Although

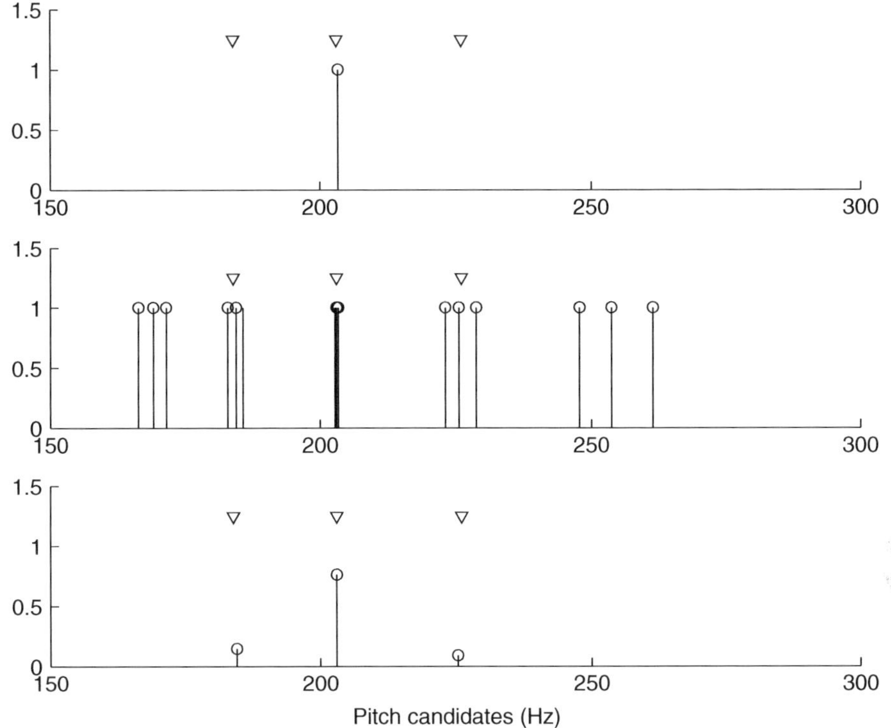

Figure 1.23 Predicted pitch percepts according to models of Walliser (a), Terhardt (b), and Goldstein (c), given a complex tone with partials at 1830, 2030, and 2230 Hz. The three arrows displayed in each part indicate frequencies at which subjects report pitch values; the pitch at 203 Hz is most strong, but percepts also occur at 184 and 226 Hz. Walliser's model gives only a single pitch prediction; Terhardt's computes subharmonics of the partials and predicts pitches for frequencies where subharmonics (nearly) coincide; Golstein's finds best-matching subharmonics for the observed partials. The amplitudes of the three predicted pitches in Goldstein's model are scaled proportionally to the inverse minimum mean-square error of that pitch value with reference to the observed partials; the sum of amplitudes equals 1, such that individual amplitudes can be interpreted as probabilities

such a shift does not alter the frequency spacing of the partials, the perceived pitch does shift. Schouten [240] found that a complex consisting of frequencies 1830, 2030, and 2230 Hz elicits a pitch of about 203 Hz; additionally, pitch matches around 184 and 226 Hz are also obtained. We investigate what pitch predictions the three pitch models will yield, with illustration thereof in Fig. 1.23. Fig. 1.23 (a) shows that the prediction by Walliser's model correctly finds a pitch of 203 Hz; it fails, however, to predict the alternate pitch values. Note that in all three parts the three arrows at frequencies 184, 203, and 226 Hz indicate the subjectively obtained pitch matches. Terhardt's model, shown in Fig. 1.23 (b), finds the most likely pitch at 203 Hz, because the subharmonics cluster most closely around that value. Pitch matches at 184 and 226 Hz are also correctly obtained, but with less likelihood, as the clustering is not as tight around these values. Fig. 1.23 (c) shows predictions from Goldstein's model. In this case, the amplitudes of the pitch matches indicate the probability of subjectively finding that particular pitch on any given occasion. For this, we have used an ad hoc metric that relates the probability of a pitch match to the inverse of the minimum mean-square error for that pitch match, given the observed harmonics. Goldstein's model also correctly matches the 203-Hz pitch, by assuming the harmonic numbers to be 9, 10, and 11; if the harmonic numbers are overestimated by 1, the pitch match at 184 Hz is predicted, and if the harmonic numbers are underestimated by 1, the pitch match at 226 Hz is predicted. Note that the pitch ambiguity that is observed in this example is primarily due to the absence of lower harmonics, which, if present, would give far less ambiguous predictions in either Terhardt's or Goldtein's model. This is also subjectively observed, and, in fact, it is believed that low-order harmonics (number $\approx$3–6) are dominant with respect to pitch determination, even if the fundamental is physically present (Plomp [209], Ritsma [227]).

Periodicity theories of pitch perception In the periodicity, or temporal, theory the locations of the BM that are excited are not important (although the tonotopic organization of the cochlea *per se* is not disputed); rather, periodicities in neural activity are used to derive pitch information. Periodicities in neural activity are caused by the fact that neural response occurs preferentially during a specific phase of the BMM waveform (phase locking). The original temporal theory is mainly due to Schouten [239], who devised an ingenious theory that combined peripheral frequency analysis and central periodicity detection. According to his theory, the lower components of a harmonic complex are spectrally resolved in the cochlea (see Sec. 1.4.3) and each map into their own pitch. The higher components, which are not resolved, create a periodic interference pattern that reflects the periodicity of the waveform. This periodicity is detected by higher neural centers and maps into a sensation of (fundamental) pitch. This gave rise to the term residue pitch, because, according to Schouten, it results from the residue of spectral components that the cochlea fails to resolve (Boff *et al.* [232]). The actual pitch that is assigned to the sound is that pitch to which attention is mainly drawn; for complex tones this is generally the residue pitch. Note that the temporal theory can also account for ambiguous pitch of inharmonically related partials, as in the example of Fig. 1.23.

Numerous experiments have shown support for both theories. It appears that neither theory alone can account for all conditions, and as such it seems likely that both place and timing information can and are used for pitch perception. Moore [177] presents a qualitative model that incorporates both place and timing information and can account for all experimental data.

1.4.6 TIMBRE

Timbre is defined by the American Standards Association [20] as follows.

Definition 9 *Timbre: that attribute of an auditory sensation in terms of which a listener can judge that two sounds similarly presented and having the same loudness and pitch are dissimilar.*

This is, as has also been noted by others, a better description of what timbre *is not* than what it *is*. It is problematic to define exactly what timbre is because it does not appear to be a one-dimensional quantity. Timbre is known to depend on short-term power spectrum (or more properly, the excitation pattern), amplitude envelope (in particular, attack and decay time), and phase spectrum. Plomp and Steeneken [210] carried out a clever experiment to investigate the relative influence of power spectrum versus phase spectrum on timbre. They used signals $s(t)$ of the form

$$s(t) = \sum_{n=1}^{N} a_n \sin(2\pi f_0 t) + b_n \cos(2\pi f_0 t), \tag{1.94}$$

and found (using triadic comparisons and multidimensional scaling) that the largest timbral difference occurred between signals where all a_n (or b_n) were zero and signals where the a_n and b_n were alternately zero. They proceeded to investigate what changes in power spectrum would yield the same magnitude of timbral difference as was found for the phase effect. The a_n and b_n were chosen to decay by a fixed amount in dB per octave, varied between -4.5 and -7.5 dB, in 0.5-dB steps. The maximum influence of phase spectrum as discussed above could be matched by a change in the power spectrum slope of 2 dB, at $f_0 = 146.2$ Hz. The influence of phase spectrum decreases for increasing f_0, being matched by a 0.7-dB change in the power spectrum slope for $f_0 = 584.4$ Hz. Also, it appeared that phase spectrum and power spectrum influences were independent of one another. In conclusion, this experiment showed that phase spectrum does influence timbre, but less so than the power spectrum. In practice, when listening to signals over loudspeakers, the relative phases of frequency components become randomized because of room reflections, which means that the signal's original phase spectrum will be modified. It thus appears that it is not practical to try to control the phase spectrum, unless listening occurs via headphones. In BWE algorithms, timbre can thus be adjusted by modifying harmonic amplitudes, for example, through filtering.

Timbre of a sound is usually qualitatively described using several descriptors. One such descriptor that seems to be well linked to an objective parameter of the signal is 'brightness'; it is related to the relative amount of high frequencies versus low frequencies contained in the signal. Brightness can be quantatively described by the spectral centroid C_S, or power spectral center of gravity, as

$$C_S = \frac{\int f\, 10 \log S^2(f)\, \mathrm{d}f}{\int 10 \log S^2(f)\, \mathrm{d}f}, \tag{1.95}$$

given a signal with power spectrum[11] $S^2(f)$. If, for example,, the relative amplitudes of the harmonics of a complex tone are modified (as most of the BWE algorithms do), C_S will change, while the pitch remains the same. Equation 1.95 is only a first-order approximation of what the spectral centre of gravity is for the internal representation of

[11] It is more proper to use a compressed power spectrum (e.g. by taking the cube root), which corresponds better to the BMM. For simplicity, we keep using $S(f)$ in the remainder of the book. It should be noted that if spectral centroid is quantatively used for analysis purposes, a perceptually much more accurate version must be employed.

the signal; refinements can be made that incorporate knowledge of auditory processing, but that is beyond the scope of this chapter.

The temporal envelope, in particular, attack and decay time, has a large influence on timbre, but as BWE algorithms do not greatly modify the temporal structure of signals, we do not need to consider this in depth.

1.4.7 AUDITORY SCENE ANALYSIS

Auditory scene analysis (ASA) is a relatively new area of interest, but has experienced increasing attention since Bregman's [38] seminal book. Bregman defines ASA as the study of how the auditory system uses sensory information to form a mental representation of the world around us. The task of forming such a representation can be subdivided into many sub-tasks: how many sound sources are present, which frequency components belong together, what is the relative positioning of the sound sources? The relevance of such questions for BWE is that BWE algorithms typically take one part of an audio signal, process it, and add it back to the other part. After the final addition of signals, the auditory system should perceive the result as deriving from one sound source. This is not an academic problem; for instance, a short time delay between the onsets of two frequency bands of a signal can be enough to perceptually separate the two bands. We will introduce some terminology here and discuss a few of the relevant ASA principles that are thought to be important for BWE applications, following Bregman [38].

Bregman uses the following concepts:

- *(Auditory) stream*: Perceptual grouping of the parts of the neural spectrogram that go together; the perceptual unit that represents a single happening. A stream can consist of more than one sound, for example, a soprano singing with a piano accompaniment.
- *Grouping*: Formation of a stream from separate sensory elements. Grouping can occur across time (sequential integration) and across frequency (simultaneous integration). The opposite of grouping is segregation, where two or more streams are formed from the sensory elements.
- *Belongingness*: A sensory element has to belong to a stream.
- *Exclusive allocation*: A sensory element can only belong to one stream; it can not be used in more than one description at the same time.
- *Closure*: Perceived continuity of a stream even if the sensory elements are interrupted. For closure to occur, it is necessary that the interruption is 'plausible'. In hearing, a plausible interruption could be masking noise, whereas a mere silent gap would not constitute a plausible interruption, and thus would not yield a continuous stream.

As explained above, BWE algorithms could potentially create two streams if an audio signal is not processed properly; in that case, the processed part appears segregated from the unprocessed part. Because this concerns the grouping of simultaneous frequency components, we will investigate the factors influencing simultaneous integration and how they might apply to BWE processing. Sequential integration is of less concern, as BWE processing does not alter the temporal structure of the processed signal, and thus should not influence grouping along the temporal dimension.

Factors influencing simultaneous integration are:

- *Correspondence over time*: If a complex spectrum consists, in part, of a simpler spectrum that was present previously, the simple spectrum would appear to continue. The 'difference' spectrum would appear as a new tone ('old-plus-new' heuristic).
- *Harmonicity*: Components with a harmonic frequency relationship are grouped, and assigned a fundamental pitch.
- *Common fate*: Components that have correlated changes in frequency (frequency modulation, FM) or amplitude (amplitude modulation, AM) are grouped. Naturally produced speech and music sounds often exhibit micromodulation (slight modulations on small time scales) as well as modulations on a larger time scale. An extreme example of amplitude modulation is an onset or offset. The common AM during an onset is an especially strong grouping cue.
- *Spatial direction*: Components arriving from the same direction are grouped; in contrast, components arriving from different directions are segregated. This cue can be quite strong, although it can be acoustically ambiguous, which is why it breaks down easily if in conflict with any of the other factors mentioned earlier.

The tendency for simultaneous frequency components to group depends on how many of the factors mentioned here favor grouping versus segregation. The various factors reinforce each other, and grouping (or segregation) will be strongest if many or all of the factors are in 'agreement'.

For BWE applications, harmonicity and common fate principles seem of greatest importance. Music and speech signals abound with harmonic signals (although noise-like signals also occur), and this harmonic structure should be maintained as much as possible. Some of the BWE algorithms actively produce harmonics of input signal components, which promotes grouping. Because the production of harmonics occurs through non-linear processing, inharmonic (distortion) components can, in some cases, also be generated. Depending on the relative energy of harmonic versus distortion products, the distortion will be audible. The distortion will covary in amplitude with the harmonic components, and may therefore be grouped into the same stream. Of course, perceptible distortion should be avoided as much as possible, regardless of grouping or segregation. Another important result is that the processed signals should not be excessively delayed with respect to the unprocessed signal: the common fate principle implies that the delayed onset of the processed signal could segregate it from the unprocessed signal. Zera and Green [304] found that in some cases, delays on the order of milliseconds can lead to discriminable changes in perception (though this does not necessarily imply segregation). This will be discussed in more detail in Sec. 2.3.3.3, and solutions will be given to avoid such potential problems.

1.4.8 PERCEPTUAL MODELLING – AUDITORY IMAGE MODEL

The auditory image model (AIM) can be used to visualize internal representations ('auditory images') of sounds, and is described elaborately in Patterson *et al.* [203, 202]. Both functional and physiological modules can be selected that model the use of fine-grain

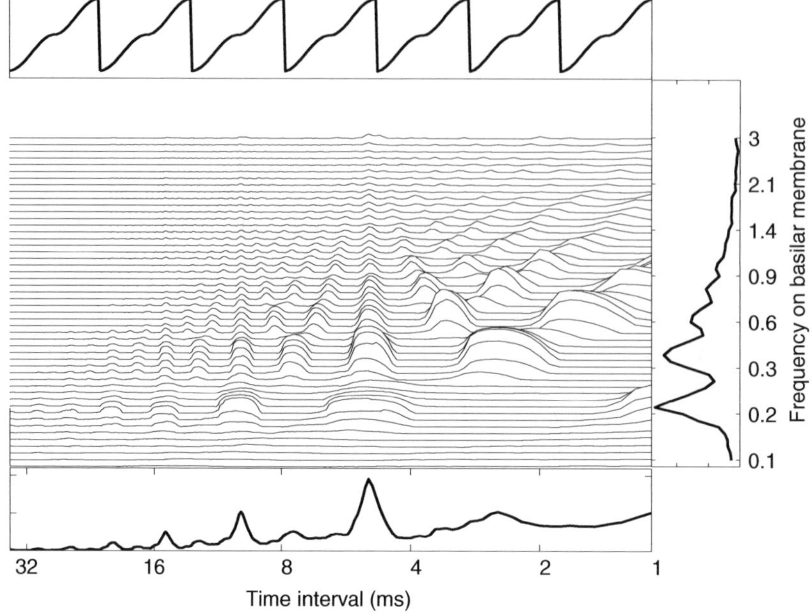

Figure 1.24 AIM calculation for a complex tone with $f_0 = 200\,\text{Hz}$ and a decaying harmonic spectrum (waveform shown in upper panel). The main panel displays the auditory image, as a function of time lag (ordinate) and frequency (abscissa), and is explained in the text. The lower panel is obtained by collapsing the auditory image over frequency, giving a lag-domain representation. The largest peak occurring at a lag of 5 ms indicates the perceived pitch at 200 Hz. The right panel is obtained by collapsing the auditory image over the lags, giving a frequency-domain representation. This clearly shows the harmonic structure of the signal, and also shows that low harmonics are better resolved than higher harmonics

timing information by the auditory system. In the following chapters, we will use AIM as a tool for predicting pitch percepts of complex tones.

The MatLab version of AIM was used (AIM-MAT); in all calculations, the functional (instead of physiological) modules of the package were used. The following processing stages are included:

- *Middle ear filtering*: The filtering described in Glasberg and Moore [90] is used.
- *Spectral analysis*: Here the response on the BM is computed in terms of BMM. The gammatone auditory filters (Eqn. 1.89 in Sec. 1.4.3) are used for this purpose.
- *Neural encoding*: First, there is a global compression of the BMM to allow a large dynamic range of sounds to be processed, as also occurs in the auditory system. Second, there is fast-acting expansion of the larger peaks in the compressed BMM response, which serves to enhance these presumably more important peaks. The last stage is a two-dimensional (over time and frequency) adaptive thresholding that serves to sharpen the resulting neural activity pattern (NAP).

- *Time-interval stabilization*: Because periodic sounds give rise to static, rather than oscillating, percepts, it is assumed that a temporal integration is performed on the NAP. To preserve the fine structure present in the timing of the NAP, a so-called 'strobed' temporal integration occurs. The integration commences when a peak in the NAP is encountered, after which the NAP input to the integrator decays in time.

The result of these processing steps is the auditory image, an example of which is shown in Fig. 1.24, where a complex tone with a 200-Hz fundamental was used. The auditory image can be collapsed over the time lags, to obtain an internal *spectral* representation of the signal (right panel of Fig. 1.24). Alternatively, a collapse over frequency provides a *temporal* representation of the signal (lower panel). Owing to the nature of the strobed temporal integration, this temporal representation is interpreted as a *lag* with respect to the last strobe: repetitions in the NAP show up as peaks in the auditory image at the appropriate lag values. Therefore, a signal with a temporal periodicity will exhibit peaks in the auditory image, which occur at a lag value equal to the periodicity interval; thereby predicting a pitch percept for the given signal at a frequency, which is the inverse of the lag. In Fig. 1.24, we find that the largest peak in the lag domain occurs at 5 ms, as we would expect on the basis of the 200-Hz fundamental frequency. We interpret the pitch strength as corresponding to the width of the peak, with wider peaks corresponding to the weaker pitch. If multiple peaks occur, the likelihood of pitch matches corresponds to the relative height of peaks. On the other hand, multiple peaks could indicate that the given signal segregates into two (or more) streams, each with their own pitch. As AIM does not incorporate auditory scene analysis principles, it does not predict which of these two possibilities applies to the given signal.

2

Psychoacoustic Bandwidth Extension for Low Frequencies

2.1 INTRODUCTION

All loudspeakers have a limited frequency range in which they can radiate sound energy at a more or less uniform level. The radiated sound pressure level can be expressed as a function of frequency through the loudspeaker's magnitude response (usually specified for an on-axis measurement). Hi-Fi enthusiasts know that a flat frequency response is very desirable, and it has been shown that the degree of 'flatness' correlates well with perceived quality (Gabrielsson and Lindström [82], Toole [274, 275]). Another desirable feature is that this frequency response is maintained for off-axis radiation, but we will not be concerned with loudspeaker directivity. Whether the loudspeaker's response is flat or not, at low and high frequencies the efficiency always decreases, leading to a *low* (f_l) and *high cut-off frequency* (f_h), usually defined as those frequencies in which the response falls 3 dB below the response at some intermediate reference frequency. Focusing on the low-frequency cut-off point, we can easily derive how the loudspeaker parameters influence its value. By rewriting the expressions derived in Sec. 1.3, we find that the efficiency η in the normal operating range, and f_l, are given by

$$\eta \sim \left(\frac{S}{m}\right)^2, \tag{2.1}$$

$$f_l = \frac{1}{2\pi}\sqrt{\frac{k_t}{m}}, \tag{2.2}$$

f_l being determined by the resonance frequency (usually denoted as f_0) of the mass-spring system that the loudspeaker is. A high efficiency η necessitates a large cone area S; a low f_l requires a low compliance k_t ('total' compliance: combined suspension and cabinet influence) and/or a large mass m. A low total compliance would necessitate a large cabinet volume; but a large mass greatly decreases the efficiency. For example, to lower the cut-off frequency of an octave by quadrupling the mass, the efficiency would decrease by a factor of 16 (12 dB). Such a measure is not in line with good loudspeaker design (high

Audio Bandwidth Extension E. Larsen and R. M. Aarts
 ISBN 0-470-85864-8

voltage sensitivity and high power efficiency). For small loudspeakers, in particular, the situation is troublesome: a small cone area, a small mass, and a high compliance lead to high values for the low-frequency cut-off point, and low efficiency. We have already found that lowering f_l by increasing m is not a viable option; also, lowering k_t is not feasible because it would necessitate a large cabinet volume (which contradicts the loudspeaker being small). The fundamental problem is that good low-frequency sound reproduction requires a large volume velocity, which is very hard to achieve for a small loudspeaker. Beyond this problem of physical origin, the perception of low-frequency sound (lower than, say, 100 Hz) is markedly worse than for intermediate frequencies (see Fig. 1.18). This means that to reproduce a low-frequency tone at equal sensation level relative to a higher frequency tone, the SPL will have to be higher. Some typical loudspeaker responses are shown in Fig. 4.13.

Nevertheless, in many applications, small loudspeakers are unavoidable, because of size and/or cost constraints. In fact, loudspeakers that are large enough to reproduce the lowest audible frequencies (around 20 Hz) at a sufficient level are huge in size and very expensive. Even a more modest goal of good reproduction at 50 Hz is difficult to achieve within the constraints usually encountered in consumer electronics, such as (flat) TV and laptop computers, and also (portable) audio and (in-ear) headphones. Another challenging case is in telephony, in which very small loudspeakers are employed.

The need for higher acoustic output has always existed, especially at low frequencies, ever since the invention of the electrodynamic loudspeaker. Improvements in loudspeaker design have yielded better low-frequency characteristics, the most popular option being vented designs. The vent introduces an additional resonance below the loudspeaker – cabinet resonance, thus extending the low-frequency response. The drawback is that the response falls off twice as fast below the new cut-off frequency, and the temporal behaviour is degraded. Even though this 'bass-reflex' design has a more extended low-frequency response than a conventional loudspeaker and cabinet, Eqns. 2.1–2.2 still hold. As mentioned previously, the fundamental problem is the limited volume velocity that is achievable with a small loudspeaker, and the physical limit cannot be overcome with purely physical modifications of the design. A partial solution to the low-frequency problem has come from BWE and the psychology of hearing. The material presented in this chapter will explain how BWE can be used to improve the bass response of small loudspeakers, and basic algorithms are presented. First, however, we discuss the traditional option of low-frequency emphasis by linear amplification of the bass portion ('bass boosting').

The loudspeaker response can theoretically be inverted using a preceding filter with the inverse of this response, as in Fig. 2.1. In practice, the limiting factors are finite cone excursion and finite power-handling capacity of the loudspeaker. Therefore, this method can only enhance frequencies at or slightly below f_l (BWE methods can enhance reproduction several octaves below f_l). At high output levels, distortion or even damage and ultimately destruction of the loudspeaker may occur. Also, this solution is very energy inefficient, because of the loudspeaker's intrinsic low efficiency at low frequencies (important for portable devices such as portable audio, cd players, or PDAs). The advantages of this approach are its simplicity and linearity.

Another solution was proposed by Long and Wickersham [164] ('ELF' system). The design specifically drives the loudspeaker below its resonance frequency, by using two

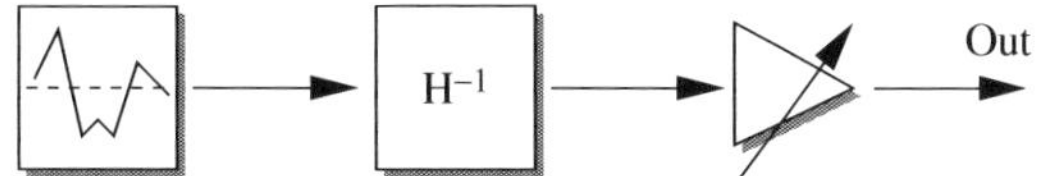

Figure 2.1 A simple circuit for 'inverting' a loudspeaker response. The signal is filtered by H^{-1}, the 'inverse' of the loudspeaker response, scaled and applied to the loudspeaker. Usually the filtering does not use the real inverse of the loudspeaker response, but simply a low-pass filter to boost low-frequency sounds. The subsequent scaling may be manually adjustable or signal dependent

integrators preceding the loudspeaker terminals. The integrators invert the high-pass characteristic of the loudspeaker, and the method is said to work better than 'traditional' amplification. But again, cone excursion and inefficiency are concerns for such an approach.

2.2 PSYCHOACOUSTIC EFFECTS FOR LOW-FREQUENCY ENHANCEMENT OF SMALL LOUDSPEAKER REPRODUCTION

Given the fact that on physical grounds it is impossible to have a good low-frequency loudspeaker response (with small loudspeakers), it is pertinent to ask whether other options are available. One option is to use BWE, with the 'extension' taking part in the auditory system, instead of extending the actual physical bandwidth of the signal. In fact, we sometimes have to reduce the bandwidth of the signal to prevent very low frequencies from entering the loudspeaker, as these cannot be reproduced anyway. If we modify the low-frequency part of the signal in such a way that the auditory system 'fills in' the part that the loudspeaker cannot reproduce, then we have achieved *psychoacoustic* BWE.

2.2.1 PITCH (HARMONIC STRUCTURE)

There are a number of psychoacoustic effects that can possibly be used for this kind of BWE, shown in Fig. 2.2. As before, f_l represents the low-frequency cut-off of the loudspeaker. These options are:

- *Frequency doubling*: By a frequency doubling, we can shift components at frequency f, for which $f_l/2 \le f \le f_l$, to frequency f', which is $f_l \le f' \le 2f_l$, as in panel (ii) of Fig. 2.2. Because frequency components are now above f_l, they can be reproduced at a higher level, and since the audibility curve slopes downward at low frequencies, the ear is more sensitive in this frequency range as well (Fig. 1.18). It may be expected that the resulting signal will have an increased loudness in the low-frequency range. Frequency doubling is attractive because of its extreme simplicity: a full-wave or half-wave rectification suffices, which is trivial to implement in digital signal processing, or in analog components. The drawback is that the waveform is seriously distorted, and also, the pitch has changed.

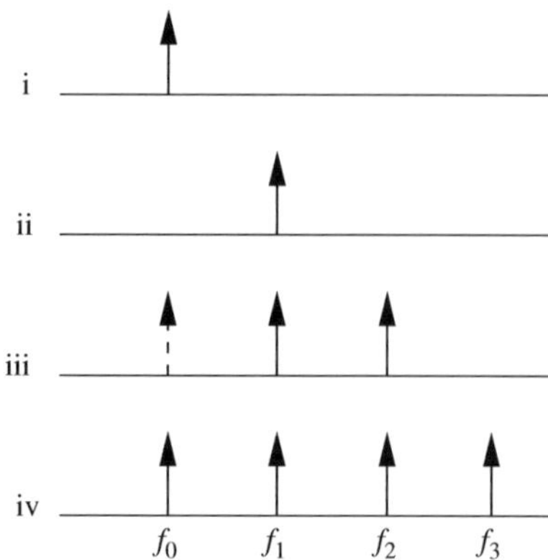

Figure 2.2 Panel (i) shows the frequency representation of a pure-tone signal at f_0 Hz. If $f_0 < f_l$, then the signal can be substituted by the following. Panel (ii): the double frequency at $2f_0$; the pitch has changed, but the loudspeaker will reproduce the signal more efficiently at $2f_0$. Panel (iii): components at $2f_0$ and $3f_0$, which will produce the cubic combination tone, and the difference tone at high levels also , in the cochlea, at f_0. Panel (iv): components at $2f_0$, $3f_0$, $4f_0$ and so on, which will produce a residue pitch at f_0. In panels (iii) and (iv), the dashed arrows are not physically radiated by the loudspeaker, but a pitch corresponding to that frequency is perceived

- *Combination tones*: As was explained in Sec. 1.4.2, non-linearities of the cochlear response generate combination tones (CT) when presented with two-tone stimuli. If pure tones with frequencies f_1 and f_2 enter the cochlea, the generated CT frequencies $f(n)$ correspond to those given in Eqn. 1.87. To elicit a low-frequency pitch at f_0, we need $f_0 = f(1) = 2f_1 - f_2$; this corresponds to the cubic CT, which is highest in level of all CTs. Probably the best option is to have $f_1 = 2f_0$ and $f_2 = 3f_0$. Although the ratio $f_1/f_2 = 1.5$ is unfavourable (the level of the cubic CT at this ratio is very low), at least these two components are harmonically related to f_0. Some advantage might be obtained because the difference tone (DT) (also described in Sec. 1.4.2) $f_2 - f_1$ now coincides with the cubic CT, which might increase the loudness of the f_0 component. The frequencies are shown in panel (iii) of Fig. 2.2. Also, choosing f_1 and f_2 in the manner described above will aid in the perception of virtual pitch, to be described next.
- *Virtual pitch*: Perhaps the most attractive option is to make use of the 'missing fundamental' effect: a special case of residue pitch, also known as virtual pitch. In Sec. 1.4.5, it is shown how the auditory system creates a pitch percept at f_0 if presented with a harmonic series, that is, a tone complex of several frequency components, which have a common fundamental frequency f_0. For this, it is not necessary that the f_0 component is actually present (nor the second, third, etc.).

  For low-frequency psychoacoustic BWE applications, we can substitute an $f < f_l$ by a series kf, $k \geq 2$, to evoke the residue pitch of f, while the loudspeaker does not radiate energy at frequency f. There are many non-linear operations that can be used to generate a harmonics series that will serve this purpose, as will be presented later in this chapter. Note that at high sound pressure levels, residue pitch and distortion products may occur simultaneously. The spectrum of a harmonic complex with fundamental at f_0 is shown in panel (iv) of Fig. 2.2.

Thus, there are three options to increase (apparent) bass reproduction below a loudspeaker's cut-off frequency. Note that the original low-frequency components below f_l can be either removed by appropriate filtering or retained.

2.2.2 *TIMBRE (SPECTRAL ENVELOPE)*

In addition to a correct pitch, the extended frequency components (harmonics of f_0) should have a timbre that is close to what would be perceived over an ideal loudspeaker. In Sec. 1.4.6, it was discussed that timbre depends on magnitude spectrum more than on phase spectrum, and the spectral centroid C_S was introduced as an objective metric to represent the subjective quality of 'brightness' of a sound.

For an analysis of low-frequency psychoacoustic BWE, we apply the simple wideband formulation as in Eqn. 1.95 to compute the spectral centroid $C_{S,0}$ for an input pure tone of frequency f_0 and amplitude a_0 and the spectral centroid $C_{S,1}$ for a synthetic signal generated by BWE, that consists, say, of fundamental and harmonics 1–5 (as discussed in Sec. 2.2.1), with amplitudes $a_0 \ldots a_5$:

$$C_{S,0} = (f_0 a_0^2)/a_0^2 = f_0, \tag{2.3}$$

$$C_{S,1} = \left(\sum_{i=0}^{5} f_i a_i^2\right) / \sum_{i=0}^{5} a_i^2 = f_0 \times \left(\sum_{i=0}^{5} i a_i^2\right) / \sum_{i=0}^{5} a_i^2 = f_0 \times \alpha, \quad 1 \le \alpha \le 6. \tag{2.4}$$

The conclusion is that $C_{S,0} \neq C_{S,1} \forall \alpha > 1$, that is, the brightness of the input signal will never be equal to the brightness of the output signal, unless all harmonic amplitudes are zero. One can easily verify this by listening to the two above signals and concluding that the pitch of both will always be equal, but the timbre will never be. We can deduce from Eqn. 2.4 that the timbre of the harmonics signal will be closest to that of the pure-tone input if the low-order harmonics are relatively larger in amplitude. In the limit that $a_0 \ll a_i$ ($i = 1 \ldots 5$), the two timbres will be indistinguishable (neglecting other factors, which may influence timbre, such as phase spectrum), but in that case, there is no bandwidth extension taking place. In practice, this means that there has to be a compromise between a large BWE effect (weakly decaying harmonics spectrum) and a good timbre match (strongly decaying harmonics spectrum). How this compromise is achieved is discussed later.

2.2.3 *LOUDNESS (AMPLITUDE) AND TONE DURATION*

After pitch and timbre, the last perceptual variable to control is loudness, measured in phones (Sec. 1.4.4). To keep matters simple, we will only consider the influence of intensity and frequency on loudness here. For this, we refer to the equal-loudness contours by Fletcher and Munson [73] (Fig. 1.18). For signals below about 500 Hz, we can state that equal-intensity signals will sound louder if the frequency content is higher. For low-frequency psychoacoustic BWE applications, this is a favourable circumstance, as the algorithms typically replace low-frequency signals with higher harmonics. The equal-loudness contours were measured for steady-state pure tones, so to assess the loudness of time-varying complex signals (such as music or speech), it would be better to use more sophisticated models, such as those by Stevens [257], Paulus and Zwicker [205], Zwicker [310], or Glasberg and Moore [90] (discussed in Sec. 1.4.4.2). Using any of these methods, we could compute the loudness of a pure-tone signal s_0 of frequency f_0 as perceived through a 'perfect' loudspeaker, as $L(s_0)$. We must also take into account the other frequencies' components present in the signal, which we represent by signal s_m. The loudness of the entire signal, perceived through a perfect loudspeaker, would then be $l_0 = L(s_0 + s_m)$. BWE processing is applied to s_0, which creates a signal s_0', and the loudness of the total signal would be $l_0' = L(gs_0' + s_m)$, where s_0 is scaled by factor g. However, this signal is reproduced over a non-ideal loudspeaker that has an average response of h_0' in the frequency region of s_0' and a response of 1 in the frequency region of s_m. Then the perceived loudness will be $l_\mathrm{h}' = L(gh_0's_0' + s_m)$. So, we should have

$$L(s_0 + s_m) = L(gh_0's_0' + s_m). \tag{2.5}$$

The increase in loudness ΔL that the processed signal would have relative to the unprocessed signal, played over the same loudspeaker, would be

$$\begin{aligned}\Delta L = L(gh_0's_0' + s_m) - L(h_0s_0 + h_ms_m) &\approx L(gh_0's_0' + s_m) - L(h_ms_m),\\ \text{if } h_0s_0 \ll h_ms_m; &\end{aligned} \tag{2.6}$$

h_0 is the loudspeaker response in the frequency region of s_0, which, being below the loudspeaker's resonance frequency, is assumed to be negligible. Because L is a complicated non-linear function, we cannot give a closed-form expression for g (Eqn. 2.5) in terms of the other variables, nor for ΔL (Eqn. 2.6). This analysis does show us that beside a model for L (loudness perception of complex tones), we also need to know the characteristics of the loudspeaker, at least in the bass frequency range. A few attempts have been made to create appropriate loudness models and algorithms for low-frequency psychoacoustic BWE applications, which will be discussed in Sec. 2.3.4.

The frequency dependence of both the equal-loudness contours and the loudspeaker response can also influence the perceived duration of tones. This will be illustrated with respect to Fig. 2.3. A 50-Hz decaying tone is shown in part (a), as produced by, for example, a bass drum. The amplitude of the signal is determined by the loudspeaker's response at 50 Hz. The dashed line indicates the minimum audible field at 50 Hz. The perceived duration of the tone is indicated by the horizontal line, and equals T_1 s. Say 50 Hz lies below the cut-off frequency of the loudspeaker and a low-frequency psychoacoustic BWE algorithm is applied to enhance the bass response, resulting in the signal shown in the Fig. 2.3 (b). The BWE algorithm has created harmonics (say, 100 and 150 Hz) at which frequencies the loudspeaker will be more efficient. This is shown by the increased

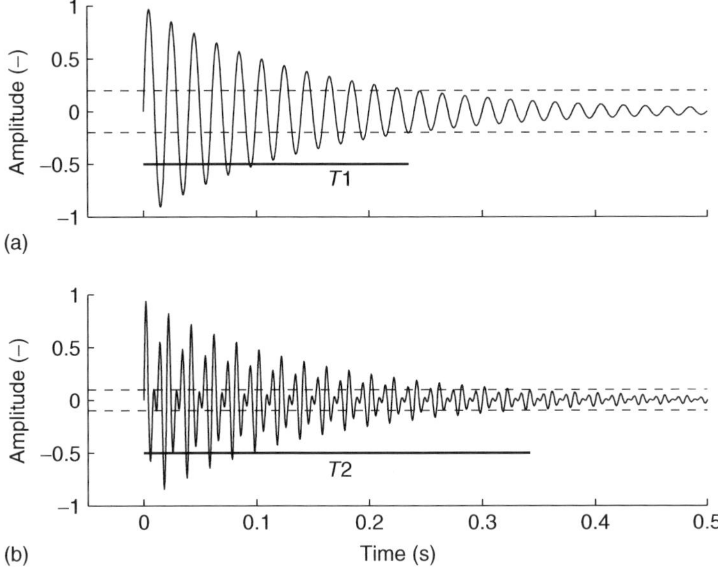

Figure 2.3 (a) Shows a decaying signal at 50 Hz. The loudspeaker characteristics and the minimum audible field at this frequency cause signals below the dashed line to be inaudible; thus, the tone has a duration T_1 of about 0.23 s. (b) Shows the signal with the 50-Hz component replaced by components at 100 and 150 Hz. The higher efficiency of the loudspeaker and the lower value of the minimum audible field at these frequencies cause the signal level where the tone becomes inaudible to be lower; thus, the tone has a duration of T_2 of about 0.34 s now

amplitude of the signal. As the equal-loudness contours slope downward at low frequencies, the minimum audible field has decreased[1]. Together, these two effects increase the perceived duration of the tone from T_1 s to T_2 s, in the given example from 0.23 to 0.34 s (of course, the loudness of the tone also increases greatly). This increase in tone duration can, for some repertoire, sound artificial. In fact, it merely shows that the BWE extension is doing its job well, as the same repertoire reproduced on a high-quality subwoofer has the same long duration bass notes. The artificial aspect of the perception on a small loudspeaker system with BWE is probably due to the fact that one does not expect good bass reproduction for such systems.

Careful inspection of the equal-loudness contours will reveal that the spacing of the contours is not constant. Rather, the contours are more 'compressed' for very low frequencies than for higher frequencies, if we restrict our attention to the bass frequency range. The consequence is that if we vary the level of two pure tones of unequal frequency by the same amount, then the loudness variation of the two will be unequal. The lower

[1] In most cases, the lowest audible intensity is not determined by the minimum audible field, but rather by masking effects of ambient noise. Therefore, the increased sensitivity of the ear at higher frequencies would not usually influence tone duration.

frequency tone will appear to have the greater loudness variation. For low-frequency psychoacoustic BWE, this could imply that if very low-frequency components are replaced by higher frequency components, loudness variations decrease. Gan *et al.* [83] have taken this into account in their BWE algorithm, as will be discussed in Sec. 2.3.4.

2.3 LOW-FREQUENCY PSYCHOACOUSTIC BANDWIDTH EXTENSION ALGORITHMS

2.3.1 OVERVIEW

We will discuss low-frequency psychoacoustic BWE algorithms using the general structure shown in Fig. 2.4; compare this structure to the general BWE framework introduced in Chapter I.3 as Fig. I.2. The implementation is in the time domain, which has the benefit of computational efficiency. Frequency-domain algorithms would be possible, but suffer the drawback that it would be difficult to achieve the desired frequency resolution while at the same time keeping the analysis window sufficiently short to satisfy stationarity of the input signal.

The essential element of the structure in Fig. 2.4 is the *non-linear device* (NLD), which converts frequencies below the loudspeaker cut-off frequency f_l to frequencies above f_l. The NLD will be chosen such that the pitch of the input signal is preserved, which will be the case if the output frequencies are harmonically related to the input frequencies (Sec. 2.2.1). In Sec. 2.3.2, we will present several options for the NLD. The filters FIL1 and FIL2, placed respectively before and after the NLD, serve two functions. FIL1 ensures that only frequencies below f_l enter the NLD; it is assumed that higher frequencies are reproduced properly by the loudspeaker, and therefore should not be modified. The filter characteristics therefore depend mainly on f_l. FIL2 does the spectral envelope shaping of the complex signal produced by the NLD. Its characteristics do not depend heavily on the loudspeaker, but on the implementation of the NLD, in particular, the relative

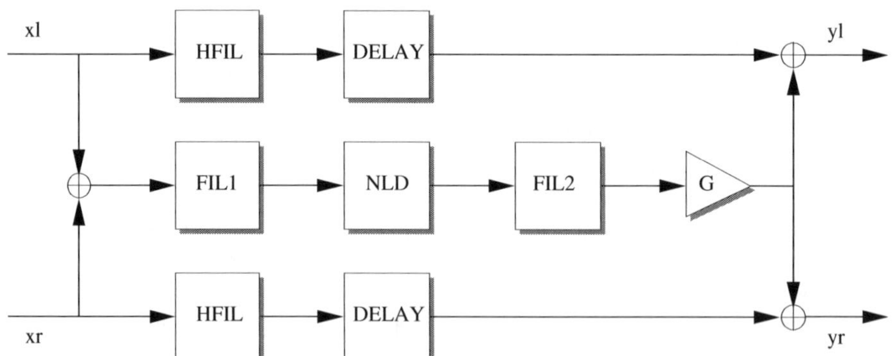

Figure 2.4 Overview of low-frequency psychoacoustic BWE in a time-domain algorithm. The harmonics are generated by the non-linear device (NLD), with appropriate filtering by FIL1 and FIL2. After scaling, the extended signal is added back to the input signals, which is delayed and possibly high-pass filtered (HFIL)

amplitudes of generated harmonics. FIL2 attempts to control the timbre of the synthetic bass signal. In Sec. 2.3.3, we will go into more detail regarding the characteristics of FIL1 and FIL2. Finally, the harmonics signal must be scaled such that an appropriate loudness is achieved, after which it is added back to the input signal. The gain may be fixed or, in more complex algorithms, adaptively determined by characteristics of the input and output signals, as will be discussed in Sec. 2.3.4. The higher frequencies of the input signals are usually passed straight through to the output, although a high-pass filter (with cut-off frequency of approximately f_l) may be applied to eliminate very low-frequency components. The rationale for this is that these very low components are not audible anyway (due to the poor loudspeaker response at those frequencies and the high audibility threshold), but do contribute to cone excursion of the loudspeaker. By removing these components, the cone excursion is decreased, which can be beneficial for the quality of the reproduced signal.

The structure of Fig. 2.4 shows that for a stereo input signal, processing is done on the summed input. This is because low-frequency content is usually identical in both channels. Also, localization is quite poor at very low frequencies, wherefore the actual distribution in left and right output channels is irrelevant.

2.3.2 NON-LINEAR DEVICE

The essential element of low-frequency psychoacoustic BWE algorithms is the non-linear device (NLD), but non-linearity is a very general property, and there are many kinds of non-linear functions. For any type of BWE, we usually require amplitude linearity, such that the relation between input and output signals is independent of level. Thus, in Vaidyanathan's [278] terminology (see Sec. 1.1), we prefer NLDs to be homogeneous systems. In this section, we review several NLD implementations that can be useful for low-frequency psychoacoustic BWE.

In discussing the various NLDs and their characteristics, we shall make reference to the auditory image model (Patterson *et al.* [203]) discussed in Sec. 1.4.8, which we use to assess the pitch of complex tones.

2.3.2.1 Multiplier

Spectral characteristics Figure 2.5 shows an NLD whereby the input signal is repeatedly multiplied with itself, producing a harmonic series. Although this system is not homogeneous, it has the advantage that one can control at the outset the number of harmonics created, and their relative amplitudes. Because the output spectrum is under direct control, a shaping filter (FIL2 in Fig. 2.4) is not necessary. The problem that this multiplying NLD is non-linear in amplitude can be solved by using an automatic gain control (AGC) before the NLD, which will scale the signal level to a reference value. After the NLD, the inverse gain will be applied to restore the signal to its original level. In this way, the whole system has effectively become homogeneous.

We begin the analysis by assuming a pure-tone input signal x of frequency f_0 at the reference level (defined to be 1). Thus,

$$x(t) = \sin(2\pi f_0 t). \tag{2.7}$$

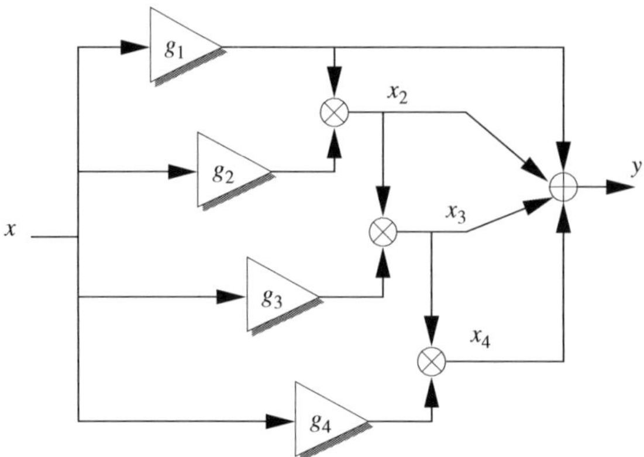

Figure 2.5 Harmonics generation by multiplication. The input signal x is repeatedly multiplied by itself to generate the harmonics. For a pure-tone input, the weights g_i can be chosen such that the output signal has a prespecified spectrum, according to Eqn. 2.11. The example shown here generates four harmonics, but more or less numbers is also possible

After multiplying a scaled x with a scaled replica, we get x_2, as

$$x_2(t) = \frac{g_1 g_2}{2}[1 - \cos(2 \times 2\pi f_0 t)]. \tag{2.8}$$

The frequency doubling is apparent; g_1, g_2 are the scaling factors. If the heterodyning option of Sec. 2.2.1 is chosen, the second harmonic is all we need and this concludes the processing of the NLD (but in this case a more effective NLD is a rectifier, which will be discussed later in this section). If, however, we propose to use the virtual pitch option of Sec. 2.2.1, we will need to generate at least an additional two to three harmonics. By multiplying x_2 with another scaled replica of x, the third harmonic x_3 can be created, and so on. The output signal y, assuming that three harmonics above f_0 are generated, will become

$$y(t) = h_0 + \sum_{i=1}^{2} \left[h_{2i-1} \sin((2i-1) \times 2\pi f_0 t) + h_{2i} \cos(2i \times 2\pi f_0 t) \right] \tag{2.9}$$

the h_i being the scale factors given as

$$\begin{aligned} h_0 &= \frac{g_1 g_2}{2} \left[1 + \frac{1}{4} g_3 g_4 \right], \\ h_1 &= g_1 \left[1 + \frac{3}{4} g_2 g_3 \right], \\ h_2 &= -\frac{g_1 g_2}{2} \left[1 + g_3 g_4 \right], \end{aligned} \tag{2.10}$$

$$h_3 = -\frac{g_1 g_2 g_3}{4},$$
$$h_4 = \frac{g_1 g_2 g_3 g_4}{8},$$

If we prespecify what the amplitudes h_i should be (the amplitude of h_0 is the uninteresting DC term, which will be filtered out later, thus we do not care about its value), then we must choose the scaling factors g_i such that

$$\begin{aligned} g_1 &= h_1 + 3h_3, \\ g_2 &= -2\frac{h_2 + 4h_4}{h_1 + 3h_3}, \quad h_1 \neq -3h_3, \\ g_3 &= \frac{2h_3}{h_2 + 4h_4}, \quad h_2 \neq -4h_4 \\ g_4 &= -2\frac{h_4}{h_3}, \quad h_3 \neq 0 \end{aligned} \tag{2.11}$$

As long as none of the numerators of Eqn. 2.11 are zero, the scale factors are well behaved and can be chosen to yield any desired harmonics spectrum. If, for example, all harmonics amplitudes are to be +1, then we have $g_1 = 4$, $g_2 = -10/4$, $g_3 = 2/5$, and $g_4 = -2$. Note that the even harmonics will be $\pi/2$ rad out of phase with the odd harmonics (Eqn. 2.9). For the multiplying NLD, we do not give an example of an AIM simulation (Sec. 1.4.8), as the number and amplitudes of harmonics is not fixed in this case: they both depend on the g_i and the level of multiplication used.

Intermodulation distortion Non-linear devices exhibit the so-called intermodulation distortion: the presence of frequency components in the output that are not harmonically related to frequency components in the input. It is the interaction of the input frequency components that produces these intermodulation distortion products. The frequencies at which they occur are the sum and difference frequencies of the input components, but the amplitudes depend on the kind of non-linearity and also on the amplitudes of the input frequency components.

Assume a signal s with two frequency components, at f_1 and f_2 (which are not themselves harmonically related), with amplitudes 1 and $0 \leq a \leq 1$. The magnitude of the Fourier representation is given by delta functions, which are

$$s \overset{|\mathcal{F}|}{\rightarrow} [f_1] + a[f_2]. \tag{2.12}$$

Here, we have used a shorthand notation on the right-hand side of the arrow, where $a[f_2]$ indicates a frequency component at frequency f_2 with amplitude a, and so on. Multiplying s with itself yields

$$\begin{aligned} s^2 \overset{|\mathcal{F}|}{\rightarrow} (*) \quad & [2f_1] + a^2[2f_2] \\ (\dagger) \quad & 2a\left([f_1 - f_2] + [f_1 + f_2]\right). \end{aligned} \tag{2.13}$$

The harmonic components are indicated by (∗) and the intermodulation distortion components by (†). We can define a harmonic-to-distortion energy ratio ς (akin to signal-to-noise ratio)

$$\varsigma = 10\log\frac{\sum f_{(*)}^2}{\sum f_{(\dagger)}^2} \tag{2.14}$$

which, for s^2 would be

$$\begin{aligned}\varsigma_2 &= 10\log\left[\frac{1^2+(a^2)^2}{2\times(2a)^2}\right] = 10\log\left[\frac{1+a^4}{8a^2}\right] \\ &= -10\log 8 + 20\log\frac{1}{a} \quad \text{for } a \ll 1.\end{aligned} \tag{2.15}$$

High ς_2 can be achieved for $a \ll 1$; the minimum value (worst-case distortion) of $\varsigma_2 = -6.0\,\text{dB}$ occurs for $a = 1$. Continuing along the same lines, we now multiply s with itself twice, and obtain

$$\begin{aligned} s^3 \overset{|\mathcal{F}|}{\rightarrow} (*) \quad & \frac{1}{4}((3+6a^2)[f_1] + [3f_1] + 3a(2+a^2)[f_2] + a^3[3f_3]) \\ (\dagger) \quad & \frac{3a}{4}(a([f_1 - 2f_2] + [f1 + 2f_2]) + [2f_1 - f_2] + [2f_1 + f_2]).\end{aligned} \tag{2.16}$$

For s^3, the harmonic-to-distortion energy ratio ς_3 becomes

$$\begin{aligned}\varsigma_3 &= 10\log\left[\frac{23a^6 + 18a^4 + 36a^2 + 5}{9a^2(a^2+1)}\right] \\ &= -10\log\frac{9}{5} + 20\log\frac{1}{a} \quad \text{for } a \ll 1.\end{aligned} \tag{2.17}$$

The lowest value $\varsigma_3 = 6.4\,\text{dB}$ occurs for $a = 0.832$; compared to the minimum value of ς_2, this is a much better situation. For s^4, we compute ς_4 directly as

$$\begin{aligned}\varsigma_4 &= 10\log\left[\frac{17a^8 + 288a^4 + 17}{a^2(176a^4 + 36a^2 + 176)}\right] \\ &= -10\log\frac{176}{17} + 20\log\frac{1}{a} \quad \text{for } a \ll 1.\end{aligned} \tag{2.18}$$

The worst-case distortion value is now $\varsigma_4 = -1.5\,\text{dB}$, at $a = 0.545$. This analysis can be continued for all necessary levels of multiplication (depending on how many harmonics are desired), but the point should be adequately illustrated. The 'grand total' ς_t, that is, that of the weighted sum of the s^i, with weighting factors g_i, is not a simple combination of the ς_i, but has to be recomputed by adding all harmonic and distortion energies and taking the ratio. The final result will be a function of a and the g_i, and may serve to choose a particular combination of g_i that will maximize ς_t, subject to some constraints on the

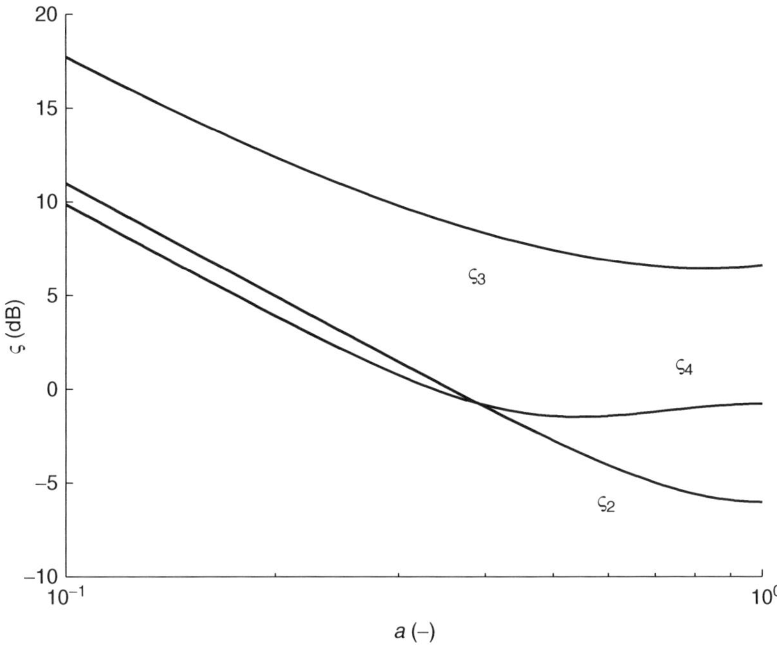

Figure 2.6 The harmonic-to-distortion energy ratio ς for the output signal of a multiplying non-linearity, given a two-tone input with amplitudes 1 and a, as a function of a. The subscripts indicate the highest harmonic number at the output, for example one ς_2 applied to one level of multiplication, where the double frequency is generated

desired harmonic amplitudes h_i (Eqns. 2.10 and 2.11). In Fig. 2.6, $\varsigma_2 - \varsigma_4$ are plotted as a function of a.

The above analysis is of course incomplete in the sense that situations with more than two input frequency components can occur in practice. The analysis becomes very tedious for such involved cases, though. Also, the metric ς can only be regarded as a very crude approximation to subjective quality, for it ignores that some of the weaker components may be masked by the stronger components. As the distortion components are generally smaller than the harmonic components, one could argue that ς overestimates the effects of distortion: in practice, some of the distortion components will be masked. In other words, a high value for ς is always good, but a low value is not necessarily very bad. It would be interesting to study this with a psychoacoustic model, using for example the two-component signal illustrated in the above analysis.

2.3.2.2 Rectifier

Spectral characteristics A very efficient method of harmonics generation is by rectification; either half-wave or full-wave. Both analog and digital implementations are trivial, and another favourable aspect is that rectification is a homogeneous operation. Of course, as a whole the system is non-linear, and the output frequency components are mostly double those at the input. To compute the harmonics spectrum exactly, we apply a pure-tone input signal of frequency f_0, and compute the Fourier series b_k of the full-wave-rectified signal

$$b_k = f_0 \int_0^{1/f_0} |\sin(2\pi f_0 t)| \, e^{-2\pi i k f_0 t} \, dt = \begin{cases} \frac{2}{\pi(1-k^2)} & \text{for even } k, \\ 0 & \text{for odd } k. \end{cases} \quad (2.19)$$

The resulting spectrum consists of only the even harmonics of f_0, which implies that the fundamental frequency of the output signal is now $2f_0$. Perceptually, this means that the synthetic bass sounds an octave too high, compared to the input signal. However, the increase in bass perception using this kind of low-frequency psychoacoustic BWE can still be attractive (mainly because of the efficient implementation). The harmonics spectrum decays quite rapidly, at -12 dB per octave.

Figure 2.7 shows the AIM calculations for a 200-Hz rectified signal, with the input component added back. Thus, the signal contains 200, 400, 800, 1200 Hz, and so on.

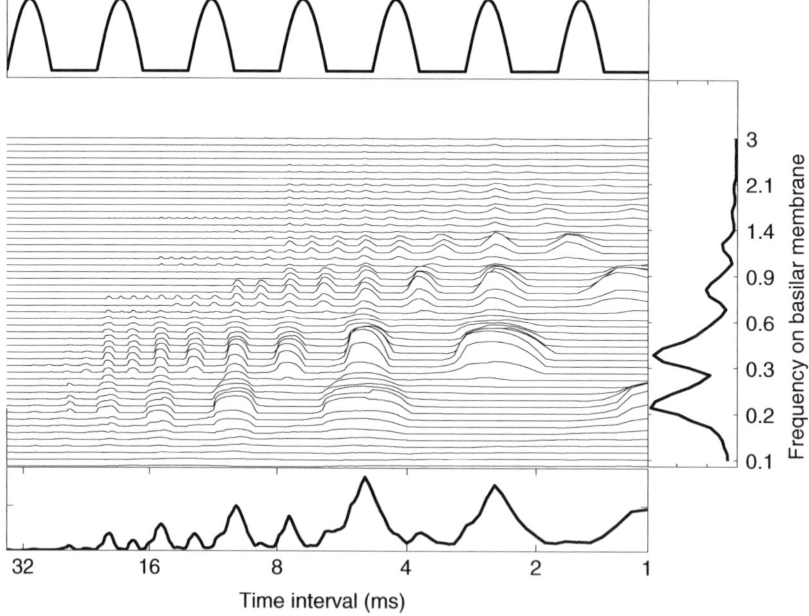

Figure 2.7 AIM calculations for a rectified pure tone with frequency $f_0 = 200$ Hz; the pure tone is added back to the rectified signal, giving a complex tone that includes the fundamental and all even harmonics. The strongest peak in the frequency-collapsed plot (lower panel) occurs at a lag of 5 ms (200 Hz). However, there is also a strong peak at a lag of 2.5 ms (400 Hz). This might indicate an ambiguous pitch or a signal that is segregated into two percepts

Although the 200-Hz component would not normally be present in the output of the rectifying NLD, it is added in this analysis, as without it the pitch percept would be unambiguously at 400 Hz. This situation might occur when using a small loudspeaker that cannot reproduce 200 Hz at a sufficient level. It is of interest to see what the effect of adding back the original 200-Hz component might be on the perceived pitch. The lower panel shows a dominant peak at a lag of 5 ms, corresponding to 200 Hz. There is another peak of almost equal amplitude, however, at a lag of 2.5 ms, corresponding to 400 Hz. This may imply an ambiguous pitch percept, or a failure of the 200-Hz component to group with the harmonic series, leading to two auditory objects with different pitches. If it is indeed a grouping problem, then common amplitude modulation of all components (as would occur in practice) could increase the likelihood that one auditory object is perceived, instead of two.

Temporal characteristics Beside frequency characteristics, temporal behaviour is important as well. In particular, it is desirable that the temporal envelope of the signal remains as close to the original as possible. If, for example, the attack time of an impulsive sound is increased, this can be very noticeable. The temporal behaviour of the rectifying NLD is satisfactory, refer to Fig. 2.8. A 5-cycle 50-Hz tone burst is shown in (a). The rectifying

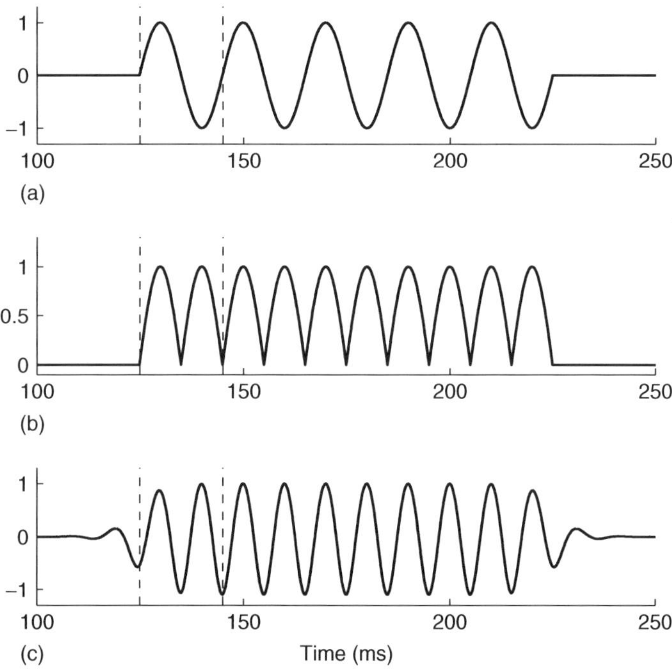

Figure 2.8 (a) Shows 5 cycles of a 50-Hz tone as input signal to a rectifying NLD, the output of which is shown in (b). (c) Shows the result after bandpass filtering between 70 and 150 Hz. The filtered output signal reaches full amplitude in the first cycle, which is beneficial for perceptual quality

NLD produces the signal shown in (b), after which bandpass filtering between 70 and 150 Hz is done (linear phase, with delay compensation), shown in (c). The filtered output reaches maximum amplitude within the first cycle.

Intermodulation distortion The robustness of this NLD to intermodulation distortion when presented with two-tone stimuli, or more complex input spectra, can be assessed using expressions derived in the appendix of this chapter, which are taken from Larsen and Aarts [156]. Although the rectifier is a non-linear system, the full output spectrum can be computed conveniently for arbitrary periodic input signals. Consider a signal $f(t)$, with Fourier series coefficients a_k. The rectified output signal has Fourier series coefficients b_k, which are given by

$$b_k = (2t_0 - 1)a_k - \sum_{n \neq k} \frac{na_n}{i\pi k(k-n)}[1 - e^{i2\pi(n-k)t_0}] \quad (n \in \mathbb{Z},\ k \neq 0), \tag{2.20}$$

where it is assumed that the input signal has period 1, with a single zero crossing at t_0. Note that for a pure tone ($a_1 = 1/2i$, $a_{-1} = -1/2i$, $t_0 = 1/2$) the result of Eqn. 2.19 is obtained. The more general case, in which the period of the signal is arbitrary and with an arbitrary number of zero crossings per period, is presented in the appendix at the end of this chapter. There, it is also shown that for large k the b_k are mainly determined by the slope of f at its zero crossings.

To quantify the amount of harmonic energy versus intermodulation distortion energy, we can again use the metric ς (Eqn. 2.14). A complicating factor in evaluating Eqn. 2.20, or its more general form as given in the appendix (Eqn. 2.93), is that the period, or rather the fundamental frequency f_0, of the input signal is a function of the input frequency components. Assume an input signal with frequencies f_1 and f_2 and amplitudes 1 and $0 \leq a \leq 1$; f_0 is the greatest common divisor of f_1 and f_2, for example, 10 Hz for input frequencies of 50 and 70 Hz, which are then the fifth and seventh harmonics: $a_{\pm 5} = \pm 1/(2i)$ and $a_{\pm 7} = \pm a/(2i)$ in Eqn. 2.93. In accounting the harmonic energy of the output signal, we sum all $b_{\pm 5n}^2$ and $b_{\pm 7n}^2$, $n \in \mathbb{Z}$. We numerically compute ς for a number of representative cases. Hence, given f_1 and f_2 (not themselves harmonically related) with amplitudes 1 and a, we first compute all zero crossings in (0, 1] and then apply Eqn. 2.93. The harmonic energy is computed as stated above, and all excess energy is considered as originating from intermodulation distortion. Figure 2.9 shows ς for various $f1/f2$ ratios; $\varsigma(x, y)$ indicates the harmonic-to-distortion energy ratio for frequencies x and y (and multiples thereof, e.g. $\varsigma(2, 3)$ would be valid for frequencies 40 and 60 Hz, and 50 and 75 Hz, etc.). It appears that ς does not depend heavily on the ratio f_1/f_2. As ς is quite large for small a, the rectifier will perform well if there is one frequency component that dominates all others; however, if there are two (or more) components of comparable amplitude, then distortion will be severe.

2.3.2.3 Integrator

Spectral characteristics Another efficient method of generating harmonics is by integrating the rectified input signal, and resetting the output to zero after each second zero crossing. A discrete-time algorithm would thus process an input signal $x(n)$ into output

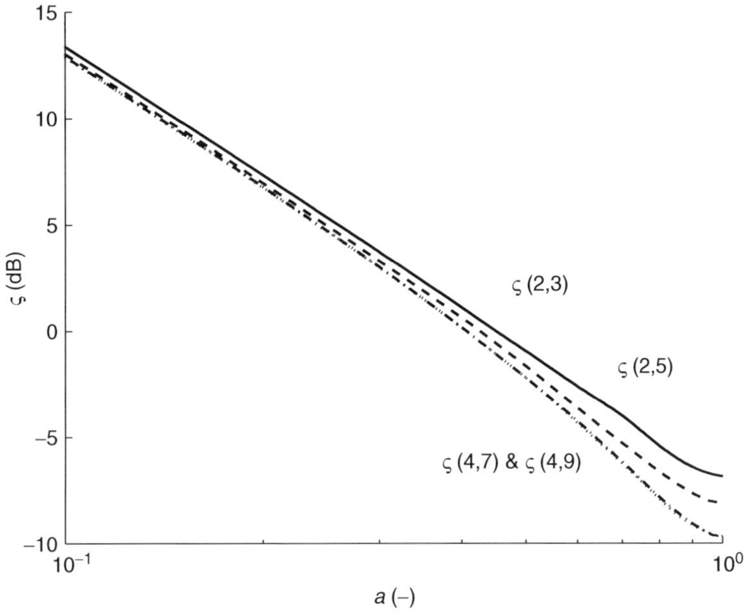

Figure 2.9 For a two-tone input with frequency ratios (f_1, f_2), and amplitudes 1 and a, Eqn. 2.93 can be used to compute the harmonic-to-intermodulation distortion energy ratio ς, for the rectifying NLD

$y(n)$ according to

$$y(n) = \begin{cases} 0 & \text{if} \quad z(x(n)) = 1 \text{ and } x(n) - x(n-1) > 0, \\ y(n-1) + c|x(n)| & \text{otherwise.} \end{cases} \tag{2.21}$$

c is a constant of integration and z is a function that detects zero crossings. In Eqn. 2.21, the output is reset to zero at positive-going (positive derivative of $x(n)$) zero crossings, but negative-going is also possible. The output signal will have the same fundamental frequency as the input signal, and for a pure-tone input will resemble a saw-tooth waveform. The integration has a low-pass filtering effect, but the discontinuities in the output due to the resetting create a strong harmonics spectrum. As is true for the rectifier, the integrator is a homogenous system, that is, input and output amplitudes are linearly related. Assuming a pure tone of frequency 1, the output signal will be (continuous-time)

$$y(t) = \begin{cases} \frac{2}{\pi}(1 - \cos(2\pi t)) & \text{for} \quad t \leq \frac{1}{2}, \\ \frac{2}{\pi}(3 + \cos(2\pi t)) & \text{for} \quad t > \frac{1}{2}. \end{cases} \tag{2.22}$$

Here we define $t \in [0, 1)$, which is the periodicity interval of $y(t)$. The Fourier series coefficients c_k then follow as

$$c_k = \frac{2k^2 + (-1)^k - 1}{i \times 2k\pi(k^2 - 1)}. \tag{2.23}$$

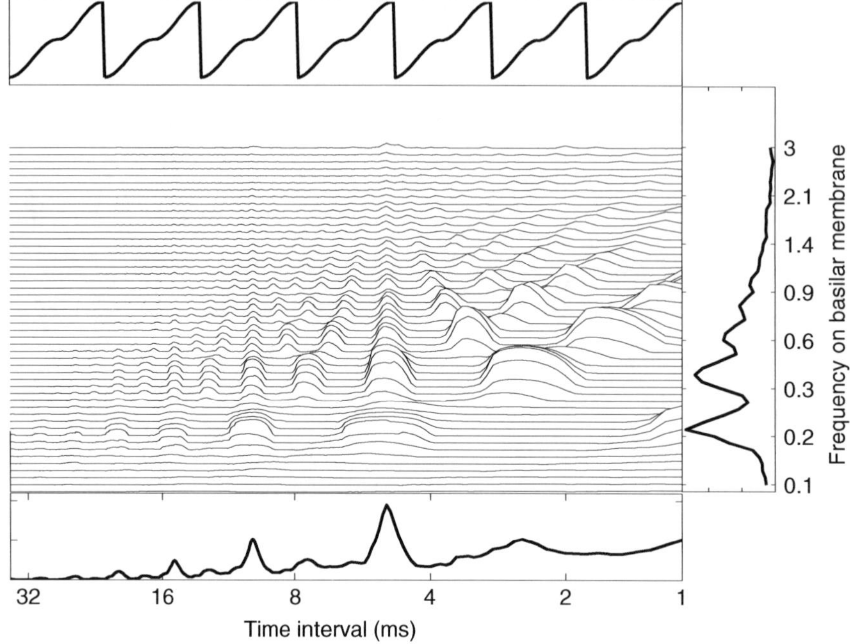

Figure 2.10 AIM calculation for the output of the integrating NLD for a 200-Hz pure-tone input. There is a single large peak at a lag of 5 ms, indicating a strong pitch percept at 200 Hz

Thus, the harmonics spectrum comprises all (even and odd) harmonics, decaying at a rate of −6 dB per octave for large k, and should give rise to a strong pitch at the fundamental. Therefore, we might expect the integrator to be a good algorithm to serve as NLD for low-frequency psychoacoustic BWE. This is confirmed by Fig. 2.10, which shows an AIM calculation for the integrating NLD output of a 200-Hz pure-tone input. The presence of all harmonics in the output signal yields one dominant peak at a lag of $\tau = 5$ ms (200-Hz pitch). Even if the fundamental is removed by filtering, or by a poor loudspeaker response at low frequencies, the 200-Hz pitch remains strong and unambiguous, as is shown in Fig. 2.11. The AIM calculation was done on a high-pass filtered (300 Hz cut-off) version of the integrated 200-Hz pure tone.

Temporal characteristics An analysis of the temporal characteristics shows that the output lags the input, refer to Fig. 2.12. A 5-cycle 50-Hz tone burst is shown in (a). The integrating NLD produces the signal shown in (b), after which bandpass filtering between 70 and 150 Hz is done, shown in (c). This filtered output signal remains small during the first cycle, and reaches a significant amplitude only at the end of the first cycle (where the integrator resets the output to zero). The combined effect is that the attack time of the input signal is increased, and also that the onset is delayed. This could be particularly detrimental if higher harmonics of the 50-Hz fundamental are present. More will be said

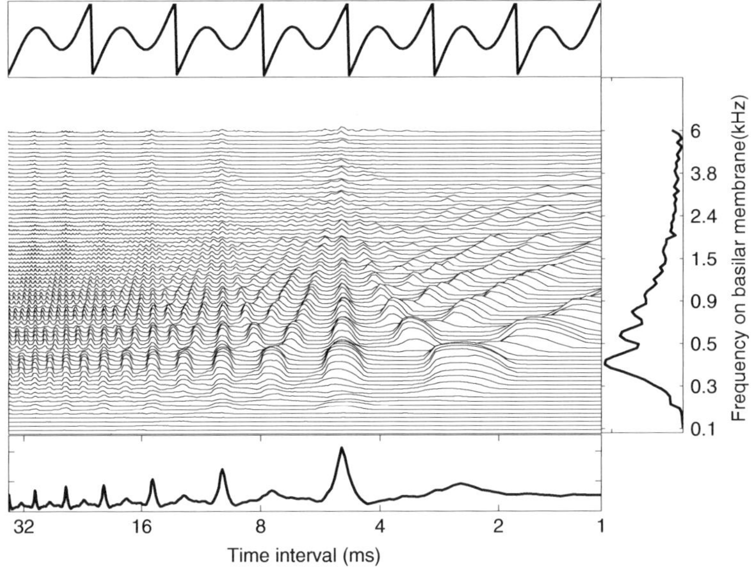

Figure 2.11 AIM calculation for the output of the integrating NLD for a 200-Hz pure-tone input, in which the 200-Hz component is filtered out (simulating the transfer function of a small loudspeaker). Even though the fundamental is not present in the spectrum, the remaining harmonics yield a single large peak at a lag of 5 ms, indicating a strong pitch percept at 200 Hz

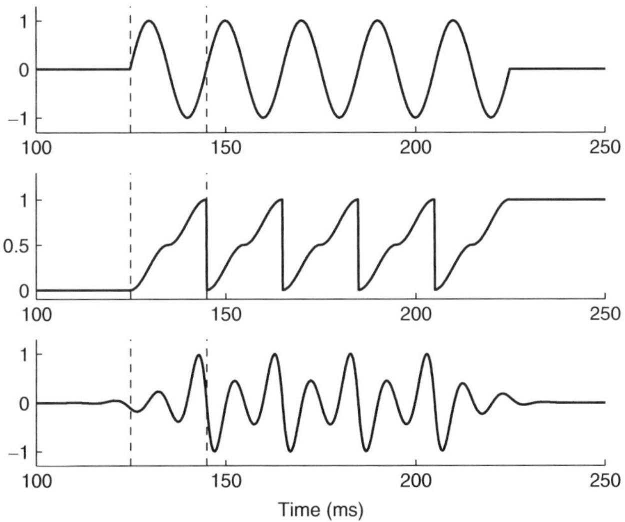

Figure 2.12 (a) Shows 5 cycles of a 50-Hz tone as input signal to an integrating NLD, the output of which is shown in (b). (c) Shows the result after bandpass filtering between 70 and 150 Hz. The filtered output signal rises more slowly than the input, and only reaches a significant amplitude at the end of the first cycle. This can lead to a degraded percept of the signal

about temporal characteristics when discussing the phase characteristics of the filters of low-frequency psychoacoustic BWE, in Sec. 2.3.3.3.

The percept is that signals with fast attacks and/or decays sound less 'tight'.

Intermodulation distortion As for the rectifier, the output signal spectrum for arbitrary periodic input signals can be computed conveniently; the derivation is presented in the appendix. The resulting Fourier series c_k for a given input Fourier series a_k is

$$c_0 = \int_0^1 (1-t)|f(t)|\,\mathrm{d}t, \tag{2.24}$$

$$c_k = \frac{b_k - \alpha_0}{i2\pi k}, \quad k \neq 0, \tag{2.25}$$

for the special case that we assume $f_0 = 1$ and there is one zero crossing in the interval [0, 1]. The magnitude of the integrator output just before resetting is α_0; the b_k are the Fourier series coefficients of the rectified output signal, and are given in Sec. 2.3.2.2. The b_k decay as $1/k$, so for large k, c_k will be proportional to α_0/k. Equation 2.108 (appendix) can be used to assess the relative amount of intermodulation distortion energy given a two-tone input signal (with possibly multiple zero crossings in the periodicity interval). An analytic solution is not available as the parameter α_0 depends on the frequencies and amplitudes of the input frequency components in a complicated non-linear way. Thus, we resort to numerical methods to compute ς, the harmonic-to-intermodulation distortion energy ratio, as in Sec. 2.3.2.2. Results are shown in Fig. 2.13 for a few f_1, f_2 combinations, where the amplitude of f_1 is always 1, and the amplitude of f_2 is $0 \leq a \leq 1$. In comparison to Fig. 2.9, which plots ς for various f_1/f_2 ratios using the rectifying NLD, the integrating NLD is seen to be significantly more robust against intermodulation distortion, as the ς's are considerably larger. In fact, $\varsigma > 0$ for almost all a. The graphs in Fig. 2.13 display a number of 'knee points', where ς suddenly decays more rapidly with increasing amplitude a of the f_2 component. This occurs because at particular values of a, additional zero crossings are created in the periodicity interval, which cause large changes in the output spectrum (see also Eqn. 2.108 in the appendix).

2.3.2.4 Clipper

Spectral characteristics A convenient way to generate a harmonics signal with only odd harmonics is by means of a limiter or a clipper. The limiter output signal g_l in response to an input f is

$$g_l(t) = \begin{cases} 1 & \text{if} \quad f(t) \geq 0, \\ -1 & \text{if} \quad f(t) < 0. \end{cases} \tag{2.26}$$

For the clipper, the output signal g_c is

$$g_c(t) = \begin{cases} f(t) & \text{if} \quad |f(t)| \leq l_c \\ l_c & \text{if} \quad f(t) > l_c, \\ -l_c & \text{if} \quad f(t) < -l_c, \end{cases} \tag{2.27}$$

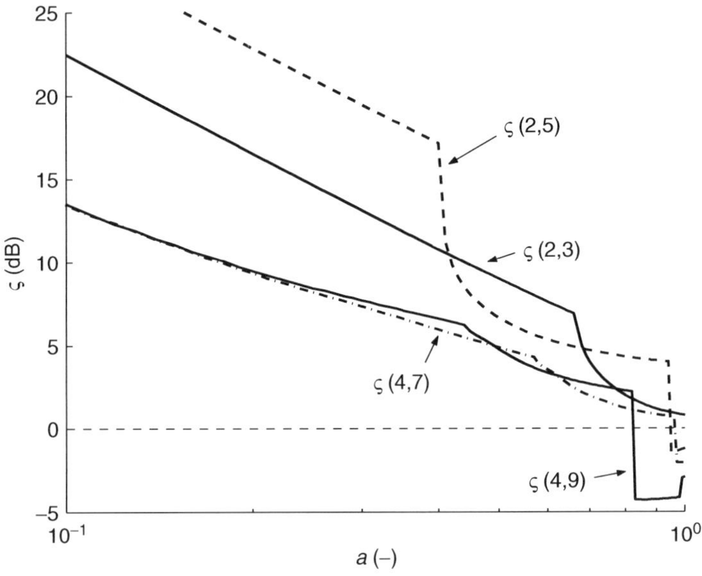

Figure 2.13 For a two-tone input with frequency ratios (f_1, f_2), and amplitudes 1 and a, Eqn. 2.108 can be used to compute the harmonic-to-intermodulation distortion energy ratio ς, for the integrating NLD

where l_c is the *clipping level*, here taken to be symmetrical around zero. One could also define different clipping levels for positive and negative signal values. Both the limiter and the clipper will generate odd harmonics of a pure-tone input signal, but are not directly suitable for BWE applications, as they are not homogeneous systems. For the limiter, this is because the output level is always ± 1, a highly non-linear characteristic. This can be overcome by detecting the envelope of the input signal and scaling the limited signal appropriately. For the clipper, the situation is a little bit more complicated. For low input levels, there may be no clipping at all, if $|f|$ does not exceed l_c, thus $g_c = f$. At intermediate input levels, moderate clipping will occur, with the desired harmonics generation. At very high input levels, such that mostly $|f| \gg l_c$, the clipper becomes a limiter (with output $\pm l_c$). The characteristics of a clipper vary significantly as the input level varies. Again, this can be overcome, or at least reduced, by scaling l_c in response to the level of f. In fact, by doing this in a special way, the clipper has demonstrated very good subjective results in the low-frequency psychoacoustic BWE application – we will elaborate on this later. As the subjective performance of the clipper is generally superior to that of the limiter, we will focus on the clipper in the remainder of this section. In Chapter 5, on high-frequency BWE of audio, we will introduce the 'soft' clipper, an operation that does not have a 'hard' threshold above which the output signal is not allowed to rise, but rather a mild compression of the input as the input level increases. The clipping as discussed in this section is hard clipping.

The spectral characteristics of a clipped sine depend greatly on the clipping level l_c. As a special case of the more general situation described in the appendix (Sec. 2.6.4), the

Fourier series coefficients of the clipped sine are, for the fundamental

$$a_1 = \frac{(2t_1 + \sin 2t_1)}{\pi}, \tag{2.28}$$

(using $t_1 = \sin^{-1} l_c$), and for the odd harmonics (even harmonics are zero because of the symmetry)

$$a_{2n+1} = \frac{\sin 2(n+1)t_1}{\pi(n+1)(2n+1)} + \frac{\sin 2nt_1}{\pi n(2n+1)} \tag{2.29}$$

As l_c approaches 0, we find as limiting case (for all values of n)

$$a_n \approx \frac{4l_c}{\pi n}, \tag{2.30}$$

which is also the result for the limiter (at level l_c). Figure 2.14 shows the Fourier series coefficients according to Eqns. 2.28 and 2.29 for $0 <= l_c <= 1$.

Figure 2.15 shows an AIM calculation for a clipped 200-Hz pure tone. There is a clear peak at $\tau = 5$ ms, corresponding to a pitch percept of 200 Hz. There are two rather large and broad peaks at smaller lags, and these become very prominent if the fundamental frequency is removed, for example, by high-pass filtering at 300 Hz, as shown in Fig. 2.16. The sharpest peak still occurs at $\tau = 5$ ms (200-Hz pitch), but the other peaks (around 3 and 2 ms) are very large as well. The reason these peaks occur is that whereas only a 200-Hz fundamental 'fits' the given harmonic series perfectly, there are other possible

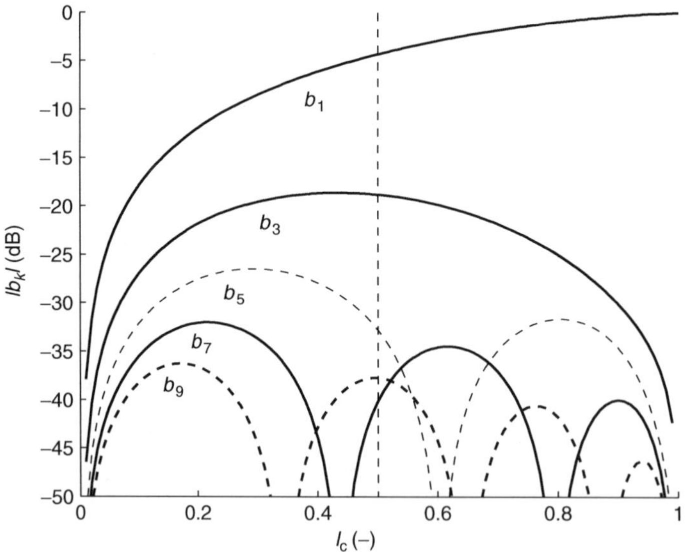

Figure 2.14 Magnitudes of harmonics of a clipped sine; clipping level l_c. Harmonic k is indicated as b_k; note that for $k \geq 5$, magnitudes may be zero for some l_c. The dashed vertical line indicates the commonly used value of 0.5 for the clipping level

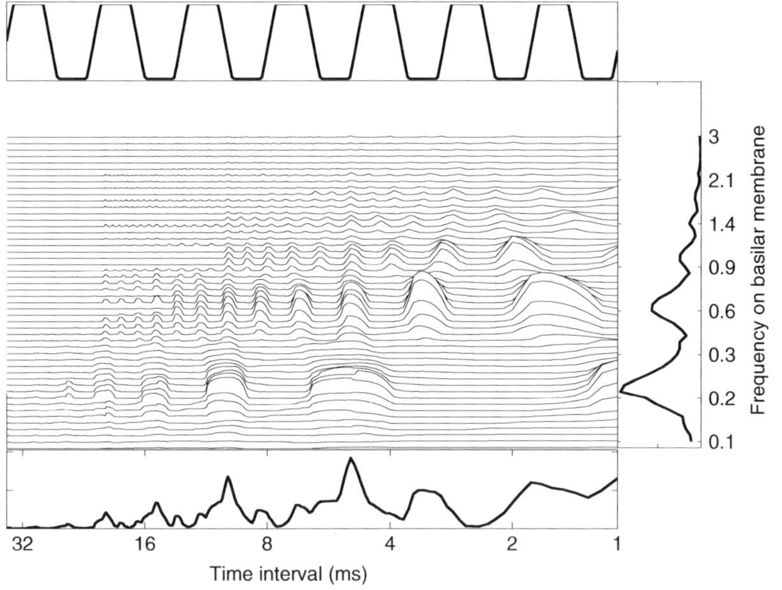

Figure 2.15 AIM calculations for a complex tone consisting of f_0 with odd harmonics (3, 5, and 7)

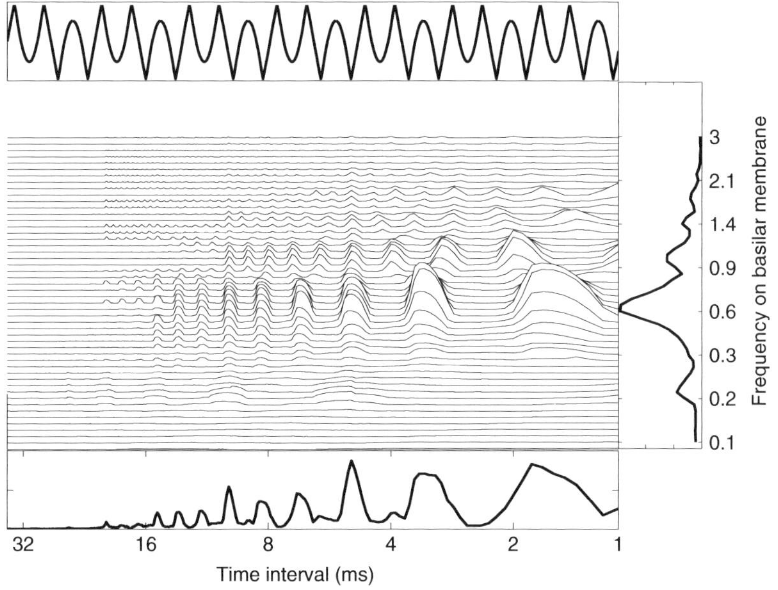

Figure 2.16 AIM calculations for a complex tone consisting of odd harmonics (3, 5, and 7), without fundamental f_0

fundamentals that fit 'reasonably' well. Especially, if we only consider the odd harmonics 3, 5, and 7 (which are in the dominance region for pitch perception (Ritsma [227]), then a 333.3- or a 500-Hz fundamental fits the harmonic at 1000 Hz exactly and roughly matches the 600 and 1400 Hz harmonics. Because the fit is not exact, pitches corresponding to either 333.3 or 500 Hz would be quite vague. But it would seem that the signal as a whole would not have a well-defined pitch, as would for example, the output signal of an integrator (Sec. 2.3.2.3 and Figs. 2.10–2.11). Thus, when a clipping NLD is used in low-frequency psychoacoustic BWE, and the fundamental frequency is not reproduced, then the resulting pitch may not be very strong at the original fundamental.

Temporal characteristics Figure 2.17 shows a 50-Hz signal (a), which clipped at $l_c = 0.5$; the resulting signal is shown in (b) (amplitude normalized). The clipped signal is filtered between 70 and 150 Hz, as shown in (c). The filtered output reaches maximum amplitude within the first cycle, thus the temporal characteristics are good.

Intermodulation distortion The general expression for the output spectrum Fourier coefficients b_n of a clipped periodic input signal with Fourier coefficients a_n is given in the appendix, Sec. 2.6.4. The result for a signal with period 2π is

$$b_n = \frac{1}{2\pi} \sum_{k=1}^{K} \sum_{m=-\infty}^{\infty} m a_m \frac{e^{i(m-n)\beta_k} - e^{i(m-n)\alpha_k}}{i(m-n)}, \tag{2.31}$$

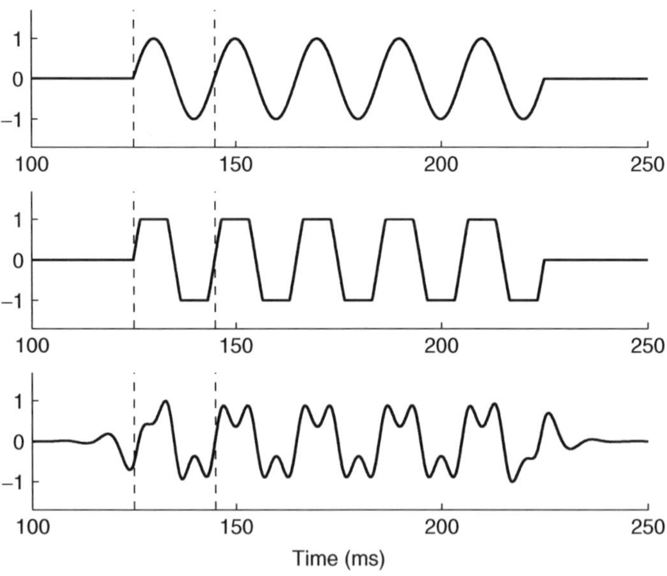

Figure 2.17 (a) This shows 5 cycles of a 50 Hz tone as input signal to a clipping NLD, the output of which is shown in (b). (c) This shows the result after bandpass filtering between 70 and 150 Hz. The filtered output reaches maximum amplitude in the first cycle, which is beneficial to perceived quality

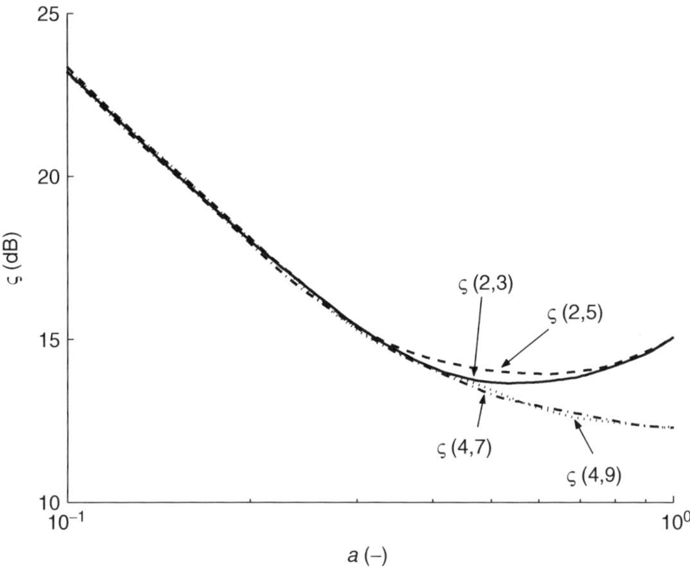

Figure 2.18 For a two-tone input with frequency ratios (f_1, f_2), and amplitudes 1 and a, the harmonic-to-intermodulation distortion energy ratio ς for the clipping non-linearity, at a clipping level of 0.5 (signal is normalized to the range [-1, +1])

with the α_k, β_k ($k = 1 \ldots K$) determining the intervals [α_k, β_k] where the signal is not clipped; these will depend on the clipping level l_c and the amplitude of the signal.

This analysis was performed for a two-tone input signal, with frequencies f_1 and f_2, amplitudes 1 and a = [0, 1]. Because a clipping non-linearity is not a homogeneous system, all signals were normalized to the range [−1, +1], and then clipped at a level of 0.5. Without this normalization ς, values are about 2 dB lower for $a \approx 1$. For the normalized clipped signals, ς is shown in Fig. 2.18. It is obvious that the values are significantly higher than for any of the preceding non-linearities discussed (multiplier, rectifier, integrator). Presumably, this is due to the fact that during portions in which the signals are clipped, the output remains fixed at the clipping level, and the influence of the interfering frequency components is thus minimized. The effect of clipping level l_c (for a fixed two-tone input signal) is not large for $l_c < 0.5$: ς drops by only a few dB as $l_c \downarrow 0$ (indicating that a limiting non-linearity performs slightly worse than a clipping non-linearity with respect to intermodulation distortion). As $\lim_{l_c \uparrow 1}$, $\lim_{\varsigma \to \infty}$.

Input-level-dependent clipping level It was already mentioned that the clipping level l_c should be scaled according to the envelope of its input signal f. Here, we shall discuss how this scaling can be implemented, following a method proposed by C. Polisset. The basic idea is to follow the envelope of f with different time constants during the attack and decay of the waveform. We define the nominal clipping level $l_{\mathrm{N},c}$ such that for stationary input signals

$$l_c = l_{\mathrm{N},c} \max |f(t)|, \tag{2.32}$$

for example, $l_{\mathrm{N},c} = 1/2$ would be a typical choice, such that the clipping level is half of the maximum absolute value of f. The time dependence of l_c can then be defined as (assuming a sample rate of $1/T_{\mathrm{s}}$)

$$l_c(t) = \begin{cases} a l_c(t - T_{\mathrm{s}}) & \text{if} \quad |f(t)| \leq l_c(t)/l_{\mathrm{N},c}, \\ l_{\mathrm{N},c}|f(t)| & \text{if} \quad |f(t)| > l_c(t)/l_{\mathrm{N},c}. \end{cases} \tag{2.33}$$

Such a dependence will cause l_c to follow without delay, any increase in amplitude of f, but to decrease at a maximum rate given by the parameter a of Eqn. 2.33. For stability reasons, $0 < a < 1$. To achieve a specified decay time $\tau_{1/2}$ (duration in which l_c will halve in value), a is given by

$$a = \mathrm{e}^{-\ln 2 \frac{\tau_{1/2}}{T_{\mathrm{s}}}}. \tag{2.34}$$

The limited rate of decrease of l_c is purposefully chosen such that typical musical signals decay much faster. An illustration hereof is given in the upper panel of Fig. 2.19. The solid line shows a decaying musical signal, and the dashed line is the associated clipping level. During the initial phase of the signal, where the envelope rises fast, the clipping level is increased to half of the envelope. During the next phase, the level of the signal is sustained, causing a stationary envelope. The clipping level is also stationary. The harmonics spectra during the various phases are shown schematically in the accompanying insets of Fig. 2.19. The last part of the signal is the decay, where the clipping level decays at a much slower rate than the signal envelope. Therefore, before the signal reaches zero amplitude, the clipping level has exceeded the signal level. From that point on, no harmonics will be generated. Effectively, after the sustained period of the signal, the harmonics spectrum slowly changes from its maximum strength to complete absence (no harmonics).

The time constant for the decay of l_c is usually taken in the order of a few seconds. This will present a problem if the audio signal decreases its overall level rapidly, because no harmonics generation will occur until l_c has decayed sufficiently. Such cases do not often occur though, in particular not in modern music, which typically has a very low dynamic range. The clipping level usually does not vary over a great range.

The perceptual effect of the varying harmonics strength seems to be generally beneficial to the low-frequency psychoacoustic BWE application. Some reasons for this observed benefit might be:

- Most instruments have time-varying spectral characteristics, and the level-dependent clipper might emulate these characteristics better than other NLDs.
- During the final part of the signal, there is no harmonics generation at all, and thus the output of the level-dependent clipper is equal to its input. This means that the latter part of the output signal will have a much lower audibility than if a full harmonics spectrum were generated. The 'tone-lengthening' effect mentioned in Sec. 2.2.3 might thus be decreased, or prevented altogether. This means the audibility of the output signal has been greatly increased relative to the input, but tone duration has not markedly changed. One could assume that this is preferred by listeners.

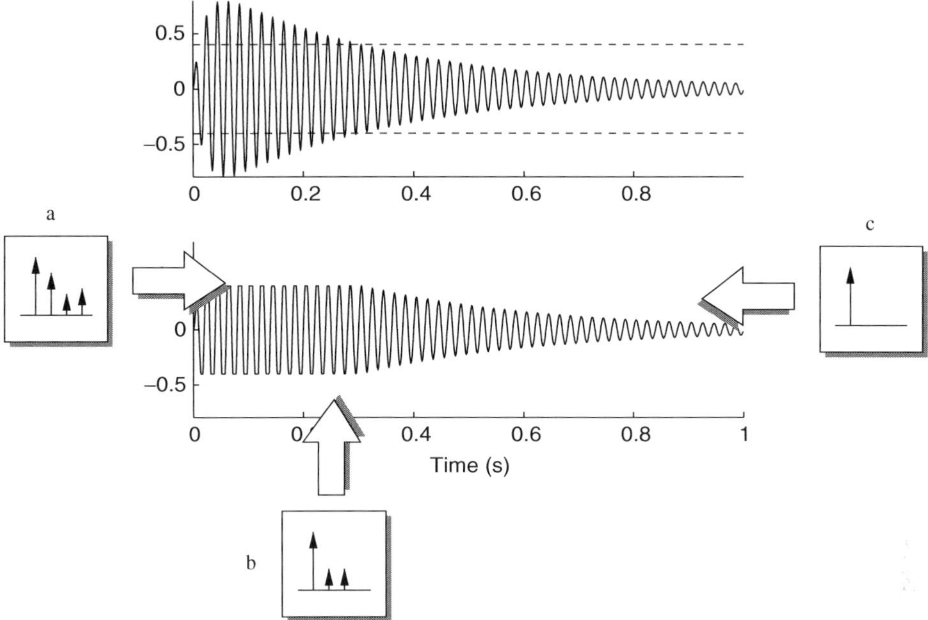

Figure 2.19 A transient signal and associated clipping level (dashed line). As the signal level decreases, the clipping level decreases at a much slower rate (it is shown constant in this figure). This causes the harmonics spectrum to change during the lifetime of the signal, as is shown in the accompanying insets. Inset a shows the harmonics at maximum strength at the beginning of the signal; inset b shows a moderate strength signal when the clipping level becomes relatively high; inset c shows the harmonics spectrum during the decay, where the signal level is below the clipping level. The only component left is the fundamental frequency

- Because the latter parts of the output signal have a weak or non-existent harmonics spectrum, the overall timbre remains closer to that of the input. Even though this comes at a cost of reduced loudness, the closer-matching timbre might be preferred by listeners.

The time-varying harmonics spectrum is a natural effect of a clipping non-linearity, and the method of varying the clipping level according to Eqn. 2.33 gives an appropriate spectro-temporal characteristic to the clipper. It would be much harder to implement something similar for a rectifying (Sec. 2.3.2.2) or integrating (Sec. 2.3.2.3) non-linearity. These NLDs produce a harmonics spectrum that is independent of level, and the only way to create a time-varying harmonics spectrum would involve modifying some basic property of the NLD.

Amplitude non-linearities on various time scales As presented here, a point has been made that the level-*dependent* nature of the clipper is beneficial to subjective quality.

However, on several earlier occasions it has been emphasized that a level-*independent* NLD is the best for BWE applications (NLDs should be homogeneous systems). Thus, there appears to be a contradiction. The resolution of this (apparent) contradiction is the fact that the notion 'level dependent' can be viewed on different time scales. The clipper with varying l_c (Eqn. 2.33) is designed with the aim to provide a level-dependent clipping on a small time scale, that of a single note. If the entire level of the audio signal changes on a larger time scale, the clipping level will adjust appropriately. Therefore, the level-dependent clipper is level independent (linear in amplitude) if one considers a 'large-enough' temporal window. In conclusion, a more precise statement might be that BWE algorithms should be homogeneous on a 'large' time scale, in which 'large' means considerably longer than a typical musical note. On a 'small' time scale (approximately the duration of a tone), the algorithm may be non-linear in amplitude.

2.3.2.5 Discussion of Non-linear Devices

All of the NLDs discussed previously in this section have distinct advantages and disadvantages. For each NLD, an analysis was presented of spectral, temporal, and intermodulation distortion characteristics, and a summary of these is shown in Table 2.1. Apart from such an objective point of view, a subjective rating is ultimately more important; of course the objective analysis helps to understand the subjective impressions.

Subjective experiments will be discussed in Sec. 2.5, in which the rectifying and integrating NLDs were included. The result of that experiment was that both these NLDs were rated approximately equal, with a slight advantage for the rectifying NLD, which may seem surprising given the better spectral characteristic of the integrator. The clipping NLD was not included in this test as it was, at the time of the experiment, not fully developed. Subsequent subjective evaluations have usually favoured the clipping NLD over all others, although no formal experiments have been conducted to confirm this. Much more on the subjective evaluation of low-frequency psychoacoustic BWE will follow in Sec. 2.5.

Table 2.1 Summary of objective features of the various NLDs. The last row describes a frequency-tracking NLD, to be discussed in Sec. 2.4

Characteristic	Amp.-linear	Spectral	Temporal	Interm. dist.
Multiplier	No, needs signal level scaling	Flexible	Good	Variable, depends on harmonic number
Rectifier	Yes	Even harmonics (pitch doubling)	Good	Moderate-poor depends on input
Integrator	Yes	All harmonics	Poor, slow attack/decay	Good
Clipper	Long-time: yes Short-time: no	Odd harmonics, is ambiguous without f_0	Good	Very good
Freq. track.	Yes	Flexible	Good	Excellent

2.3.3 FILTERING

Whichever NLD is used in the low-frequency psychoacoustic BWE system (Figs. I.2 and 2.4), it must be supplied with an appropriate input signal. Also, its output usually needs some filtering to yield a pleasant timbre. The characteristics of these two filters will be discussed in this section. Also, for low-frequency psychoacoustic BWE, it is important that the filters are linear phase, as will be demonstrated in Sec. 2.3.3.3.

2.3.3.1 Filter 1

Filter 1 precedes the NLD and its function is to pass only those frequencies that need to be enhanced. The use of this filter is one of the essential differences between the use of controlled distortion for low-frequency psychoacoustic BWE applications versus uncontrolled distortion as may occur in amplifiers or loudspeakers. In the discussions of various NLDs in Sec. 2.3.2, we found that introducing more than one frequency component to the NLD leads to intermodulation distortion, which should be avoided. Thus, the bandwidth of filter 1 should not be too large. If necessary, filter 1 could be replaced by a filterbank, spanning the same frequency range as the original filter, with each filter connected to an identical NLD, the outputs of which will be summed at the end. In such an arrangement, each filter has a very narrow bandwidth, and intermodulation distortion will be minimized, at the expense of increased algorithmic complexity. However, on the basis of our experience, the use of one single filter does not cause excessive intermodulation distortion, and therefore the use of the just-mentioned filterbank does not seem necessary.

Filter 1 should not pass frequencies above the low-frequency cut-off, f_l, of the loudspeaker, as these components should be adequately reproduced without processing. Therefore, the upper limit of filter 1 will be at most f_l. In most applications, this value will vary somewhere between 70 and 200 Hz. In principle, the lower limit of filter 1 should be approximately 20 Hz, as this is around the minimum audible frequency. But, if the upper limit of filter 1 is very high, it may be better to increase this lower limit somewhat; a bandwidth of two octaves should suffice. Note that the lower limit of musical pitch lies around 40 Hz (Guttman and Pruzansky [102]), so it might be argued that including frequency components below this limit is of questionable value. Nevertheless, frequencies below 40 Hz do occur in music (albeit very rarely), and as it is the aim of low-frequency psychoacoustic BWE to enhance bass perception, we will advocate the use of 20 Hz as the lower limit. If the limiting frequency is set even lower, then any energy below 20 Hz (if present in the audio signal for whatever reason), will be reproduced at the correct fundamental frequency, which, being so low, is heard as an amplitude modulation instead of a unified low-pitch percept. The effect on artificially generated tones of frequency lower than 20 Hz does not sound good, and therefore, the lower cut-off frequency of filter 1 should not be lower than 20 Hz.

The order of the filter does not seem to have too great an effect on quality. Low- and high-pass flanks of second order (−12 dB per octave) seem to be sufficient for adequate separation of the bass frequencies. Alternatively, a stopband attenuation of 20 dB will suffice. For the passband ripple, a value of 1 dB seems good enough; it is hard to perceive any deleterious effects of this ripple if one is presented only with the BWE-processed signals.

The phase response of filter 1 should be linear, the reason for which will be demonstrated in Sec. 2.3.3.3.

2.3.3.2 Filter 2

The second filter, placed behind the NLD, filters the harmonics spectrum such that a reasonable timbre results. This is necessary as the timbre of the harmonics directly out of the NLD usually sounds too 'sharp', which is caused by the harmonic amplitudes being too large. By low-pass filtering, a more pleasant timbre can be achieved. Again, a second-order filter (12 dB per octave) usually suffices. Note that a low-pass flank of −12 dB per octave does not mean that successive harmonics will be attenuated by 12 dB relative to each other. Because the fundamental frequency of the filtered signal is usually quite low, there may be several harmonics present in a single octave at the low-pass flank of the filter. This is illustrated in Fig. 2.20.

The fundamental frequency is usually attenuated by filter 2, because it is either not desired in the output signal, and if it is, it is available directly from the original audio signal. Thus, filter 2 employs a high-pass flank, of moderate order, with a cut-off frequency that is (roughly) equal to the higher frequency limit of filter 1. Thus, filter 2 has a bandpass characteristic, with a bandwidth of about 1 to 1.5 octaves wide.

Filter 2 is preferentially implemented as a linear-phase filter, for reasons that will be explained in the next section.

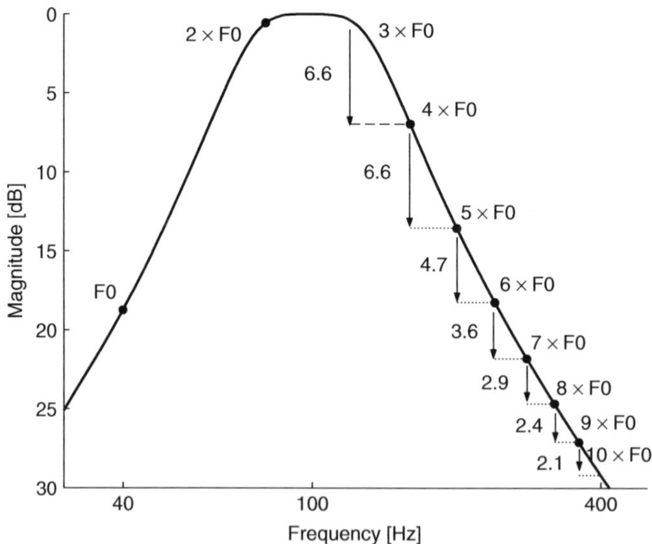

Figure 2.20 A second-order Butterworth bandpass filter with cut-off frequencies of 70 and 140 Hz. A harmonics signal with 40-Hz fundamental has harmonics as indicated by the filled circles: the filter attenuation in dB of successive harmonics is indicated on the right flank of the filter. Note that this attenuation depends both on the filter order *and* on the fundamental frequency value

2.3.3.3 Linear versus Non-linear Phase Filters

The topic of using linear-phase versus non-linear-phase filters in audio processing can sometimes be a controversial one. Although there is little scientific evidence that people are sensitive to phase distortion – excluding some special cases – some would claim that linear-phase systems sound far better than their non-linear-phase counterparts. The issue of linear or non-linear phase for low-frequency psychoacoustic BWE filtering can be analysed objectively, and the conclusion is that it *is* better to employ linear-phase filtering, for reasons to be explained next.

Because in low-frequency psychoacoustic BWE the filter bandwidths and cut-off frequencies are usually orders of magnitude smaller than the system sample rate, IIR implementations are more efficient computationally than a direct FIR implementation. For a modest sample rate of 10 kHz, a frequency resolution of 10 Hz (which would be required to design a filter with a passband of about 100 Hz) would necessitate 1000 taps. In contrast, an IIR filter of ten or less coefficients will probably achieve the desired requirements as well. A drawback of IIR filters is that the phase is non-linear. Lower-order FIR filters are possible if the signal is downsampled before NLD processing. Because NLD implementations are computationally trivial (rectification, clipping, integration), there is not much to gain from downsampling from a computational point of view, and the required anti-alias filters will probably negate the advantage of processing at a lower sample rate. Another option to use FIR filters at reduced complexity is through frequency warping (Härmä *et al.* [104]). With this technique, it is possible to trade high-frequency resolution for low-frequency resolution, which would allow lower-order FIR filters to be used. The concept has not been evaluated for low-frequency psychoacoustic BWE, though.

To be explicit, the problems with non-linear-phase filters in low-frequency psychoacoustic BWE are:

- Interference of synthetic harmonics signal with other signal components. The output of the NLD consists of harmonics, and sometimes the fundamental frequency component as well. After filtering by filter 2, these are added back to the main signal. Because the main signal also contains the fundamental, and possibly some harmonics, interference will occur. As the phase relationships of the original fundamental and its harmonics and the synthetic BWE signal are impossible to predict a priori, the nature of this interference (constructive or destructive) is unknown. We can examine this issue to some degree by using a method devised by C. Polisset (private communication), in which we use a pure-tone input and compute the steady-state output signal for a BWE algorithm. If we then compare the energy of this output signal to the input signal, it will be apparent if interference occurs. We will denote this frequency ratio by $h(f, g)$, akin to a transfer function, with frequency f and harmonics gain g as parameters. Note that $h(f, g)$ is not a transfer function in the common sense of the term, because BWE systems are not linear (and sometimes not time-invariant either, in the case of level-dependent clipping). We have

$$h(f, g) = \frac{[\sin(2\pi ft) + g\phi(\sin(2\pi ft))]_{\mathrm{rms}}}{[\sin(2\pi ft)]_{\mathrm{rms}}}. \tag{2.35}$$

The function ϕ indicates the BWE processing. We can also compute $h(f, g)$ for the ratio of rms value of harmonics signal (without adding the input signal) to the rms value of the input signal. By means of example, we use a BWE algorithm with elliptic IIR filters (FIL1 from 20–70 Hz, both flanks of second order; FIL2 from 70–140 Hz, both flanks of second order) and a clipping non-linearity at 50% clipping level. The results are shown in Fig. 2.21. Panels a and c show values of $h(f, g)$, with harmonics-to-input energy ratio in panel a and harmonics + input-to-input energy ratio in panel c. Note that significant gain is obtained in a narrow frequency interval, and that destructive interference occurs in panel c for frequency values slightly below 100 Hz. This would mean that the output energy of the BWE system would actually be smaller than the input energy. In contrast, panels b and d show $h(f, g)$ in the same conditions, but using linear-phase implementations of the same filter characteristics (to be discussed below). Note that gain is obtained over a much broader frequency interval, and that variations of h as functions of f and g are much smoother than in panels a and c. The general features shown in the four panels of Fig. 2.21 are not dependent on the use of elliptic filters or the specific frequency bands used.

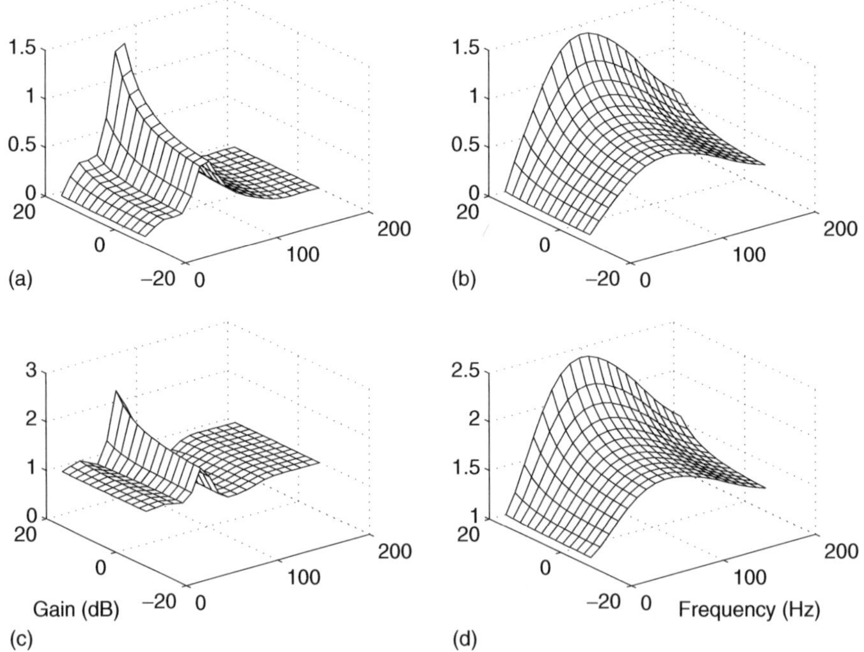

Figure 2.21 (a): h (see Eqn. 2.35) considering BWE with *non-linear* phase filters; input frequency and harmonics gain are variables. The ratio of harmonics-to-input energy is shown as the third dimension. (b): same, but for a *linear* phase filter with the same spectral specifications, and otherwise similar BWE processing. (c): h as in (a), but the value shown is of harmonics+input-to-input energy. (d): same, but for a *linear* phase filter. Note that in both cases the linear-phase filter implementations of the BWE processing give much smoother characteristics

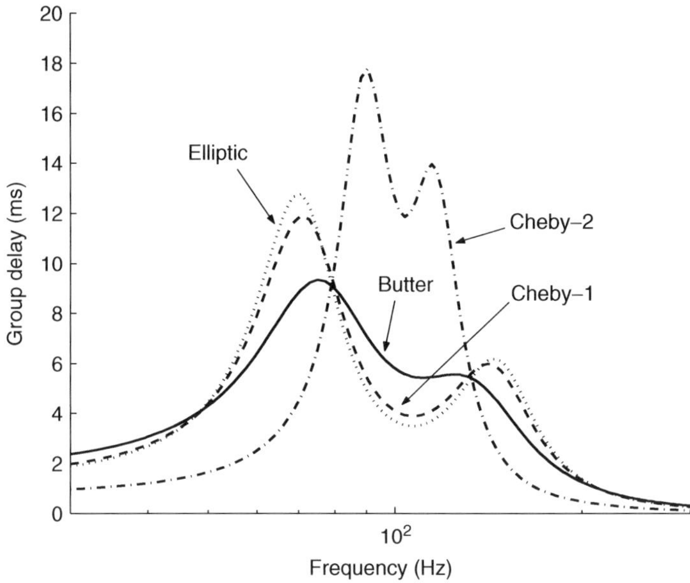

Figure 2.22 Group delay of four IIR filters (Butterworth, Chebyshev type 1, Chebyshev type 2, and elliptic, with bandpass 70–140 Hz, second-order low-pass and high-pass, $f_s = 44.1$ kHz). Every filter shows large variations in group delay

- Frequency-dependent delay of harmonics signal. Because the group delay of a filter is equal to the derivate of the phase response, a non-linear-phase filter will have a non-constant group delay. This means that different frequency components will be delayed by different amounts of time. For low-frequency psychoacoustic BWE applications, this variation in delay can be significant even for successive harmonics. To see this, we computed group delay for four common IIR filter types in Fig. 2.22, where all filters were bandpass filters with cut-off frequencies of 70 and 150 Hz, and both flanks were of second order. The total group delay varies with filter type, but reaches 10 to 15 ms in the passband of the filters, and almost 20 ms for the Chebyshev type 2 filter; this is around 1–2 signal periods for the frequencies of interest. The *variation* in group delay with frequency, being more important, is around 5 ms for most filter types. This may seem a small amount, but results from a study by Zera and Green [304] indicate that such delays may be audible. They investigated the audibility of onset asynchronies of various harmonics of a multi-tone complex, for a variety of onset times, and found that delays of 2 ms are audible. Also note that the BWE algorithm uses two filters, and, thus, group delay variations will be approximately twice the value as indicated in Fig. 2.22: in the order of 10 ms. They also found that thresholds for offset asynchronies are significantly larger than for the onset asynchronies. Thus, we may expect that the delay variations caused by non-linear-phase filters in BWE are audible at the onset of tones with a fast attack. Another effect that may be important is that common onset of individual frequency components is a strong grouping cue (Bregman [38]). Conversely, an asynchrony in onset across frequency may cause segregation of some harmonics from the bulk of the harmonic complex, causing two (or more) tones to be heard in the BWE-processed signal. We have some anecdotal evidence that this indeed occurs, as musically trained listeners have sometimes commented that bass tones sound delayed with respect to the rest of the music, when listening to low-frequency psychoacoustic BWE processing (with IIR filters).

Low-frequency psychoacoustic BWE sounds better with linear-phase filters than with non-linear-phase filters, and it is plausible that the two reasons discussed above are responsible for this. It is not clear how much either effect (interference and group delay variations) deteriorates the quality by itself, but in any case, there is sufficient motivation to use linear-phase filtering in low-frequency psychoacoustic BWE algorithms. A useful method of implementing linear-phase IIR filters was devised by Powell and Chau [213]. Basically, the method involves a double filtering of the signal; once in 'forward' time and once in 'backward' time (reversing the order of the samples). The backward-time filtering exactly compensates the frequency-dependent delay of the forward-time filtering; the magnitude characteristic imposed on the signal is the square of the filter when applied once, but this can be accounted for in the design of the filter. In a real-time system, finite blocks of data must be used, and for a good block connection, the overlap-add method (OLA) is used (Allen [18]). In this way, the low computational complexity of the IIR filter is maintained, although the OLA requires each sample to be effectively processed four times. Still, this will be much more efficient than a direct FIR implementation.

The linear-phase filtering by filter 1 and filter 2 should be accompanied by a corresponding delay of the main signal in the unprocessed signal branch (Fig. 2.4). If the main signal is high-pass filtered (to remove frequencies below f_l), this is best done with a linear-phase filter as well. The net effect is then a delay of the entire signal.

2.3.4 GAIN OF HARMONICS SIGNAL

The final step before adding the generated harmonics signal back to the main signal is scaling. In Sec. 2.2.3, three points were noted:

- Frequency dependence of loudness: This implies that loudness of equal-level (sound pressure level) harmonics will be higher than that of the fundamental.
- Frequency dependence of 'loudness growth': This implies that equal variations in sound pressure level will lead to smaller loudness variations for the harmonics than for the fundamental.
- 'Tone-lengthening' effect: The combination of increased loudness and better loudspeaker response at frequencies of harmonics (vs fundamental) leads to tones that sound longer when BWE-processed.

In Sec. 2.3.4.2, a method is presented to adaptively vary the gain of the harmonics signal, based on the equal-loudness contours. In Sec. 2.3.4.3, a method that varies the gain according to the total output level of the BWE signal is presented.

2.3.4.1 Fixed Gain

The simplest solution is to simply ignore the loudness variations with frequency, and apply a fixed gain to the harmonics signal. Loudness variations for various frequencies are generally not huge, and, therefore, a fixed gain can be a suitable solution for a simple low-frequency psychoacoustic BWE system. The gain value will depend on the NLD used and the characteristics of the loudspeaker. From a manufacturer's point of view, maximum bass loudness is usually desired. The gain value can be set as high as is desired, the only limitation being cone excursion and power-handling capacity of the loudspeaker (Sec. 1.3.2).

2.3.4.2 Frequency-adaptive Gain

Another method is due to Gan *et al.* [83] ('Virtual Bass'). They consider that the 'transfer function' of pressure amplitude to loudness (SPL to phones) is similar to that of a downwards expander, with a frequency-dependent expansion ratio E_r. In other words, if the level at the input is lower by x dB, the loudness will decrease by $E_r(f) \times x$ dB. For example, for 40 Hz the expansion ratio is 1.52, while for 100 Hz the expansion ratio is 1.24. A further assumption is that the expansion ratio is nearly independent of absolute loudness in the range 20–80 phones, for frequencies of 110–1000 Hz[2]. With respect to log-frequency, a simple relationship is found to estimate the frequency-dependent expansion rate $\hat{E}_r$, as

$$\hat{E}_r = -0.103 \ln f + 1.71, \quad f > 100 \text{ Hz}, \tag{2.36}$$

where it is acknowledged that for frequencies below 100 Hz this approximation underestimates the actual expansion ratio. Suppose now that harmonics of a fundamental frequency f_0 are generated, and consider the nth harmonic. The 'harmonics expansion ratio' $H\hat{E}_r(f_0, n)$ to be used for this harmonic is then given by

$$H\hat{E}_r(f_0, n) = \frac{\hat{E}_r(f_0)}{\hat{E}_r(nf_0)}, \tag{2.37}$$

and specifies the expansion ratio of the nth harmonic *relative* to the fundamental. For $40 < f_0 < 100$ Hz, the expansion ratio for higher harmonics is fairly independent of f_0 and is given as in Table 2.2. The 'Virtual Bass' algorithm uses a modulation technique (the original reference does not detail this procedure) to generate the individual higher harmonics, and therefore it is possible to apply the appropriate expansion ratio to each harmonic. Observing that the expansion ratio for all the harmonics are roughly 1.10, a simplification could be to apply this expansion ratio to the entire harmonics signal. In a structure as in Fig. 2.4, such an expansion may be achieved by first estimating the envelope $\tilde{f}(t)$ of the BWE harmonics signal f and scaling according to

$$f(t) = (\tilde{f})^{1.1}(t) \times f(t). \tag{2.38}$$

Envelope of a signal can be estimated by low-pass filtering the absolute value of the signal. Figure 2.23 illustrates how such an approach can be incorporated in to the general BWE structure.

[2] It is also implicitly assumed that the equal-loudness-level contours that were measured for pure tones can be used to assess loudness growth of complex tones, which is unlikely to be entirely valid.

Table 2.2 Harmonics expansion ratio as proposed by Gan *et al.* [83], using Eqn. 2.37. These values are valid for the range of fundamental frequencies Gan *et al.* considered (40–100 Hz)

Harmonic n	2	3	4	5
$HEr(f_0, n)$	1.06	1.10	1.13	1.15

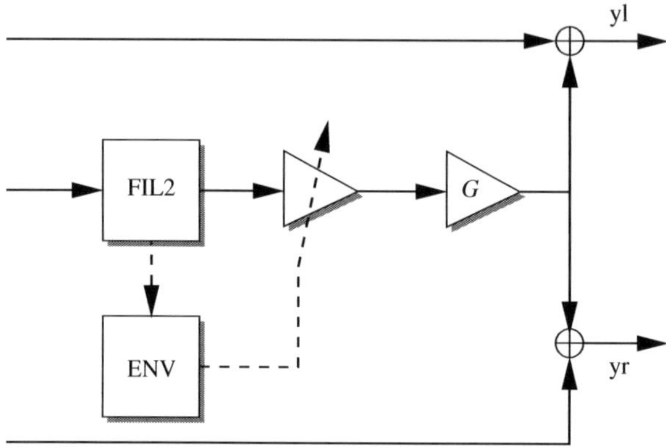

Figure 2.23 Part of a low-frequency psychoacoustic BWE algorithm, compare to Fig. 2.4. The dashed lines indicate a feedforward loop from the output of filter 2, which estimates the envelope of the harmonics signal. The envelope is then expanded by a factor of 1.1, and this value is used to scale the harmonics signal. The final scale factor G is used to bring the entire harmonics signal to an appropriate level

2.3.4.3 Output-level-adaptive Gain

In low-frequency psychoacoustic BWE, the scaled harmonics signal is added back to the main signal (Fig. 2.4) and applied to the loudspeaker terminals. The BWE processing emphasizes frequencies above the loudspeaker cut-off frequency f_l relative to frequencies below f_l, but still care must be taken to avoid overloading the loudspeaker. One could of course implement a fixed scaling of the harmonics signal such that at high reproduction levels distortion is avoided, but this may compromise performance at lower reproduction levels. Especially if a large bass response is desired, a high gain for the harmonics signal should be used at low reproduction levels, as audibility rapidly decreases at low sound pressure levels. Both loudspeaker protection and better matching of human audibility can be achieved by controlling the gain of the harmonics signal in response to the level of the output signal, as in Fig. 2.24. The feedback loop will ensure that at intermediate and low output levels, the gain is at its maximum value, but if the output level is high, the gain is adjusted appropriately. The decay time of the gain control signal must be very small, such that distortion is prevented when a sudden loud sound is reproduced. The gain should increase so slowly as to be inaudible, that is, over a period of several seconds.

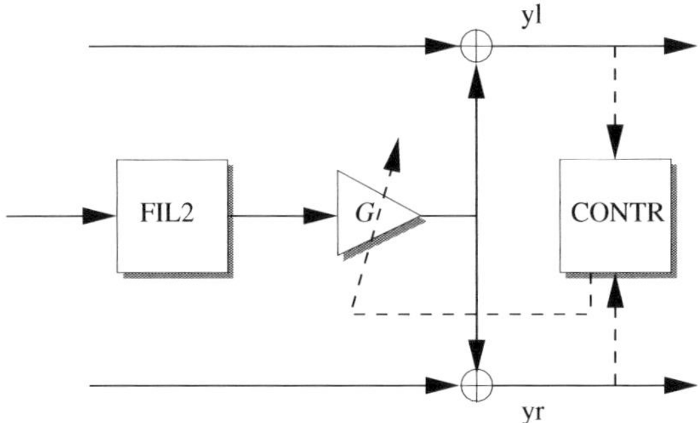

Figure 2.24 Part of a low-frequency psychoacoustic BWE algorithm, compare to Fig. 2.4. The dashed lines indicate the feedback loop from the output to a control unit that modifies the gain of the harmonics signal, such that at low and intermediate output levels the gain is maximum, but gradually decreased as the output level becomes high

2.4 LOW-FREQUENCY PSYCHOACOUSTIC BANDWIDTH EXTENSION WITH FREQUENCY TRACKING

2.4.1 NON-LINEAR DEVICE

The BWE algorithm as outlined previously does not discriminate between tonal and atonal (noise-like) signals. This is because any signal with frequency components within the passband of FIL1 will be processed by the NLD. Occasionally, this can result in annoying artifacts, if noise-like signals are processed, such as may occur in music and speech. This problem could be prevented if BWE processing is only applied to periodic signals. The scheme presented in Fig. 2.25 achieves this goal. One of the attractive features of the algorithm is that it does not explicitly decide if the signal has a tonal or noise characteristic. Rather, BWE processing is implicitly faded out when noise-like signals are present.

The algorithm will be explained following Fig. 2.25. The first step is to estimate the dominant frequency ω_0 in the input signal $x(t)$ (the box labeled FT), where $x(t)$ is obtained by filtering the full-bandwidth input signal, such that only low-frequency components are retained. This frequency estimation is carried out by a recursive frequency-tracking algorithm, which updates at each new sample according to

$$\hat{r}_k = \hat{r}_{k-1} + x_{k-1}\gamma[x_k + x_{k-2} - 2x_{k-1}\hat{r}_{k-1}], \tag{2.39}$$

where $\hat{r}_k = \cos(\omega_0(k)T_\mathrm{s})$, T_s being the sample time, gives the frequency estimate. The frequency-tracking algorithm will be discussed in more detail later. The frequency estimate is used to control a harmonics generator (box labeled HG), which generates a harmonics signal $x_\mathrm{h}(t)$ as

$$x_\mathrm{h}(t) = \sum_{k=M}^{N} A_k \sin(k\omega_0 t), \tag{2.40}$$

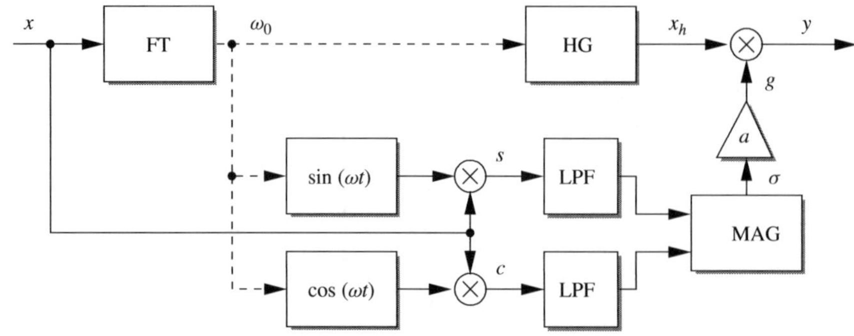

Figure 2.25 Part of a low-frequency psychoacoustic BWE scheme using frequency tracking. FT is a frequency tracker, LPFs are low-pass filters, HG is a harmonics generator, and a is a scaling factor. The output signal $y(t)$ contains harmonics of the strongest frequency component contained in $x(t)$, but only if $x(t)$ is periodic

where M and N equal the minimum and maximum harmonic numbers that are desired. These could be determined by ω_0 or simply be constant, for example, $M = 2$ and $N = 5$. Note that this method of harmonics generation prevents intermodulation distortion, which is an advantage over the NLDs discussed previously in this chapter. Also, the amplitudes of the harmonics can be arbitrarily chosen to produce a desired timbre. It would even be possible to adapt the amplitudes depending on the input signal, although this possibility has not yet been further investigated.

Next, the signal is scaled by a gain control signal $g(t)$, resulting in the output signal $y(t)$. To see how $g(t)$ is determined by $x(t)$, we first assume that $x(t) = A_0 \sin(\omega_0 t + \phi)$. We also assume that the frequency tracker (FT) correctly estimates the frequency of this signal. As shown in Fig. 2.25, the following signals are then generated

$$s(t) = A_0 \sin(\omega_0 t + \phi) \sin(\omega_0 t), \tag{2.41}$$

$$c(t) = A_0 \sin(\omega_0 t + \phi) \cos(\omega_0 t), \tag{2.42}$$

which can also be written as

$$s(t) = \frac{A_0}{2}[\cos\phi - \cos(2\,\omega_0 t + \phi)], \tag{2.43}$$

$$c(t) = \frac{A_0}{2}[\sin\phi + \sin(2\,\omega_0 t + \phi)]. \tag{2.44}$$

After averaging (by low-pass filtering), we get

$$\overline{s}(t) = \frac{A_0}{2}\cos\phi, \tag{2.45}$$

$$\overline{c}(t) = \frac{A_0}{2}\sin\phi. \tag{2.46}$$

where $\overline{s}(t)$ and $\overline{c}(t)$ are slowly time-varying signals. Taking the square root of the sum of squares, this becomes

$$\sigma(t) = \sqrt{\overline{s}^2(t) + \overline{c}^2(t)} = \frac{A_0}{2}. \tag{2.47}$$

The control signal g is then obtained as

$$g(t) = a\sigma(t) \tag{2.48}$$

if, for example $a = 2$, then $g(t) = A_0$. Thus, we see that for a sinusoidal input signal $y(t) = x_{\mathrm{h}}(t)$, that is, the output signal has maximum amplitude. Now if $x(t)$ is a noise signal, the averaged (low-pass) $s(t)$ and $c(t)$ will tend to zero if the averaging time is sufficiently long. For intermediate cases, the gain control signal $g(t)$ will be scaled $x_{\mathrm{h}}(t)$ down to an appropriately lower amplitude. It thus appears that the gain control signal varies between 0 (noise inputs) and 1 (sinusoidal input), with a gradual transition between these two extremes, depending on the periodicity of the input signal. In practice, $x(t)$ may contain multiple sinusoids and/or a sinusoid in the presence of noise. Section 2.4.2.3 shows how the initial frequency estimation is affected by such signals.

In conclusion, this alternate NLD generates a harmonics signal without intermodulation distortion, but only for periodic input signals. For noisy input signals, the output tends to zero.

2.4.2 FREQUENCY TRACKING

Here, we will elaborate on the frequency-tracking algorithm utilized in the NLD of the previous section. There is a vast literature regarding frequency tracking, owing to the many applications in, for example, astronomy, acoustics, and communications; see e.g. Quinn and Hannan [214] and Tichavsky and Nehorai [271] for a comparative study of four adaptive frequency trackers. Recently, an algorithm was devised for rapid power-line frequency monitoring (Adelson [15]), on the basis of a number of formulas presented in Adelson [14]. Most of the algorithms presented in the book of Quinn and Hannan are complex and not very suitable for real-time implementation, while for BWE algorithms, we (as usual) strive for maximum computational efficiency, by avoiding divisions, trigonometric operations such as FFTs – which also necessitate the use of buffers – and the like.

Here, we will develop an efficient frequency-tracking procedure, which uses only a few arithmetic operations, and is insensitive to the initial state of the algorithm parameters. We also analyse the convergence behaviour of the algorithm for stationary input signals, and the dynamic behaviour if there is a transition to another stationary state, the latter being considered important to assess the tracking abilities for realistic signals. The following derivations and analyses were also published in Aarts [5], and in analogous form previously for another application (cross-correlation tracking) in Aarts *et al.* [8]. In slightly modified form, the algorithm can also be used to track amplitude instead of frequency.

We shall show in Sec. 2.4.2.1 that the recursion

$$\hat{r}_k = \hat{r}_{k-1} + x_{k-1}\gamma[x_k + x_{k-2} - 2x_{k-1}\hat{r}_{k-1}], \tag{2.49}$$

estimates to a good approximation the frequency of a signal given by

$$r_k = \cos(\omega_0(k)T_s), \tag{2.50}$$

where $\omega_0(k)$ is the frequency of the input signal x to be determined, k is the time index, and $f_s = 1/T_s$ is the sampling frequency. The parameter γ determines the convergence speed, and hence determines the tracking behaviour of $\hat{r}$, but not the actual value of $\lim_{k\to\infty} \hat{r}_k$ in the stationary case. Equation 2.49 is the basis for our approach of recursively tracking the frequency. In Sec. 2.4.2.2, we shall analyse the solution of Eqn. 2.49, starting from an initial value $\hat{r}_0$ at $k = 0$, when $\gamma \downarrow 0$, and we shall indicate conditions under which

$$\lim_{\gamma\downarrow 0}[\lim_{k\to\infty} \hat{r}_k] = \cos(\omega_0 T_s). \tag{2.51}$$

The analysis is similar to that of an algorithm (Aarts *et al.* [8]) to track correlation coefficients, and can be facilitated considerably by switching from difference equations, as in Eqn. 2.49, to differential equations.

In Sec. 2.4.2.3, we consider the case of a sinusoidal input signal x, and we compute explicitly the left-hand side of Eqn. 2.51 for the solution of Eqn. 2.49. It turns out that the recursion Eqn. 2.49 yields the correct value r for the left-hand side of Eqn. 2.51.

2.4.2.1 Derivation of Tracking Formulas

Here, we consider r as defined in Eqn. 2.50, and we show that r satisfies to a good approximation (when γ is small) the recursion in Eqn. 2.49.

We start with Adelson's [15] Eqn. 1

$$r = \frac{\sum_{j=1}^{n-1} x_j(x_{j-1} + x_{j+1})}{2\sum_{j=1}^{n-1} x_j^2}. \tag{2.52}$$

In order to make this formula suitable for tracking purposes, it is modified into

$$r_k = \frac{\sum_{j=1}^{n-1} x_{k-j}(x_{k-j-1} + x_{k-j+1})}{2\sum_{j=1}^{n-1} x_{k-j}^2}. \tag{2.53}$$

Now r_k depends on $n - 1$ samples from the past, and the current sample x_k. However, it is not optimal for tracking purposes, since it suffers from the fact that it requires many operations and may lead to numerical difficulties in the case of a small denominator in Eqn. 2.53. Therefore, a second modification is made by using – instead of a rectangular window and an averaging over $2n$ $x_i x_{i+1}$ products – an exponential window. In order to minimize the number of operations, we select $n = 2$. Now, we define

$$r_k = \frac{S_n}{S_d}, \tag{2.54}$$

where

$$S_n(k) = \sum_{l=0}^{\infty} c\,\mathrm{e}^{-\eta l} x_{k-l-1}(x_{k-l} + x_{k-l-2}), \tag{2.55}$$

$$S_d(k) = \sum_{l=0}^{\infty} 2c\,\mathrm{e}^{-\eta l} x_{k-l-1}^2, \tag{2.56}$$

$$c = 1 - \mathrm{e}^{-\eta}, \tag{2.57}$$

with η is a small but positive number that should be adjusted to the particular circumstances for which tracking of the frequency is required. We now show that r of Eqs. 2.54–2.57 satisfies to a good approximation the recursion in Eqn. 2.49. To this end, we note that

$$S_n(k) = \mathrm{e}^{-\eta} S_n(k-1) + c x_{k-1}(x_k + x_{k-2}), \tag{2.58}$$

and

$$S_d(k) = \mathrm{e}^{-\eta} S_d(k-1) + 2c x_{k-1}^2. \tag{2.59}$$

Hence, from the definition in Eqn. 2.54,

$$r(k) = \frac{S_n(k-1) + c\,\mathrm{e}^{\eta} x_{k-1}(x_k + x_{k-2})}{S_d(k-1) + 2c\,\mathrm{e}^{\eta} x_{k-1}^2}. \tag{2.60}$$

Since we consider small values of η, $c = 1 - \mathrm{e}^{-\eta}$ is small as well. Expanding the right-hand side of Eqn. 2.60 in powers of c and retaining only the constant and the linear term, we get after some calculations

$$r(k) = r(k-1) + \frac{c\,\mathrm{e}^{\eta}}{S_d(k-1)} x_{k-1}[x_k + x_{k-2} - 2r(k-1)x_{k-1}] + O(c^2). \tag{2.61}$$

Then, deleting the $O(c^2)$ term, we obtain the recursion in Eqn. 2.49 when we identify

$$x_{\mathrm{rms}}^2 = S_d(k), \tag{2.62}$$

for a sufficiently large k, and

$$\gamma = \frac{c\,\mathrm{e}^{\eta}}{x_{\mathrm{rms}}^2}, \tag{2.63}$$

which is a constant for a stationary signal $x(t)$.

We observe at this point that we have obtained the recursion in Eqn. 2.49 by applying certain approximations (as in Eqn. 2.62) and neglecting higher-order terms. Therefore, it is not immediately obvious that the actual r of Eqn. 2.50 and the solution of $\hat{r}$ of the recursion in Eqn. 2.49 have the same value, in particular for large k. However, next we shall show that $\hat{r}$ shares some important properties with the veridical r.

2.4.2.2 Analysis of the Solution of the Basic Recursion

Now we consider the basic recursion in Eqn. 2.49, and we analyse its solution $\hat{r}(k)$, given an initial value $\hat{r}_0$ at $k = 0$, when $\gamma \downarrow 0$. It is convenient to introduce the new variables

$$\beta_k = 2x_{k-1}^2, \tag{2.64}$$

and

$$\delta_k = x_{k-1}(x_k + x_{k-2}), \tag{2.65}$$

to remain compatible with the notation in Aarts *et al.* [8] and Aarts [5]. Thus, we shall consider the recursion in Eqn. 2.49, which we rewrite as

$$\hat{r}(k) = (1 - \gamma\beta_k)\hat{r}(k-1) + \gamma\delta_k \tag{2.66}$$

for $k = 1, 2, \ldots$ with γ a small positive number and δ_k, β_k bounded sequences with $0 \le \beta_k \le 1$.

In Aarts *et al.* [8], it was shown how to obtain the limiting behaviour of $\hat{r}(k)$ as $k \to \infty$ when $\gamma > 0$ is small. This was done under an assumption (slightly stronger than required) that the mean values (denoted by $M[.]$)

$$\begin{aligned} b_0(\gamma) &= M\left[\frac{-1}{\gamma}\log(1 - \gamma\beta_k)\right] \\ &= \lim_{K\to\infty} \frac{1}{K}\sum_{l=1}^{K} \frac{-1}{\gamma}\log(1 - \gamma\beta_l), \\ \mathrm{d}_0 = M[\delta_k] &= \lim_{K\to\infty} \frac{1}{K}\sum_{l=1}^{K} \delta_l \end{aligned} \tag{2.67}$$

for the discrete-time case and

$$\begin{aligned} b_0 = M[\beta(t)] &= \lim_{T\to\infty} \frac{1}{T}\int_0^T \beta(s)\,\mathrm{d}s, \\ \mathrm{d}_0 = M[\delta(t)] &= \lim_{T\to\infty} \frac{1}{T}\int_0^T \delta(s)\,\mathrm{d}s, \end{aligned} \tag{2.68}$$

for the corresponding continuous-time case, exist.

Since $b_0(\gamma) \to b_0$ as $\gamma \downarrow 0$, it was shown that

$$\lim_{\gamma\downarrow 0}\left[\lim_{k\to\infty} \hat{r}(k)\right] = \frac{M[\delta_k]}{M[\beta_k]} = \frac{\mathrm{d}_0}{b_0} = \frac{M[\delta(t)]}{M[\beta(t)]}, \tag{2.69}$$

and for any number $b < b_0(\gamma)$

$$\hat{r}(k) = \frac{\mathrm{d}_0}{b_0(\gamma T_\mathrm{s})} + O(\mathrm{e}^{-\gamma bkT_\mathrm{s}}),\ k \ge 0. \tag{2.70}$$

This shows that the time constant τ, that is, the time for the exponential term to drop to e^{-1} of its original value, for the tracking behaviour is given by

$$\tau = \frac{T_s}{\gamma b_0(\gamma T_s)}. \tag{2.71}$$

We finally observe that $b_0(\gamma) \to b_0$ as $\gamma \downarrow 0$. In the next section, we shall work this out for sinusoidal input signals.

2.4.2.3 Sinusoidal Input Signals

In this section, we test the algorithms derived in Sec. 2.4.2.1, and analysed in Sec. 2.4.2.2, with respect to their steady-state behaviour, for sinusoidal input signals. Hence we take

$$x_k = A_0 \sin(\omega_0 k T_s + \phi), \tag{2.72}$$

with arbitrary A_0 and ϕ. Calculating δ and β with Eqs. 2.65–2.64, and using Eqn. 2.69, it is easy to find

$$\lim_{\gamma\downarrow 0}\left[\lim_{k\to\infty} \hat{r}(k)\right] = \cos\omega_0 T_s; \tag{2.73}$$

compare with Eqn. 2.50. This limit obviously does not depend on A_0, nor on ϕ. If Eqn. 2.72 and $r_{k-1} = \cos\omega_0 T_s$ are substituted into Eqn. 2.49, then we get $r_k = r_{k-1}$, independent of γ, indicating that r has converged to a constant value. Using Eqn. 2.71 and $b_0 = A_0^2$, it appears that the time constant τ_d of the tracking behaviour is equal to

$$\tau_d = T_s/(\gamma A_0^2). \tag{2.74}$$

Now consider the case that the signal x consists of two sinusoids (with unequal frequencies), where the latter can represent a disturbance signal or as harmonic distortion of the first sinusoid. Thus,

$$x_k = A_0 \sin(\omega_0 k) + A_1 \sin(\omega_1 k), \tag{2.75}$$

and by using Eqns. 2.64, 2.65, and 2.69 we get

$$\lim_{\gamma\downarrow 0}\left[\lim_{k\to\infty} \hat{r}(k)\right] = \frac{A_0^2\cos\omega_0 T_s + A_1^2\cos\omega_1 T_s}{A_0^2 + A_1^2}. \tag{2.76}$$

Alternatively, consider the case that the signal x consists of a sinusoid with additional noise $n(k)$ (with autocorrelation function $R_n(k)$). Then

$$x_k = A_0 \sin(\omega_0 k) + n(k), \tag{2.77}$$

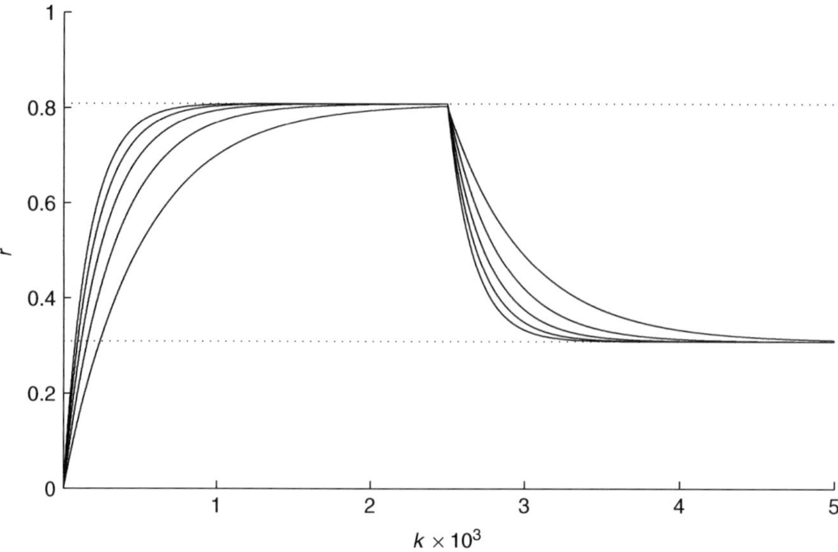

Figure 2.26 Step response of Eqn. 2.49 for a sinusoidal input signal with amplitude $A_0 = 1$, $\hat{r}_0 = 0$, and $\gamma = 2 \cdot 10^{-3}$–$6 \cdot 10^{-3}$ in three increments, and making a step from $\omega_0 T_s = \pi/5$ ($r = 0.81$) to $\omega_0 T_s = 2\pi/5$ ($r = 0.31$). The dotted lines are the final values given by Eqn. 2.73

and we get

$$\lim_{\gamma \downarrow 0}[\lim_{k \to \infty} \hat{r}(k)] = \frac{A_0{}^2 \cos \omega_0 T_s + 2R_n(T_s)}{A_0{}^2 + 2R_n(0)}. \tag{2.78}$$

Equation 2.78 shows that if $R_n(T_s)$ and $R_n(0)$ are known, or can be estimated, the estimation of $\hat{r}$ can be easily improved.

To demonstrate the tracking behaviour of Eqn. 2.49, in Fig. 2.26 the step response is plotted for a sinusoidal input signal, making a change in frequency, for various values of γ. It appears that the time constants correspond well with the values predicted by Eqn. 2.74. The values of γ used in Fig. 2.26 are just for illustration purposes, and in practice they could be much larger. To obtain stability, we need $|1 - \beta_k \gamma| < 1$. Practical values for sinusoidal input signals are $0 < A_0 \gamma < 0.5$.

Using the same procedure as for tracking frequency, we can track the amplitude A_0 of the input signal as well. To that end, β and δ are modified into

$$\beta'_k = 1 - r_k^2, \tag{2.79}$$

and

$$\delta'_k = x_{k-1}^2 - x_k x_{k-2}. \tag{2.80}$$

Using Eqn. 2.66, we get

$$A(k) = (1 - \gamma\beta'_k)A(k-1) + \gamma\delta'_k, \tag{2.81}$$

and, finally, we get $A_0 = \sqrt{A(k)}$.

2.5 SUBJECTIVE PERFORMANCE OF LOW-FREQUENCY PSYCHOACOUSTIC BANDWIDTH EXTENSION ALGORITHMS

There is little published data on the subjective quality of low-frequency psychoacoustic BWE systems, although from informal listening it is known that well-designed systems can yield good-quality sound. Here, we shall discuss the results of a formal listening test, parts of which was also published in Larsen and Aarts [156]. First we present the results from two other studies.

2.5.1 'VIRTUAL BASS'

Performance of 'Virtual Bass' system In Gan *et al.* [83], a low-frequency psychoacoustic BWE algorithm is presented, called 'Virtual Bass'. They also report results from human subject testing, which was performed in the following manner. Ten naive subjects were used in an age group of 24 to 35 years old. Three test signals were employed: a sequence of gunshots, a bass guitar soundtrack, and classical music. Three loudspeakers were used: a high-quality two-way monitor (5" cone; 68–20,000 Hz; 30 W), a multi-media speaker (3" cone; 110–15,000 Hz; 6 W), and a flat-panel speaker (150–20,000 Hz; 3.6 W). The authors did not mention whether the subjects could see the speaker that was being used; if so, this might have biased the results. For each signal that was tested, subjects first heard the unprocessed signal as a reference (presumably using the monitor speaker), followed by the processed signal using either the 'Virtual Bass' algorithm or a commercially available system (unspecified). The subjects were then asked to grade both bass quality and signal impairment on a five-grade scale as in Table 2.3. Bass quality was judged as good for both the 'Virtual Bass' and the commercial bass system, with a slight advantage for the 'Virtual Bass' system. The impairment ratings were a bit lower for the 'Virtual Bass' than for the commercial bass system, however. The average impairment was 3.67 for 'Virtual Bass' and 4 for the commercial bass system. The impairment of the 'Virtual

Table 2.3 Five-grade quality and impairment scale used by Gan *et al.* [83]

Grade	Quality	Impairment
5	Excellent	Imperceptible
4	Good	Perceptible, not annoying
3	Fair	Slightly annoying
2	Poor	Annoying
1	Bad	Very annoying

Bass' signals were reported as a humming pitch, attributed to the method of harmonics generation (modulating function). These artefacts were more audible in the better-quality monitor speakers than in the other two reproduction systems. This is favourable considering that low-frequency psychoacoustic BWE will typically not be used in good-quality speakers with extended low-frequency response.

Cross-talk cancellation application Tan *et al.* [261] use the 'Virtual Bass' algorithm of Gan *et al.* [83] in a cross-talk cancellation method. In cross-talk cancellation, the objective is to eliminate sound from the left loudspeaker reaching the right ear and sound from the right loudspeaker reaching the left ear. This is important for virtual audio applications using loudspeakers. The use of headphones in virtual audio applications would not require the use of cross-talk cancellation.

Tan *et al.* argue that the low interaural level difference (ILD) for low-frequency sounds makes cross-talk cancellation difficult, because it requires inversion of an ill-conditioned matrix. Even if cross-talk cancellation is possible, the required boosting of low frequencies will cause problems in the loudspeakers because of large cone excursion and power-handling capacity. They propose to circumvent these problems by using a low-frequency psychoacoustic BWE system to replace very low frequencies by higher harmonics, for which it is easier to cancel the cross-talk. A subjective test was performed with two different signals, at two different phantom source azimuths (45 and 90° relative to straight ahead); ten subjects were used. The quality was determined by ranking on a five-grade scale (different from the one used in Table 2.3). The phantom sources at 90° received slightly higher scores (about 0.4–0.5 points). We performed a t-test for the two samples (different azimuths) of each signal and found no significant difference between the means at the 10% significance level, however. In fact, the reference condition, which consisted of a cross-talk system without the 'Virtual Bass' processing, did not differ significantly at the 5% significance level from any of the conditions tested with the 'Virtual Bass' system (only one of the signals at 90° azimuth had a significantly different mean at the 10% level), as was determined by t-tests for each condition.

2.5.2 *'ULTRA BASS'*

In Larsen and Aarts [156], a discussion was presented on the results of a listening test that was conducted to assess the subjective quality of two low-frequency psychoacoustic BWE systems ('Ultra Bass'). Here, some of this discussion is repeated, together with some new analysis.

The experiment had three objectives:

1. To rank order preference for unprocessed, linearly amplified (bass only), and BWE-processed musical signals.
2. To evaluate if preferences vary per subject.
3. To evaluate if preferences vary per repertoire.

Algorithms tested In the following text, we will refer to four different algorithms, which are as follows:

1. Unprocessed signal, which was included as reference against which the processed signals would be compared.
2. Linear amplification, which is considered to yield 'baseline' performance for bass enhancement. The quality of the two BWE systems should at least match but preferably exceed the quality of the linear system.
3. Low-frequency psychoacoustic BWE system with rectifier as NLD.
4. Low-frequency psychoacoustic BWE system with integrator as NLD. This and the previous algorithm were chosen because from informal listening it was observed that the quality of the processed signals is quite different for the two cases (which is not surprising given the analysis in Secs. 2.3.2.2 and 2.3.2.3).

The linear amplification was done with commercial sound-editing software, using a graphic EQ in 1/2-octave bands. The amplification was 6 dB (44 Hz), 9 dB (62.5 Hz), 9 dB (88 Hz), and 6 dB (125 Hz); these values were chosen to give maximum bass boost without creating audible distortion at the reproduction level used in the experiment. The processing was identical for the two BWE systems, except for the implementation of the NLD. Filter 1 was implemented as a second-order Chebyshev-type I IIR filter; passband ripple was 1 dB, and the passband was 20–70 Hz. Filter 2 was implemented as a third-order elliptic IIR filter (also non-linear phase), passband ripple of 3 dB, stopband attenuation of 30 dB and passband of 70–140 Hz. The gain value for the harmonics signal was fixed at 15 dB, for both BWE systems. The signal in the main path was not processed (no high-pass filter, no delay). The implementation of the BWE systems as used in the test is now known to be suboptimal; particularly, the use of non-linear-phase IIR filters would be avoided in favour of using linear-phase filters.

Music selection, signal generation, and reproduction Music was selected according to genre and an a priori evaluation of subjective quality. Genres were pop and rock, and subjective quality criteria were that the bass content of the signals should be 'difficult' to reproduce well on a small loudspeaker system. This approach was taken so that the obtained results would indicate performance of some of the most demanding signals. Excerpts ($\approx$10 s duration) from the following four tracks were used:

1. 'Bad' by Michael Jackson. This track contains a typical pop bass line, which was known to give good subjective performance after BWE processing. It was included to contrast the other, more demanding, signals.
2. 'My Father's Eyes' by Eric Clapton. A very deep and strong bass line accompanies the music on this track, which may sound too imposing if the reproduced bass is not well balanced.
3. 'Hotel California' by The Eagles (live version). The excerpt was from the start of the track, which consists of a bass drum only (and some audience noise). This makes it easier to focus on the bass quality. The difficulty in reproduction lies in the low frequency and very fast attack of the drum. Also, the decay is very gradual and should not sound unnatural.
4. 'Twist and Shout' by Salt n' Peppa. In this track, the bass follows a tight beat, the main difficulty being to preserve the tight temporal envelope.

These four signals were processed by each of the three algorithms as described previously, and the test included the unprocessed signals as well. Prior to processing, all four test signals were scaled to obtain approximately equal loudness.

Reproduction was on a commercially available medium-sized Hi-Fi system. The low cut-off frequency was about 140 Hz. Listeners were seated at a distance of about 1 m in the median plane between two loudspeakers. Reproduction was at a comfortable listening level, and was fixed prior to the start of the experiment, being the same for all subjects.

Human subjects Fifteen unpaid volunteers (eleven male, four female) participated in the experiment. The age range was 25–30 years old. All had self-reported good hearing, varying degrees of experience in listening tests, and varying degrees of interest in music. Subjects were asked to indicate their preferred genres of music, which were pop and rock.

Experimental procedure A direct ranking of the various processed signals would be difficult, and the paired comparison paradigm was chosen because it is known to yield good results when used to compare several perceptually close signals (David [56]). Thus, a pair of signals (same repertoire, different processing) would be presented, and listeners were instructed to choose the version with the best bass quality. Although this allows the possibility that different listeners use different criteria in their selection, this was done to obtain general preference ratings; furthermore, one of the objectives was to find out if there would be differences in preferences among subjects. Instructing listeners to choose on the basis of the 'best bass quality' should meet both these objectives. Subjects could listen to the pair of the signals as long as was required to make a selection. Because each repertoire had four versions, six pairs were presented to the listener. After the six presentations, the next repertoire was used, until all four repertoire were completed. There were no repetitions, as in most cases the signal pair presented on any trial differed enough to be distinguishable, and we did not expect listeners to change their preference during the course of the experiment. Some listeners had prior exposure to signals processed by the BWE system.

The responses were recorded in preference matrices $\mathcal{P}$ (one for each repertoire); $\mathcal{P}$ is an anti-symmetric 4×4 matrix with elements $p_{ij} = \{0, 1\}$, a 1 indicating that the column element i is preferred over the row element j, and vice versa. The diagonal elements are not used. The preference matrix can be summarized in a score vector $\mathbf{s}$, which is a column vector, the elements of which are the sum of the rows of $\mathcal{P}$. The ranking of the different algorithms then follows directly from $\mathbf{s}$.

Results Table 2.4 gives the score vectors for all subjects, for each repertoire (numbered as indicated previously). Also, the number of circular triads is shown (CT) (Levelt *et al.* [159]). A circular triad occurs if, for example, version 2 is preferred over 3, 3 is preferred over 4, and 4 is preferred over 2; this indicates an inconsistency in the subject responses. A high value for CT probably indicates that the task is confusing, and would necessitate caution in interpretation of the results. As Table 2.4 shows, for most subjects CT is zero or one, which is normal.

For a preliminary analysis of the results, we plotted the normalized score of each algorithm, averaged over the four repertoire, for each subject; see Fig. 2.27. The normalized

Table 2.4 Score vectors obtained in the listening test. The fifteen subjects are labeled A–O. CT indicates the number of circular triads

	v	A	B	C	D	E	F	G	H	I	J	K	L	M	N	O	Total
Bad	1	0	0	0	0	3	0	2	0	0	1	0	0	1	0	0	7
	2	2	2	3	1	2	1	3	1	2	2	2	2	3	1	1	28
	3	1	2	2	2	1	3	0	3	2	2	2	3	0	3	3	26
	4	3	2	1	3	0	2	1	2	2	1	2	1	2	2	2	26
Eyes	1	1	0	3	1	3	0	3	0	0	0	0	0	2	1	2	15
	2	1	2	1	2	1	1	2	1	3	2	1	2	2	2	3	26
	3	1	2	0	1	0	2	1	3	1	2	2	2	0	0	1	18
	4	3	2	2	2	2	3	0	2	2	2	3	2	2	3	0	30
Hotel	1	0	1	2	1	2	0	1	0	1	0	2	1	1	1	1	14
	2	1	2	3	3	3	1	2	1	2	2	0	2	2	3	3	30
	3	2	3	1	2	1	3	3	3	3	2	3	3	3	2	1	35
	4	3	0	0	0	0	2	0	2	0	2	1	0	0	0	1	11
Twist	1	0	0	2	0	2	0	2	0	0	2	1	0	0	0	0	9
	2	2	2	3	1	3	1	3	1	1	3	1	2	1	2	1	27
	3	2	1	0	2	0	3	1	3	3	1	1	3	3	2	3	27
	4	2	3	1	3	1	2	0	2	2	0	3	1	2	2	2	26
Total	1	1	1	7	2	10	0	8	0	1	3	3	1	4	2	3	45
	2	6	8	10	7	9	4	10	4	8	9	4	8	8	8	8	111
	3	6	8	3	7	2	11	5	12	9	7	8	11	6	7	8	106
	4	11	7	4	8	3	9	1	8	6	5	9	4	6	7	5	93
CT		2	2	0	1	0	0	0	0	1	3	2	1	1	1	1	–

score was obtained as the sum of corresponding elements of the subject's four score vectors, for example, element 1 for the unprocessed version of each signal, divided by 12. In this way, the normalized score varies between 0 and 1. The subjects have been divided into two groups, A (subjects 3, 5, and 7) and B (all others); later on, we will motivate this division. For now we merely notice that, for each algorithm, the mean score assigned by groups A and B is different. Group A rates the unprocessed and linearly amplified signals higher than both BWE-processed signals; for group B, all three processing algorithms have scored approximately the same, while the unprocessed version gets a low score. Table 2.5 gives the mean normalized score for each algorithm, for both groups as well as overall (mean over groups). On the basis of Table 2.5, for group A the rank order of the algorithms would be: (1) linear amplification, (2) no processing, (3) BWE with rectifier, and (4) BWE with integrator. For group B, the rank order would be: (1) BWE with rectifier, (2) BWE with integrator, (3) linear amplification, and (4) no processing.

Discussion Division of subjects in two groups (A and B) can be made plausible by visualizing the subjects' responses with multidimensional scaling (MDS), see App. A. Fig. 2.28 shows the resultant two-dimensional mapping obtained using as proximities the Euclidian distances between score vectors (which are four dimensional). Subjects have been divided into five clusters, S0–S4. The previously mentioned group A corresponds to

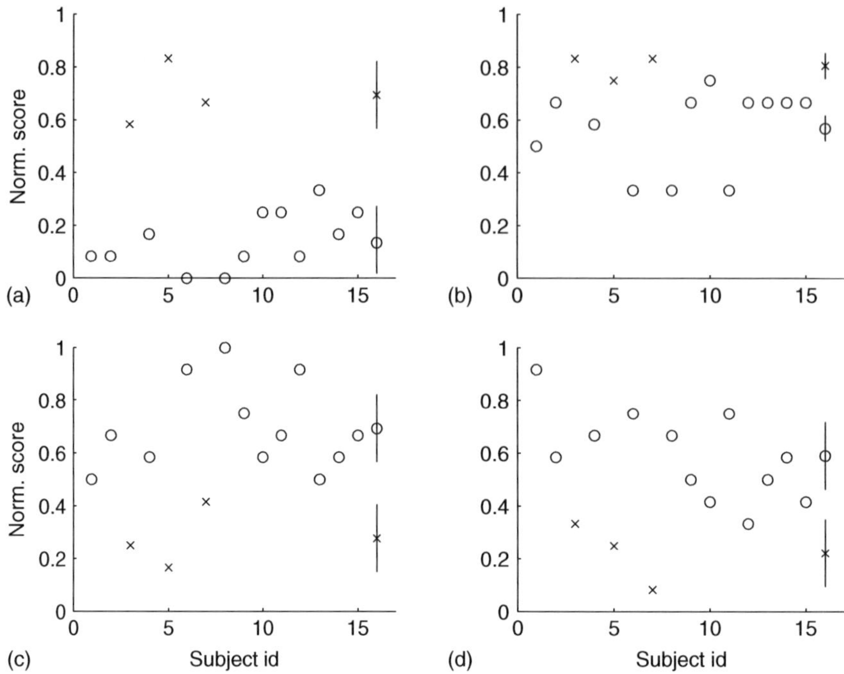

Figure 2.27 Normalized scores for each algorithm (a: unprocessed, b: linear gain, c: BWE with rectifier, and d: BWE with integrator), for each subject. The subjects were divided into two groups and plotted with different symbols (o, x). At the right-hand side of each graph, the means and standard deviation of each subject group is indicated

Table 2.5 Normalized score vectors for the four algorithms, which may be used for subjective quality ranking. Groups A and B are different subject groups, as defined in the text. The 'overall' numbers are weighted averages of the two group values

	A	B	Overall
1	0.69	0.15	0.26
2	0.81	0.57	0.62
3	0.28	0.69	0.61
4	0.22	0.59	0.52

cluster S2 of Fig. 2.28; group B corresponds to the other four clusters. The division into groups A and B is now obvious, as Fig. 2.28 shows that the MDS maps subjects in group A (cluster S2) far away from all the other subjects. By inspecting the subject's individual responses from Table 2.4, we can interpret the MDS dimensions. The horizontal dimension seems to indicate preference for BWE processing, with higher preference towards the right-hand side. The vertical dimension seems to indicate preference for NLD type, with integrator towards the top and rectifier towards the bottom. In Larsen and Aarts [156],

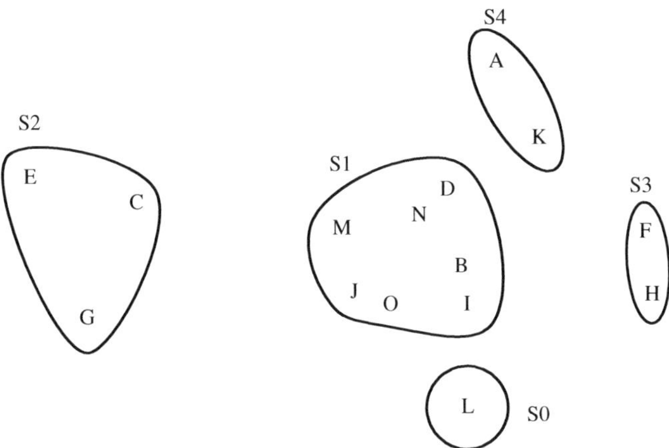

Figure 2.28 Two-dimensional scaling of subject (A–O) preferences of a subjective comparison between various bass enhancement systems. Interpretation of the two dimensions is made in the text. Subjects were grouped in five clusters. Group A mentioned in the text, and in Fig. 2.27 and Table 2.5 corresponds to cluster S2 here. Group B corresponds to the other four clusters. From Larsen and Aarts [156]

results were further analysed with biplots (Gabriel [81]), which showed that the two BWE systems were judged most similar, and the linear system versus BWE with integrator were judged most dissimilar. Conclusion It appears that there is no consistent judgement from the whole subject group regarding preference for a particular processing type. Out of 15 subjects, 3 preferred a linear bass enhancement system, while the other 12 preferred low-frequency psychoacoustic BWE processing. Within these 12 subjects, there was no clear preference for a rectifier or integrator as NLD, although the rectifier did receive a somewhat higher average appreciation. On the basis of this experiment and the response of all subjects taken as a whole, the main conclusion is that low-frequency psychoacoustic BWE can perform at least as well as linear systems. More recent developments in low-frequency psychoacoustic BWE methods, such as adaptive clipping (Sec. 2.3.2.4) or frequency tracking (Sec. 2.4) have shown superior performance in informal evaluations and may show a more conclusive benefit to linear bass enhancement systems in formal listening tests.

2.6 SPECTRAL CHARACTERISTICS OF NON-LINEAR DEVICES

In Sec. 2.3.2, intermodulation characteristics of non-linear devices were analysed. For the rectifying and integrating NLDs, expressions were given for the Fourier series coefficients of processed signals, given the Fourier series coefficients of the input signals. Sections 2.6.1 and 2.6.2 will present the full derivation of these expressions, originally published in Larsen and Aarts [156], and were largely due to A.J.E.M. Janssen. Discrete-time expressions are given in Sec. 2.6.3, and in Sec. 2.6.4 the Fourier series coefficients of a clipped sinusoid are given.

Consider a real periodic signal $f(t)$ of period $T_0 = 1/f_0$ and assume that

-
$$f(t_i) = 0, \quad i = -1, 0, \ldots, N, \tag{2.82}$$
$$f(t) \neq 0, \quad t \neq t_i, \tag{2.83}$$
thereby defining the zeros of $f(t)$ during a one-period interval; t_{-1} is defined as the beginning of the period, and t_N as the end of the period (which is identical to t_{-1} of the next period). There are $N \geq 1$ zero crossing in between t_{-1} and t_N, and $t_N - t_{-1} = T_0$. We use the shorthand notation $\mathbf{t} = (t_{-1}, \ldots, t_N)^T$. Figure 2.29 illustrates the above notation.
- $f(t)$ changes sign at every t_i, which implies that N is odd.
- $f'(t_{-1}) > 0$.
- $f(t)$ is sufficiently smooth such that its Fourier coefficients a_n decay at a rate of at least $1/n^2$. This will be satisfied if, for instance, $f(t)$ is at least twice continuously differentiable.

We have for $f(t)$ the Fourier series representation

$$f(t) = \sum_{n=-\infty}^{\infty} a_n e^{i2\pi f_0 nt}, \tag{2.84}$$

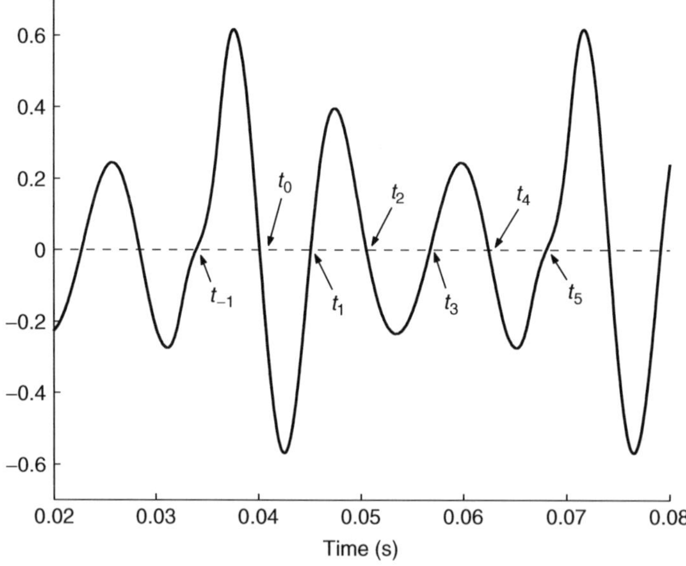

Figure 2.29 A signal with a period of 34 ms ($f_0 \approx 30$ Hz) has several zero crossings in the periodicity interval. The zero crossings for one period are indicated using the notation of this section: t_{-1} indicates the start of the period, and t_5 indicates the end of the period. There are five intermediate zero crossings. Note that $t_5 \equiv t_{-1}$ of the next period

where $a_n = a^*_{-n}$, because $f(t)$ is real. The output signal of the rectifying NLD is denoted by $g(t)$, which has the Fourier series representation

$$g(t) = \sum_{n=-\infty}^{\infty} b_n \mathrm{e}^{i2\pi f_0 n t}, \tag{2.85}$$

with of course also $b_n = b^*_{-n}$. The objective is to express the b_n in terms of the a_n.

2.6.1 *OUTPUT SPECTRUM OF A RECTIFIER*

On the periodicity interval $[t_{-1}, t_{\mathrm{N}})$, the function $g(t)$ is given by

$$g(t) = |f(t)|, \tag{2.86}$$

which can also be written as

$$|f(t)| = f(t)h(t;\ \mathbf{t}) \tag{2.87}$$

where

$$h(t;\ \mathbf{t}) = \begin{cases} 1 & \text{for} \quad t_{-1} \le t < t_0, \\ -1 & \text{for} \quad t_0 \le t < t_1, \\ \vdots & \qquad \vdots \\ (-1)^N = -1 & \text{for} \quad t_{\mathrm{N}-1} \le t < t_{\mathrm{N}}. \end{cases} \tag{2.88}$$

Let d_n be the Fourier coefficients of $h(t)$, so

$$\mathrm{d}_0 = 1 + 2f_0 \sum_{k=0}^{N} (-1)^k t_k, \tag{2.89}$$

$$\mathrm{d}_n = -\frac{1}{i\pi n} \sum_{k=0}^{N} (-1)^k \mathrm{e}^{-i2\pi f_0 n t_k}, \quad n \neq 0. \tag{2.90}$$

From the foregoing

$$g(t) = \sum_{n=-\infty}^{\infty} a_n \mathrm{e}^{i2\pi f_0 n t} \times \sum_{m=-\infty}^{\infty} \mathrm{d}_m \mathrm{e}^{i2\pi f_0 m t} = \sum_{k=-\infty}^{\infty} \mathrm{e}^{i2\pi f_0 k t} \times \sum_{n+m=k} a_n \, \mathrm{d}_m, \tag{2.91}$$

and therefore

$$b_k = \sum_{n=-\infty}^{\infty} a_n \, \mathrm{d}_{k-n}. \tag{2.92}$$

Combining all the previous results

$$b_k = \left(1 + 2f_0 \sum_{m=0}^{N} (-1)^m t_m\right) a_k - \sum_{n \neq k} \frac{a_n}{i\pi(k-n)} \sum_{m=0}^{N} (-1)^m \mathrm{e}^{i2\pi f_0 (n-k) t_m}. \tag{2.93}$$

Having expressed the b_k in terms of the a_k (also using the locations of the zeros of $f(t)$), the problem is solved in principle. However, the right-hand side of Eqn. 2.93 exhibits a decay of the b_k roughly as $1/k$, while the form of $g(t)$ suggests that there should be a decay like $1/k^2$, due to the triangular singularities at the t_i. This decay of the b_k can be made explicit by properly using the condition stated in Eqns. 2.82 and 2.83. Accordingly,

$$\sum_{n=-\infty}^{\infty} a_n \mathrm{e}^{i2\pi f_0 n t_m} = 0. \tag{2.94}$$

Then the series at the far right-hand side of equation 2.93, for $k \neq 0$, becomes

$$\begin{aligned}
\sum_{n \neq k} \frac{a_n}{i\pi(k-n)} \sum_{m=0}^{N} (-1)^m \mathrm{e}^{i2\pi f_0 (n-k) t_m} &= \sum_{n \neq k} \frac{a_n}{i\pi} \left(\frac{1}{k-n} - \frac{1}{k} + \frac{1}{k}\right) \sum_{m=0}^{N} (-1)^m \mathrm{e}^{i2\pi f_0 (n-k) t_m} \\
&= \sum_{n \neq k} \frac{n a_n}{i\pi k(k-n)} \sum_{m=0}^{N} (-1)^m \mathrm{e}^{i2\pi f_0 (n-k) t_m} + \\
&\quad \frac{1}{i\pi k} \sum_{n \neq k} a_n \sum_{m=0}^{N} (-1)^m \mathrm{e}^{i2\pi f_0 (n-k) t_m}.
\end{aligned} \tag{2.95}$$

And also

$$\begin{aligned}
\sum_{n \neq k} a_n \mathrm{e}^{i2\pi f_0 (n-k) t_m} &= -a_k + \sum_{n=-\infty}^{\infty} a_n \mathrm{e}^{i2\pi f_0 (n-k) t_m} \\
&= -a_k + \mathrm{e}^{-i2\pi f_0 k t_m} \sum_{n=-\infty}^{\infty} a_n \mathrm{e}^{i2\pi f_0 n t_m} \\
&= -a_k.
\end{aligned} \tag{2.96}$$

Hence for $k \neq 0$, the second term of Eqn. 2.95 vanishes, and

$$b_0 = f_0 \int_{t_{-1}}^{t_N} |f(t)|\, \mathrm{d}t, \tag{2.97}$$

$$b_k = \left(1 - 2f_0 \sum_{m=0}^{N} (-1)^m t_m\right) a_k - \sum_{n \neq k} \frac{n a_n}{i\pi k(k-n)} \sum_{m=0}^{N} (-1)^m \mathrm{e}^{i2\pi f_0 (n-k) t_m}. \tag{2.98}$$

The right-hand side of Eqn. 2.98 does exhibit the correct $1/k^2$ behaviour that is expected from the b_k's for large k. More precisely, assuming that $a_k = 0$ for large k, this becomes (for large k)

$$\sum_{n \neq k} \frac{na_n}{i\pi k(k-n)} \sum_{m=0}^{N} (-1)^m \mathrm{e}^{i2\pi f_0 (n-k) t_m} \approx \frac{1}{k^2} \sum_{n=-\infty}^{\infty} \frac{na_n}{i\pi} \sum_{m=0}^{N} (-1)^m \mathrm{e}^{i2\pi f_0 (n-k) t_m}. \quad (2.99)$$

Since

$$f'(t) = \sum_{n=-\infty}^{\infty} i2\pi f_0 n a_n \mathrm{e}^{i2\pi f_0 n t}, \quad (2.100)$$

this can be written as

$$\begin{aligned} &\sum_{n \neq k} \frac{na_n}{i\pi k(k-n)} \sum_{m=0}^{N} (-1)^m \mathrm{e}^{i2\pi f_0 (n-k) t_m} \\ &\quad \approx \frac{1}{k^2} \frac{1}{i\pi} \frac{1}{i2\pi f_0} \sum_{n=-\infty}^{\infty} i2\pi f_0 n a_n \sum_{m=0}^{N} (-1)^m \mathrm{e}^{i2\pi f_0 (n-k) t_m} \\ &\quad = -\frac{1}{2\pi^2 f_0 k^2} \sum_{m=0}^{N} (-1)^m f'(t_m) \mathrm{e}^{-i2\pi f_0 k t_m}. \end{aligned} \quad (2.101)$$

Thus, if $a_k = 0$ for large k, then for large k

$$b_k \sim \frac{1}{2\pi^2 f_0 k^2} \sum_{m=0}^{N} (-1)^m f'(t_m) \mathrm{e}^{-i2\pi f_0 k t_m}. \quad (2.102)$$

It appears that the spectrum of $g(t) \equiv |f(t)|$ at high frequencies is mainly determined by the slope of $f(t)$ at its zero crossings.

2.6.2 OUTPUT SPECTRUM OF INTEGRATOR

Now we consider the integrating NLD; under the same assumptions as in Sec. 2.6.1, we get on the periodicity interval $[t_{-1}, t_{\mathrm{N}})$

$$g(t) = \begin{cases} \int_{t_{-1}}^{t} |f(s)|\,\mathrm{d}s, & t_{-1} \leq t < t_1, \\ -\alpha_0 + \int_{t_{-1}}^{t} |f(s)|\,\mathrm{d}s, & t_1 \leq t < t_3, \\ \vdots & \vdots \\ -[\alpha_0 + \ldots + \alpha_{(N-1)/2}] + \int_{t_{-1}}^{t} |f(s)|\,\mathrm{d}s, & t_{z-2} \leq t < t_z. \end{cases} \quad (2.103)$$

The α_i are the 'jumps' of $g(t)$ at the 'resetting moments', and are given by ($k \in \mathbb{Z}$)

$$\begin{aligned} -\alpha_0 &= -\int_{t_{-1}}^{t_1} |f(s)|\,\mathrm{ds} \quad \text{at time } t = k/f_0 + t_1, \\ -\alpha_1 &= -\int_{t_1}^{t_3} |f(s)|\,\mathrm{ds} \quad \text{at time } t = k/f_0 + t_3, \\ &\vdots \quad \vdots \\ -\alpha_{(N-1)/2} &= -\int_{t_{z-2}}^{t_z} |f(s)|\,\mathrm{ds} \quad \text{at time } t = k/f_0 + t_\mathrm{N}. \end{aligned} \tag{2.104}$$

The above notation is illustrated in Fig. 2.30. For $t \neq t_{-1}, t_1, t_3, \ldots t_z$, we have $g'(t) = |f(t)|$, thus

$$g'(t) = |f(t)| - \sum_{m=0}^{(N-1)/2} \alpha_m \sum_{n=-\infty}^{\infty} \delta(t - n/F_0 - t_{2m+1}). \tag{2.105}$$

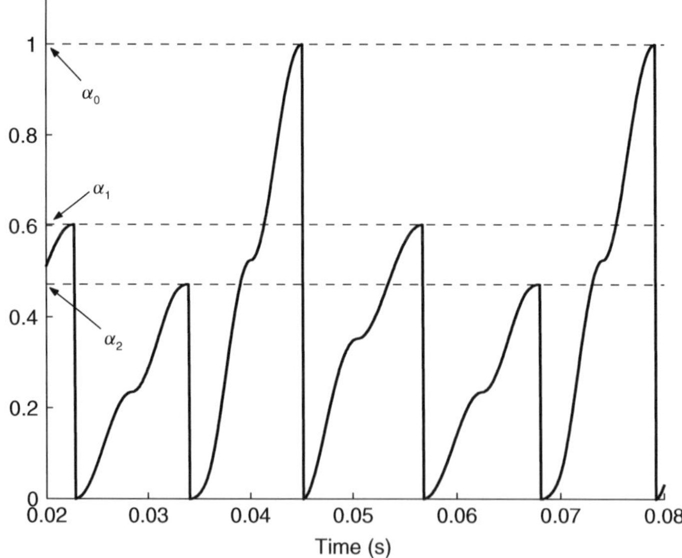

Figure 2.30 A signal with a period of 34 ms ($f_0 \approx 30$ Hz) has several zero crossings in the periodicity interval, leading to three 'resets' to 0 for one period of the output signal. The signal shown here is the result of applying the integrating non-linearity to the signal shown in Fig. 2.29. The magnitude of the resets, that is, the local maxima of the output signal, is indicated using the notation of this section

Denoting the Fourier coefficients of $|f(t)|$ by c_k, so that

$$|f(t)| = \sum_{k=-\infty}^{\infty} c_k \mathrm{e}^{i2\pi F_0 kt}, \tag{2.106}$$

and using $\sum_{n=-\infty}^{\infty} \delta(t - n/F_0 - t_{2m+1}) = \sum_{k=-\infty}^{\infty} \mathrm{e}^{i2\pi F_0 k(t-t_{2m+1})}$, we can write Eqn. 2.105 as

$$\sum_{k=-\infty}^{\infty} i2\pi F_0 \, kb_k \mathrm{e}^{i2\pi F_0 kt} = \sum_{k=-\infty}^{\infty} \left(c_k - \sum_{m=0}^{(N-1)/2} \alpha_m \mathrm{e}^{-i2\pi F_0 kt_{2m+1}} \right) \mathrm{e}^{i2\pi F_0 kt}, \tag{2.107}$$

so that

$$b_k = \frac{c_k - \sum_{m=0}^{(z-1)/2} \alpha_m \mathrm{e}^{-2\pi i \nu_0 kt_{2m+1}}}{2\pi i \nu_0 k} \qquad k \neq 0. \tag{2.108}$$

The b_k show a decay of roughly $1/k$, which is what we expect owing to the discontinuities of $g(t)$ at $t_{-1}, t_1 \ldots t_\mathrm{N}$. The c_k can be found as the b_k of Eqn. 2.98. For $k = 0$ we get, with partial integration,

$$\begin{aligned} b_0 &= \int_{t_{-1}}^{t_\mathrm{N}} g(t)\,\mathrm{d}t \\ &= [tF(t)]_{t_{-1}}^{t_\mathrm{N}} - \int_{t_{-1}}^{t_\mathrm{N}} t \left(|f(t)| - \sum_{m=0}^{(z-1)/2} \alpha_m \sum_{n=-\infty}^{\infty} \delta(t - n/G_0 - t_{2m+1}) \right) \mathrm{d}t \\ &= -\int_{t_{-1}}^{t_\mathrm{N}} t|f(t)|\,\mathrm{d}t + \sum_{m=0}^{(N-1)/2} \alpha_m t_{2m+1}. \end{aligned} \tag{2.109}$$

2.6.3 OUTPUT SPECTRA IN DISCRETE TIME

For low-frequency psychoacoustic BWE applications, the frequencies of interest are orders of magnitude lower than the sample rate, such that continuous-time expressions, as we have used until now, are good approximations to the discrete-time expressions that actually should be used. However, for other BWE applications (notably high-frequency BWE treated in Chapters 5 and 6), the frequencies of interest can be in the same order of magnitude as the sample rate, and in such cases the proper discrete-time expressions must be used. These expressions can be developed along the same lines as the continuous-time expressions, and we therefore only give results, for the clipping and integrating non-linearity.

For this section, we use square brackets to index the variables, for example, as $f[n]$ instead of $f(t)$. We assume that $f[n]$ is periodic, with a period of N samples, sampled at a rate $f_\mathrm{s} = 1/\Delta t$. Zero crossings are defined by the sequence $x[n]$ from $f[n]$ as

$$x[n] = \begin{cases} 1 & \text{for } f[n] \geq 0, \\ 0 & \text{for } f[n] < 0. \end{cases} \tag{2.110}$$

We define $I[n]$ to be the *indicator* of the event $x[n] \neq x[n-1]$; if $x[n] \neq x[n-1]$, then $I[n] = 1$, else $I[n] = 0$. Now, we will define a zero crossing in $f[n]$ to occur for $n = n'$ if $I[n'] = 1$. Further assume that

- $$I[n_{-1}] = I[n_0] = I[n_1] = \ldots = I[n_z] = 1, \tag{2.111}$$
$$I[n] = 0,\ n \neq n_{-1},\ n_0,\ n_1 \ldots n_z. \tag{2.112}$$
where all $n_{0,1\ldots z-1} \in (n_{-1}, n_z)$. Thus, $f[n]$ has z zero crossings in the interval (n_{-1}, n_z), and owing to the periodicity requirements on $f[n]$, z must be uneven; furthermore, $n_z - n_{-1} = N$ and $z \geq 1$.
- We choose $f[n_{-1}+1] - f[n_{-1}] > 0$.

We have for $f[n]$ the Fourier series representation

$$f[n] = \frac{1}{N}\sum_{k=0}^{N-1} a[k]\mathrm{e}^{2\pi ikn/N}, \tag{2.113}$$

and because $f[n]$ is real we have $a[k] = a^*[-k]$. We consider the real periodic time series $F[n]$, derived by some non-linear operation from $f[n]$. We have for $F[n]$ the following Fourier series representation

$$F[n] = \frac{1}{N}\sum_{k=0}^{N-1} b[k]\mathrm{e}^{2\pi ikn/N}, \tag{2.114}$$

and again $b[k] = b^*[-k]$. Now the problem is again to express the $b[k]$ in the $a[k]$. In the limit that the sampling frequency tends to infinity, the discrete-time expressions are expected to equal the continuous-time expressions (this is indeed the case, as can be checked by setting $\lim_{\Delta t \downarrow 0}$ and replacing sums by integrals for the given discrete-time expressions). Note that due to the assumptions, specifically the assumption of periodicity, the derived results have limited applicability. This is because a periodic signal, when sampled, is only periodic if the sample rate and the signal's fundamental frequency have a greatest common divisor (GCD), in which we allow for non-integer arguments. If the GCD exists, it determines the periodicity of the sampled signal (N as mentioned above), which can thus be much longer than the periodicity of the continuous signal. For example, a 7-Hz pure tone (periodicity 0.144 s) sampled at 19 Hz (GCD is 1), has a periodicity of 1 s. In other cases, the periodicity interval can be extremely long (or non-existent) such that practical signals are not stationary within such time intervals. Nonetheless, the derived expressions have an academic use in that they can be used to assess output spectra for specifically chosen signals that have short periodicity. We may then expect that the conclusions from these output spectra can be used more generally.

2.6.3.1 Rectifier

For the rectifier, $F[n]$ is given by

$$F[n] = |f[n]|, \tag{2.115}$$

which leads to

$$b[k] = \left(1 + \frac{2}{N}\sum_{m=0}^{z}(-1)^m n_m\right) a[k] + \tag{2.116}$$

$$\frac{1}{N}\sum_{n\neq k} a[n] \sum_{m=0}^{z}(-1)^m \mathrm{e}^{-\pi i(n_{m-1}+n_m-1)(k-n)/N} \frac{\sin\pi(n_m - n_{m-1})(k-n)/N}{\sin\pi(k-n)/N}.$$

2.6.3.2 Integrator

On the periodicity interval $[n_{-1}, n_z)$, we get

$$F[n] = \begin{cases} \Delta t \sum_{k=n_{-1}}^{n} |f[k]|, & n_{-1} \le n < n_1, \\ -\Delta t \sum_{k=n_{-1}}^{n_1-1} |f[k]| + \Delta t \sum_{k=n_{-1}}^{n} |f[k]|, & n_1 \le n < n_3, \\ \vdots & \vdots \\ -\Delta t \sum_{k=n_{-1}}^{n_{z-2}-1} |f[k]| + \Delta t \sum_{k=n_{-1}}^{n} |f[k]|, & t_{z-2} \le t < t_z. \end{cases} \tag{2.117}$$

Defining α_m as

$$\alpha_m = \Delta t \sum_{k=n_{-1+2m}}^{n_{1+2m}-1} |f[k]|, \tag{2.118}$$

we have

$$b[k] = \Delta t \frac{c[k] - \sum_{m=0}^{(z-1)/2} \alpha_m \mathrm{e}^{-2\pi i k n_{2m+1}/N}}{1 - \mathrm{e}^{-2\pi i k/N}}, \quad k \neq 0. \tag{2.119}$$

The $c[k]$ can be found as the $b[k]$ of Eqn. 2.116. For $k = 0$ we get, with partial summation[3],

$$b[0] = \sum_{k=n-1}^{n_z-1} F[k]$$

[3] Let $\sum_{n=0}^{\infty} a[n]$ be a series of which $s[n]$ ($n \in \mathbb{N}$) is the sequence of partial sums, and let $b[n]$ be a sequence. Then $\forall\ n \in \mathbb{N}$ we have that

$$\sum_{n=N_1}^{N_2} a[n]b[n] = \sum_{n=N_1}^{N_2} s[n](b[n] - b[n+1]) + s[N_2]b[N_2+1] - s[N_1-1]b[N_1].$$

$$= \sum_{k=n-1}^{n_z-1} (k+1)\{-\Delta t|f[k+1]| +$$

$$\Delta t \sum_{m=0}^{(z-1)/2} \alpha_m \sum_{l=-\infty}^{\infty} \delta[k+1-lN-n_{2m+1}]\} + \Delta t(n_z|f[n_z]| - n_{-1}|f[n_{-1}]|)$$

$$= -\Delta t \sum_{k=n-1}^{n_z-1} k|f[k]| + \Delta t \sum_{m=0}^{(z-1)/2} \alpha_m n_{2m+1}. \qquad (2.120)$$

2.6.4 OUTPUT SPECTRUM OF CLIPPER

Consider a real-valued, 2π-periodic signal $f(x)$, given in Fourier series form as

$$f(x) = \sum_{n=-\infty}^{\infty} a_n e^{inx}, \qquad (2.121)$$

with

$$a_n = \frac{1}{2\pi} \int_0^{2\pi} f(x) e^{-inx}\, dx, \qquad (2.122)$$

where the complex Fourier coefficients a_n satisfy $a_{-n} = a_n^*$, $n \in \mathbb{Z}$. Furthermore, assume a number $l_c > 0 \in \mathbb{R}$ (the clipping level), and the set

$$\{x \in [0, 2\pi] \mid |f(x)| \le l_c\} \qquad (2.123)$$

in the form

$$\bigcup_{k=1}^{K} [\alpha_k, \beta_k], \qquad (2.124)$$

where the $[\alpha_k, \beta_k] \subset [0, 2\pi]$ are pairwise disjoint intervals, as shown in an example in Fig. 2.31.

Let

$$f_{l_c}(x) = \begin{cases} f(x), & |f(x)| \le l_c, \\ l_c, & f(x) \ge l_c, \\ -l_c, & f(x) \le -l_c, \end{cases} \qquad (2.125)$$

which is the clipped version of f at clipping level l_c. We want to compute the Fourier coefficients

$$b_n = \frac{1}{2\pi} \int_0^{2\pi} f_{l_c}(x) e^{-inx}\, dx, \quad n \in \mathbb{Z}. \qquad (2.126)$$

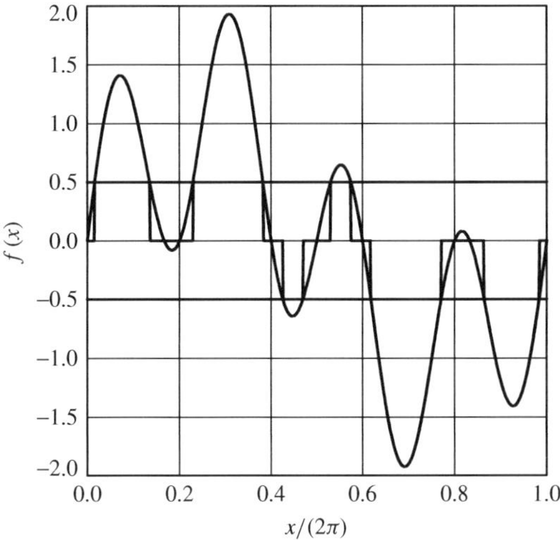

Figure 2.31 The function $f(x) = \sin(x) + \sin(4x)$ vs. x. The heavy portions of the x-axis indicate the intervals $[\alpha, \beta]$ where $|f(x)| \le l_c$, where in this case $l_c = 0.5$

Note that

$$f'_{l_c}(x) = \sum_{n=-\infty}^{\infty} b_n \mathrm{e}^{inx}, \tag{2.127}$$

hence

$$f'_{l_c}(x) = \sum_{n=-\infty}^{\infty} inb_n \mathrm{e}^{inx}. \tag{2.128}$$

On the other hand, we have

$$f'_{l_c}(x) = \begin{cases} f'(x), |f(x)| < l_c, \\ 0, |f(x)| > l_c. \end{cases} \tag{2.129}$$

Therefore, using $f(x) = \sum_{m=-\infty}^{\infty} a_m \mathrm{e}^{imx}$, we get

$$\begin{aligned} inb_n &= \frac{1}{2\pi} \int_0^{2\pi} f'_{l_c}(x) \mathrm{e}^{-inx}\,\mathrm{d}x = \frac{1}{2\pi} \sum_{k=1}^{K} \int_{\alpha_k}^{\beta_k} f'(x) \mathrm{e}^{-inx}\,\mathrm{d}x \\ &= \frac{1}{2\pi} \sum_{k=1}^{K} \int_{\alpha_k}^{\beta_k} \left(\sum_{m=-\infty}^{\infty} ima_m \mathrm{e}^{imx} \right) \mathrm{e}^{-inx}\,\mathrm{d}x \\ &= \frac{1}{2\pi} \sum_{k=1}^{K} \sum_{m=-\infty}^{\infty} ima_m \int_{\alpha_k}^{\beta_k} \mathrm{e}^{i(m-n)x}\,\mathrm{d}x \end{aligned}$$

$$= \frac{1}{2\pi} \sum_{k=1}^{K} \sum_{m=-\infty}^{\infty} i m a_m \frac{e^{i(m-n)\beta_k} - e^{i(m-n)\alpha_k}}{i(m-n)}. \tag{2.130}$$

Here, we have introduced the convention that for $\xi = 0$

$$\frac{e^{i\xi\beta} - e^{i\xi\alpha}}{i\xi} = \beta - \alpha, \tag{2.131}$$

which is correct in the limit and gives the same answer as treating the $m = n$ case separately in Eqn. 2.130. It thus follows that for $n \neq 0$

$$b_n = \frac{1}{2\pi n} \sum_{k=1}^{K} \sum_{m=-\infty}^{\infty} m a_m \frac{e^{i(m-n)\beta_k} - e^{i(m-n)\alpha_k}}{i(m-n)}. \tag{2.132}$$

For $n = 0$, we find more directly

$$b_0 = \frac{1}{2\pi} \int_0^{2\pi} f_{l_c}(x)\,dx = \frac{1}{2\pi} \sum_{k=1}^{K} \int_{\alpha_k}^{\beta_k} f(x)\,dx + \frac{l_c}{2\pi}|S_+| - \frac{l_c}{2\pi}|S_-|, \tag{2.133}$$

where $|S_+|$ and $|S_-|$ are the sizes of the sets

$$S_+ = \{x \in [0, 2\pi] \mid f(x) \geq l_c\}, \quad S_- = \{x \in [0, 2\pi] \mid f(x) \leq -l_c\}, \tag{2.134}$$

which should be available also. Note that the first number at the far right-hand side of Eqn. 2.133 can be expressed in terms of the a_n as

$$\frac{1}{2\pi} \sum_{k=1}^{K} \int_{\alpha_k}^{\beta_k} f(x)\,dx = \frac{1}{2\pi} \sum_{k=1}^{K} \int_{\alpha_k}^{\beta_k} \sum_{m=-\infty}^{\infty} a_m e^{imx}\,dx$$
$$= \frac{1}{2\pi} \sum_{k=1}^{K} \sum_{m=-\infty}^{\infty} a_m \frac{e^{im\beta_k} - e^{im\alpha_k}}{im}. \tag{2.135}$$

Example Using the preceding method, we will calculate the Fourier coefficients of a clipped sine $\sin_{l_c}(x)$. Let

$$f(x) = \sin x = \frac{e^{ix} - e^{-ix}}{2i},$$
$$a_{\pm 1} = \frac{\pm 1}{2i}, \text{ all other } a_m = 0,$$
$$l_c \in (0, 1),$$
$$\alpha = \arcsin l_c \in (0, \pi/2). \tag{2.136}$$
$$\{x \in [0, 2\pi] \mid |f(x)| \leq a\} = [0, \alpha] \cup [\pi - \alpha, \pi + \alpha] \cup [2\pi - \alpha, 2\pi]. \tag{2.137}$$

Then we get

$$b_n = \frac{1}{2\pi n} \sum_{m=-\infty}^{\infty} m a_m \left\{ \frac{e^{i(m-n)\alpha} - 1}{i(m-n)} \right.$$

$$\left. + \frac{e^{i(m-n)(\pi+\alpha)} - e^{i(m-n)(\pi-\alpha)}}{i(m-n)} + \frac{1 - e^{i(m-n)(2\pi-\alpha)}}{i(m-n)} \right\} \quad (2.138)$$

$$= \frac{1}{2\pi n} \sum_{m=-\infty}^{\infty} m a_m \left\{ \frac{e^{i(m-n)\alpha} - e^{-i(m-n)\alpha}}{i(m-n)} + (-1)^{m-n} \frac{e^{i(m-n)\alpha} - e^{-i(m-n)}}{i(m-n)} \right\} \quad (2.139)$$

$$= \frac{1}{2\pi n} \sum_{m=-\infty}^{\infty} m a_m . \frac{2\sin(m-n)\alpha}{m-n} (1 + (-1)^{m-n}). \quad (2.140)$$

Using Eqn. 2.136, we then get

$$b_n = \frac{1}{2\pi n} \left\{ \frac{\sin(1-n)\alpha}{i(1-n)} (1 + (-1)^{1-n}) + \frac{\sin(1-n)\alpha}{i(1-n)} (1 + (-1)^{-1-n}) \right\} \quad (2.141)$$

$$= \frac{1 + (-1)^{n-1}}{2\pi i n} \left(\frac{\sin(n-1)\alpha}{n-1} + \frac{\sin(n+1)\alpha}{n+1} \right), \quad (2.142)$$

and finally

$$b_n = \begin{cases} \frac{1}{\pi i(2\ell+1)} \left(\frac{\sin 2\ell\alpha}{2\ell} + \frac{\sin(2\ell+2)\alpha}{2\ell+2} \right), & n = 2\ell + 1, \ell \in Z, \\ 0, & n = 2\ell, \ell \in Z. \end{cases} \quad (2.143)$$

Using $b_{-2\ell-1} = -b_{2\ell+1}$, we thus find

$$\sin_{1_c}(x) = \sum_{\ell=-\infty}^{\infty} b_{2\ell+1} e^{i(2\ell+1)x} = \sum_{\ell=-\infty}^{\infty} (b_{2\ell+1} e^{i(2\ell+1)x} + b_{-2\ell-1} e^{-i(2\ell+1)x}) \quad (2.144)$$

$$= \sum_{\ell=-\infty}^{\infty} b_{2\ell+1} 2i \sin(2\ell+1)x = \frac{1}{\pi} \sum_{\ell=0}^{\infty} \left(\frac{\sin 2\ell\alpha}{\ell} + \frac{\sin 2(\ell+1)\alpha}{\ell+1} \right) \frac{\sin(2\ell+1)x}{2\ell+1}. \quad (2.145)$$

3

Low-frequency Physical Bandwidth Extension

3.1 INTRODUCTION

Chapter 2 discussed the situation in which the audio signal has a wider low-frequency bandwidth than the loudspeaker. The opposite situation, in which the loudspeaker has a wider bandwidth than the audio signal, can also occur (although this is probably a less common situation). Consider two possibilities:

1. The audio signal was limited in bandwidth owing to the transmission (or storage) channel. In that case, BWE processing should restore, or resynthesize, the missing frequency components as closely as possible. For speech, this occurs during transmission through the telephone network, together with a limitation in the high-frequency bandwidth. Methods to address both the high- and low-frequency limitation for speech applications will be discussed in Chapter 6. For general audio applications other than telephony, this situation is not a common one, and as such it will not further be discussed.
2. The audio signal is full bandwidth (at the low-frequency end) and the additional low frequencies are desired for enhancement only, in which case the loudspeaker must have a suitably low cut-off frequency. Applications of such methods would be, for example, (home) cinema, Hi-Fi, and automotive audio. We will devote the remainder of this chapter to a discussion of this situation.

In most cases, the algorithms for these applications physically extend the low-frequency spectrum of the signal; therefore, we refer to this kind of BWE methods as low-frequency physical BWE methods. Because the loudspeaker is able to reproduce lower frequencies than those present in the audio signal, there is a possibility to lower the perceived pitch.

3.2 PERCEPTUAL CONSIDERATIONS

Assume that a signal $x(t)$ has a low-frequency cut-off of $f_{l,x}$, and the reproducing loudspeaker has a low-frequency cut-off of $f_{l,l} < f_{l,x}$. To increase the apparent bass response

Audio Bandwidth Extension E. Larsen and R. M. Aarts
 ISBN 0-470-85864-8

of the perceived signal, we can utilize the frequency range $f_{1,1}$–$f_{1,x}$ to add frequency components related to $x(t)$. These lower-frequency components can lower the pitch of the signal, which can be used for bass enhancement.

3.2.1 PITCH (SPECTRAL FINE STRUCTURE)

Say the signal $x(t)$ has a pitch of f_0 Hz, mediated by partials at $n.f_0$ Hz, $n = 1, 2, 3 \ldots$. Reproducing $x(t)$ at a reduced pitch of, say, f_0' is most practical if the harmonics already present in $x(t)$ remain harmonically related to the added low-frequency components. For example, $f_0' = f_0/2$ would be a good choice. Adding the f_0' component to $x(t)$ to create signal $x_2(t)$ would yield a complex tone with a fundamental frequency at f_0' all *even* harmonics. This situation also occurs in low-frequency psychoacoustic BWE using a rectifier as non-linear device (NLD), see Fig. 2.7. In the related discussion, it was remarked that predictions from the auditory image model (AIM [203], see Sec. 1.4.8) indicated two pitch percepts of nearly equal strength. In the notation of this section, there is a pitch at f_0', and slightly weaker, at f_0. This could indicate that either the pitch is ambiguous or that the new fundamental at f_0' is not grouped with the harmonics, leading to two signals being perceived. Note that a common amplitude or frequency modulation of the f_0' partial and the $n.f_0$ partials should facilitate grouping of all partials into a single stream ('common fate' principle, see Sec. 1.4.7).

A less ambiguous situation would occur if instead of only adding the component $f_0' = f_0/2$, components at $(2k+1)f_0' = (k+1/2)f_0$ are also added ($k = 1, 2, 3 \ldots$). In that case, the new signal $x_2(t)$ would contain a fundamental at f_0' and *all* harmonics, akin to the situation in which an integrating NLD is used in the low-frequency psychoacoustic BWE algorithm; see Fig. 2.10 for the AIM pitch prediction of this signal. Such a signal has an unambiguous and strong pitch at f_0' (even without common amplitude and/or frequency modulation). Another possibility is to add a frequency component at $f_0'' = f_0/3$, leading to a pitch that is one-third of the original; this can be extended to a frequency division by 4, 5, and so on. We will not further consider such situations though, and concentrate on the case in which the pitch is lowered by an octave.

3.2.2 TIMBRE (SPECTRAL ENVELOPE)

Assume a complex tone with harmonic amplitudes a_i at frequencies $i.f_0/2$, with i even. As before, we simplify our modelling of timbre to include only brightness[1], modelled by the spectral centroid C_S (Eqn. 1.95), which for this signal is

$$C_S = f_0 \times \sum_i i a_i^2 / \sum_i a_i^2, \tag{3.1}$$

where the sum runs over all i for which $a_i \neq 0$. After BWE processing, harmonics are added such that the new fundamental is $f_0/2$, and the new harmonic amplitudes are $a_i + g b_i$ ($i = 1, 2, 3, \ldots$); the b_i are the synthetic frequency components and are

[1] In reality, other factors influence timbre, such as the relative phase of partials and temporal envelopes. These factors are neglected to simplify the discussion, and also because they are thought to be less important than the amplitude spectrum of the harmonics, see also Sec. 1.4.6.

determined by the algorithmic details, and g is the gain factor of the b_i. The spectral centroid of the BWE signal C'_S is

$$C'_S = f_0 \times \sum_i i(a_i + gb_i)^2 / \sum_i (a_i + gb_i)^2. \quad (3.2)$$

If the goal is to achieve $C_S = C'_S$, and say the b_i are identically zero for nonzero a_i, and vice versa, then this is achieved if

$$\frac{\sum_i ib_i^2}{\sum_i b_i^2} = \frac{\sum_i ia_i^2}{\sum_i a_i^2}, \quad (3.3)$$

that is, if the individual spectral centroids of the a_i and b_i are identical, independent of the value of g. If the a_i and b_i are nonzero for common i, it is more cumbersome to derive a relationship such that $C'_S = C_S$, in part because the relative phases of the a_i and b_i need to be accounted for as well. It could also be desirable to lower the brightness of the processed signal, such that $C'_S < C_S$. Again, to achieve this it will depend on whether the a_i and b_i are nonzero for different i, or not, and their relative phase in the latter case. Because C_S depends in such a complicated manner on the harmonic structure of the signals, as well as on the processing details, we do not further consider an analysis of these effects. In general, a mildly sloping harmonics spectrum of the b_i should maintain a similar value for C_S and, therefore, a similar timbre. Of course, if only a component at $f_0/2$ is added, without any higher harmonics, C_S will be reduced by an amount that depends on the amplitude of the $f_0/2$ component.

3.2.3 LOUDNESS (AMPLITUDE)

Many of the comments made in Sec. 2.2.3 regarding loudness effects for low-frequency psychoacoustic BWE also apply for low-frequency physical BWE methods, although the effects are in the opposite ‘direction’. It will again be useful to refer to the equal-loudness contours of Fig. 1.18. Firstly, we see that adding low-frequency components in the bass frequency range means that the added components will be less audible than the original low partials, due to the upward slope of the equal-loudness contours at low frequencies. A more proper way of analysing this exactly would be to use a loudness model, such as ISO532A or ISO532B, described in Sec. 1.4.4.2. Nonetheless, as acoustic energy is added above threshold, the loudness of the signal will increase. Furthermore, we assume that the loudspeaker has a more or less ‘flat’ response in the frequency region where the synthetic components are added, so we do need to consider the loudspeaker’s response as we did in Sec. 2.2.3 and Eqns. 2.5 and 2.6.

3.3 LOW-FREQUENCY PHYSICAL BANDWIDTH EXTENSION ALGORITHMS

In the previous chapter on low-frequency psychoacoustic BWE, we encountered several different ways to create higher harmonics from a (periodic) signal. We can borrow many of these techniques to also create subharmonics, if we make appropriate changes to the algorithms. Therefore, only a limited number of possible algorithms will be presented, as others will be obvious extensions of what was presented in Chapter 2. First, we will discuss several options on using the synthetic low-frequency components.

3.3.1 SYSTEMS WITH LOW-FREQUENCY EXTENSION

Figure 3.1 presents a low-frequency physical BWE system; comparison with Fig. 2.4 will show that there is a lot of similarity with the low-frequency psychoacoustic BWE system. However, the processing details differ. The input signal $x(t)$ is filtered by FIL1 to obtain the lowest frequency band present in the signal. Although this band will vary over time, for simplicity, a constant band of about an octave can be used. The actual bandpass region will depend on the low-frequency cut-off value f_l of the loudspeaker used, and the frequency division occurring in the non-linear device (NLD). Assuming that this frequency division is a factor of two, the bandpass region of FIL1 can be set to $2f_l$–$4f_l$, for example, for $f_l = 30\,\text{Hz}$, this would be 60–120 Hz. This bandpass signal is processed by the NLD, to create a fundamental an octave lower than the strongest frequency component at its input, and also the harmonics of this new fundamental. Depending on whether only the new fundamental or also some of its harmonics are desired in the output, filtering by FIL2 will select the desired frequency range. If only the new fundamental is desired, FIL2 will be bandpass between f_l–$2f_l$; if also harmonics are desired, the high-frequency limit of FIL2 should be increased, for example, to $4f_l$. Finally, a gain factor g is applied to create the harmonics signal $x_h(t)$.

At this point, there are several options. One option is indicated by the dash-dotted line in Fig. 3.1, where $x_h(t)$ is fed to a 'low-frequency effects' (LFE) loudspeaker (analogous to the sixth channel in a 5.1 surround sound system)[2]. The other option is to add $x_h(t)$ back to a delayed version of $x(t)$, where the delay of $x(t)$ should match the filtering delay of $x_h(t)$[3], yielding output signal $y(t)$. A standard crossover network can then be used to

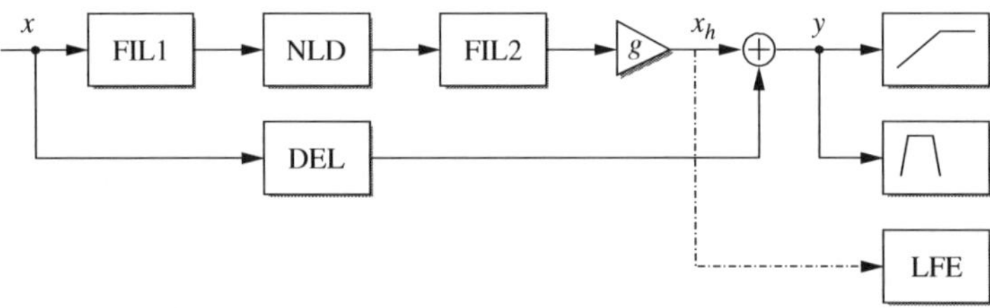

Figure 3.1 Low-frequency physical BWE system. FIL1 and FIL2 are bandpass filters, NLD is a non-linear device, g is a scaling factor. The input signal $x(t)$ is processed to yield a harmonics signal $x_h(t)$, which can be fed directly to a low-frequency effects (LFE) channel. Alternatively, it can be added back to $x(t)$ and applied to a set of loudspeakers, including a subwoofer

[2] By obvious modification to the filter bandpass regions and the NLD, extremely low frequencies that are below, say, 20 Hz can be generated. These frequencies can be used to drive *shakers* that transmit tactile vibrations.

[3] As for the low-frequency psychoacoustic BWE system, filtering for low-frequency physical BWE should preferentially be done with linear-phase filters (see Sec. 2.3.3), such that all frequency components of $x_h(t)$ will be equally delayed.

drive the available loudspeaker system, for example, a subwoofer for the frequency range 30–100 Hz, and an accompanying full-range system.

Note that we have described the low-frequency physical BWE system here as adding very low frequency components, down to 20 or 30 Hz. Proper reproduction of such frequencies requires very large and expensive loudspeaker systems. However, the frequency regions can be scaled to higher values (say to 100 Hz), such that implementation on smaller loudspeaker systems is also possible. Of course, for bass enhancement purposes, the final effects are usually more dramatic at very low frequencies.

3.3.2 NON-LINEAR DEVICE

Non-linear processing for low-frequency physical BWE applications aims to lower the frequency content of the available signal. Many of the comments made in Sec. 2.3.2 with regard to non-linear processing for low-frequency psychoacoustic BWE algorithms can be applied here as well. In particular, it is preferable to have NLDs that are level independent (homogeneous systems, see Sec. 1.1). With regard to the actual implementation of the NLDs, the various possibilities discussed in Sec. 2.3.2 can often be used here, with slight modifications such that not only higher harmonics but also the subharmonic is generated. To avoid duplicating a lot of material, we present only a brief discussion of several NLD options here.

An important difference with low-frequency psychoacoustic BWE is that for low-frequency physical BWE, we can choose to add only the halved fundamental, thereby creating a harmonic series with $f_0/2$, f_0, $2f_0$, and so on. As discussed in Sec. 3.2.1, the $f_0/2$ component may not group too well with the other partials, although common amplitude or frequency modulation of all partials will facilitate grouping (Sec. 1.4.7). On the other hand, the harmonics generated by the NLD (as discussed below) can all be added back to the main signal. In such a case, it is possible that a harmonics series of for example, $f_0/2$, f_0, $3f_0/2$, $2f_0$, and so on, is generated, in which case a strong pitch at $f_0/2$ is always perceived. The perceptual effect of low-frequency physical BWE therefore depends on how many synthetic harmonics are added back to the main signal.

3.3.2.1 Rectifier

In Sec. 2.3.2.2, a rectifier as an NLD was introduced for low-frequency psychoacoustic BWE applications. It was seen that the rectifier predominantly generates the double-frequency component of the strongest input frequency component, that its temporal response is good, and that the amount of intermodulation distortion can be quite large if more than one frequency component of comparable amplitude are contained in the input. To modify this algorithm such that it generates half the input frequency, we can simply set the output to zero for alternating periods of the input, as in Fig. 3.2, which can be done effectively by detecting zero crossings of the input signal.

3.3.2.2 Integrator

In Sec. 2.3.2.3, an integrator as an NLD was introduced. In contrast to the rectifier, all (odd and even) harmonics of the input frequency are generated. The amount of intermodulation distortion was generally low, although the temporal response was slightly worse (in terms

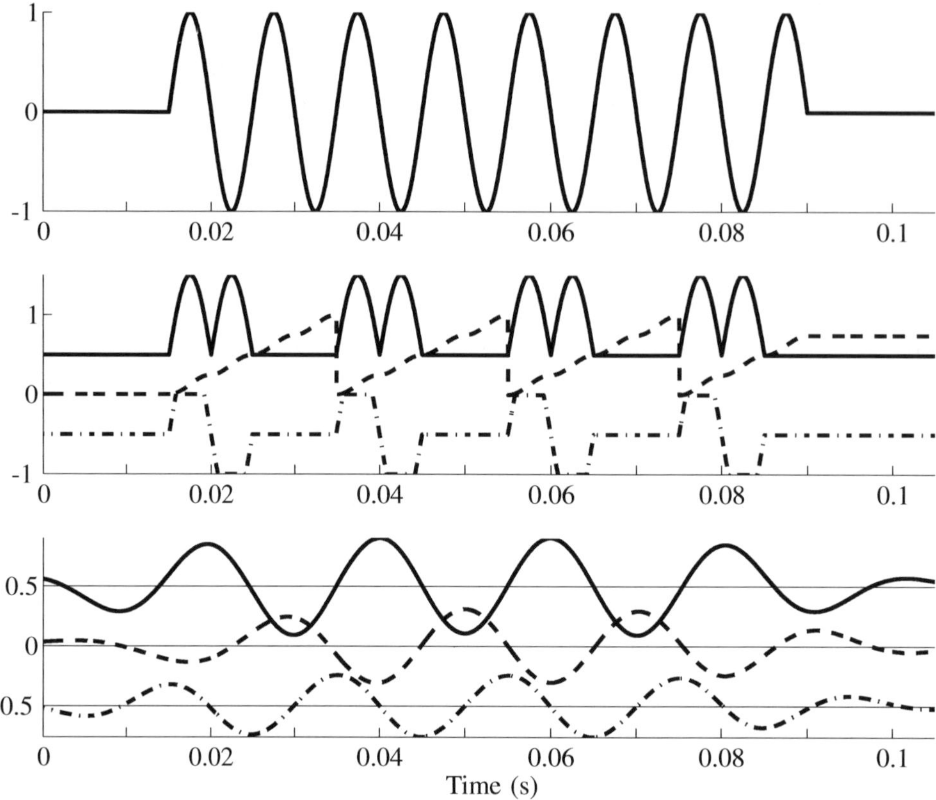

Figure 3.2 Low-frequency physical BWE processing with various NLDs. The upper panel shows a tone burst at 100 Hz. The middle panel shows the effect of rectification (solid line, offset by +0.5), integration (dashed line, no offset), and clipping (dash-dotted line, offset by −0.5). Note that period doubling occurs, because the signal is zeroed on every other period for the rectifier and clipper, while for the integrator the reset occurs after two input periods. The lower panel shows the effect of linear-phase filtering around 50 Hz, such that only the fundamental frequency is retained

of distortion of temporal envelope). To use an integrator for halving the frequency of a signal, the only necessary modification is to reset the output of the integrator to zero after two input periods have occurred. The input signal periods can again be detected by observing zero crossings. The integrated output for a pure-tone input is plotted as a dashed line in the middle panel of Fig. 3.2. Linear-phase filtering this signal around 50 Hz leads to a signal as shown by a dashed line in the lower panel of the same figure.

3.3.2.3 Clipper

In Sec. 2.3.2.4, a clipper as an NLD was introduced. This device generates odd harmonics of the input frequency. The clipper was shown to be very robust against intermodulation distortion, and to have a good temporal response. Again, a slight modification will allow the clipper to be used for low-frequency physical BWE purposes. As with the rectifier discussed previously, alternating periods of the input signal are clipped (at a clipping level of half-maximum amplitude, in this case ± 0.5), the other periods being zeroed. The resulting signal is shown as the dash-dotted line in the middle panel of Fig. 3.2. Linear-phase filtering around 50 Hz leads to the signal shown by the dash-dotted line in the lower panel of the same figure.

For low-frequency psychoacoustic BWE algorithms, it was discussed at length how to enhance performance by making the clipping level adaptive with respect to the level of the input signal. For low-frequency physical BWE, this might also have some advantage, and a similar strategy could be employed, although this has not been validated by actual listening tests.

3.3.2.4 Low-frequency Physical Bandwidth Extension with Frequency Tracking

An alternative method of generating subharmonics is by using a frequency tracker, as discussed in Sec. 2.4, refer to Fig. 2.25. This algorithm generates any desired harmonics spectrum, with a fundamental that is based on the strongest frequency component contained in the input spectrum (the tracked frequency). Advantages are that there is no intermodulation distortion and that harmonics are only generated if the input is periodic. Also, the frequency tracker is implemented in a computationally very efficient manner. The disadvantage is that the frequency tracker needs finite time to converge to the actual signal frequency, and that errors may occur if multiple input frequencies or additive noise is present. However, the particular method of frequency tracking was shown to adapt quickly and that some corrections for the tracked frequency are possible if a few statistics of any additive noise are known or can be estimated.

Assuming that the input contains frequency ω_0, and is correctly estimated by the frequency tracker, the harmonics generator (HG in Fig. 2.25) will generate a signal $x_h(t)$ as

$$x_h(t) = \sum_{k=1}^{N} A_k \sin(k\frac{\omega_0}{2}t), \tag{3.4}$$

If one desired to merely add the halved fundamental, then $N = 1$; otherwise, $N > 1$ and a desired harmonics spectrum can be generated. Because the synthetic frequency components are explicitly created, there is no intermodulation distortion.

3.3.2.5 Inclusion of Higher Harmonics

From the lower panel of Fig. 3.2, it is quite obvious that the resulting signal (representing the signal $x_h(t)$ in Fig. 3.1) does not depend a great deal on the particular choice of NLD. This is because in all cases only the halved fundamental (in this case 50 Hz) was extracted by FIL2 of Fig. 3.2. The various NLDs (rectifier, integrator, clipper) differ most in the amplitude and phase spectrum of the generated harmonics, thus if some of the harmonics are retained by FIL2, then the output signals would differ more. To illustrate this, Fig. 3.3 shows signals $x_h(t)$ as they would be obtained if FIL2 were a linear-phase filter with a bandpass of 40–160 Hz (all the other processing steps are identical to those in the first

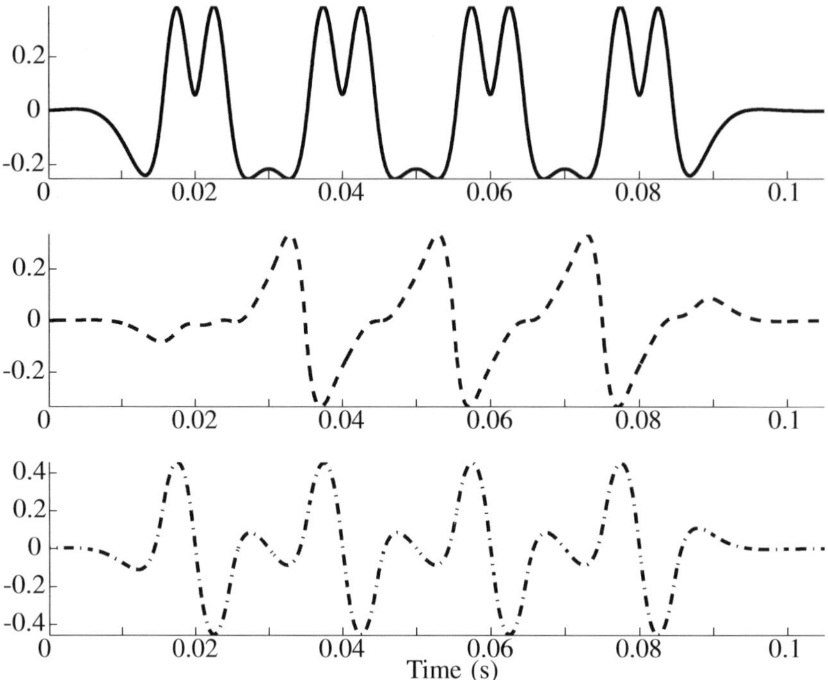

Figure 3.3 Low-frequency physical BWE processing with various NLDs; input signal is assumed to be a gated 100-Hz tone as in the upper panel of Fig. 3.2. Here, the output signal $x_h(t)$ (refer to Fig. 3.1) is shown for three different NLDs: rectifier (upper panel), integrator (middle panel), and clipper (lower panel). Owing to filtering of the spectrum generated by the NLD, only harmonics at 50, 100, and 150 Hz are appreciably present in the output signals. Note that the different spectra generated by the NLDs causes the observed differences in waveforms

example). Thus, the partials at 50, 100, and 150 Hz would be passed completely, while higher partials would be strongly attenuated by FIL2. The interference of these three partials creates the differences visible in Fig. 3.3, where signals by rectifier, integrator, and clipper are displayed as solid, dashed, and dash-dotted lines respectively.

3.3.3 FILTERING

Using the same kind of analysis as was done in Sec. 2.3.3.3 for low-frequency psychoacoustic BWE algorithms, it can be shown that it is preferable to use a linear-phase implementation for filters FIL1 and FIL2, for two reasons:

- The final output signal of the low-frequency physical BWE algorithm will be obtained by scaling $x_h(t)$ and adding it back to $x(t)$ (either electrically or acoustically, see Fig. 3.1), so common frequency components will interfere either constructively or destructively. In the example of Figs. 3.2 and 3.3, there is a synthetic 100-Hz component in $x_h(t)$ as generated by the rectifier and the integrator (but not the clipper, which generates only odd harmonics of 50 Hz), which adds to the original 100-Hz fundamental present in $x(t)$. Figure 2.21 shows examples in which synthetic and original frequency bands interfere to create an irregular amplitude spectrum in case of non-linear-phase filters. For low-frequency physical BWE algorithms, the situation would be similar, if beside the halved fundamental, higher harmonics are also added, as these would overlap in frequency range with the original signal. If only the halved fundamental is added, there is no frequency overlap between $x(t)$ and $x_h(t)$, and there should be no objection to use a non-linear-phase filter from this particular point of view.

- Because of the filtering by FIL1 and FIL2, signal $x_h(t)$ will be delayed with respect to $x(t)$, by an amount that depends on the filter orders (higher filter order giving larger delays). In Sec. 1.4.7 on auditory scene analysis, it was explained that an onset delay between frequency components can lead to segregation, that is, $x(t)$ and $x_h(t)$ being separately perceived as two different streams. This was analysed for low-frequency psychoacoustic BWE in detail in Sec. 2.3.3.3, and typical values for total group delay *variation over frequency* of components of $x_h(t)$ was shown to be on the order of 10 ms (Fig. 2.22), which combines the effect of both filters. Such delays can be assumed to be detectable in principle (Zera and Green [304]), and there is some circumstantial evidence from informal listening tests that such can indeed lead to segregation. To avoid such problems, linear-phase filters will not lead to group delay variations, and if $x(t)$ is delayed (using a delay line) by the same amount as $x_h(t)$ (the required amount of delay can be easily computed if the filters FIL1 and FIL2 are known), both signals can be added exactly in phase. Note that in this case it does not matter whether $x(t)$ and $x_h(t)$ have overlapping frequency components or not.

The standard way to implement linear-phase filters with FIR structures presents a problem for low-frequency physical BWE in that the bandpass region is usually a very small fraction of the sample rate. Also, the filter's low cut-off frequency can be very small, even smaller than for low-frequency psychoacoustic BWE algorithms. An FIR filter would have to be of very high order (many taps) to implement such a specification, leading to a high computational burden. A more efficient way is to use an IIR filter that is designed to have an amplitude spectrum that is the square root of the desired specification. The filter can be applied once in forward time, after which the filtered signal is time reversed and filtered using the same filter. This output is then time reversed again. The result is zero phase shift (as phase changes of the first filter are canceled by the second time-reversed filter) and an attenuation that is the square of the IIR filter's amplitude spectrum. This was also discussed in Sec. 2.3.3.3 and is discussed elaborately in, for example, Powell and Chau [213].

3.3.4 GAIN OF HARMONICS SIGNAL

Again, the situation is very analogous to the case of low-frequency psychoacoustic BWE applications. Gain (or scaling) $g(t)$ of $x_h(t)$ prior to addition to $x(t)$ to form the output $y(t)$ (Fig. 3.1) could be fixed, frequency adaptive, and/or output-level adaptive. For a simple implementation, $g(t)$ could simply be a constant value, such that the loudness balance of the synthetic harmonics signal and the original signal is subjectively appropriate, and does not lead to distortion at high signal levels.

A slightly more subtle scheme could be adopted analogous to Gan *et al.* [83], which takes into account the equal-loudness level contours (Sec. 2.3.4.2 and Fig. 1.18) to expand the envelope of the generated harmonics signal. While Gan *et al.* designed this scheme for low-frequency psychoacoustic BWE applications, the same procedure could be used for a low-frequency physical BWE algorithm. In a simple implementation of their scheme, the envelope of the harmonics signal $x_h(t)$ would be expanded by a factor of about 1.10 (intended for fundamental frequencies in the range 40–100 Hz; the expansion ratio varies with fundamental frequency, the quoted value is an average). A more sophisticated approach implements a different expansion ratio for each harmonic, reflecting the change in loudness growth over frequency. A scheme like this, in which each harmonic is scaled separately, is attractive if harmonics are generated individually, such as in Gan *et al.*ś modulation technique, or the method described in Sec. 2.4 and 3.3.2.4 (using a frequency tracker). In the latter case, expansion ratios can be determined even more accurately, as the fundamental frequency is known (estimated), which is not implemented in Gan *et al.*ś original algorithm.

An output-level-adaptive gain could be implemented entirely analogous to that described in Sec. 2.3.4.3, and illustrated in Fig. 2.24. This scheme employs a feedback loop from the output $y(t)$ to control $g(t)$. In normal circumstances, $g(t)$ has a fixed value, but if the level of $y(t)$ exceeds a predetermined value, $g(t)$ should be decreased very quickly to prevent distortion occurring in the loudspeaker. While $y(t)$ is below the threshold level, $g(t)$ can be slowly increased back to its nominal value. This increase should occur slowly to prevent envelope distortion of $y(t)$ (i.e. on a time scale of a few seconds). An additional advantage of such an automatic gain control (AGC) scheme is that low-level bass sounds are maximally amplified, while high-level bass sounds are attenuated. This is a good match to the audibility characteristics of low-frequency sounds, which have very low loudness at low-to-intermediate sound pressure level, but which increase in loudness very rapidly (more so than intermediate frequency sounds) with increasing sound pressure level.

3.4 LOW-FREQUENCY PHYSICAL BANDWIDTH EXTENSION COMBINED WITH LOW-FREQUENCY PSYCHOACOUSTIC BANDWIDTH EXTENSION

We have so far presented several options for physical extension of the low-frequency spectrum of an audio signal. In Chapter 2, we have presented the same for a psychoacoustic bandwidth extension. These two concepts can be combined such that the bass pitch in audio signal is lowered (usually by an octave), but in such a way that very low frequency components are not radiated nor ever physically present. This would permit application to smaller loudspeaker systems.

Beside the obvious cascade of a low-frequency physical BWE system followed by a low-frequency psychoacoustic BWE system, this concept can be more efficiently implemented by the 'standard' low-frequency physical BWE approach of Fig. 3.1. The only modification is that FIL2 should have a higher low cut-off frequency, such that the halved fundamental is not actually present in $x_h(t)$. For example, a complex tone with a 70-Hz fundamental would yield a synthetic signal after non-linear processing that has a fundamental at 35 Hz, including harmonics (the spectrum of which depends on the particular choice of the NLD). When using a small loudspeaker, with a low-frequency resonance at 100 Hz, FIL2 could be implemented as a bandpass filter between 100–200 Hz. The result is that the pitch of the bass tone has been lowered to 35 Hz, which is effected by frequency components >100 Hz. In other words, a signal without very low pitched tones has been modified such that it is perceived as having very low pitched tones, using a loudspeaker that cannot reproduce very low frequencies.

4

Special Loudspeaker Drivers for Low-frequency Bandwidth Extension

The preceding chapter dealt with low-frequency BWE exclusively through signal-processing algorithms. For the case of small loudspeakers, it was shown that non-linear processing could enhance the perception of very low frequency tones. The advantage of such an approach is that the loudspeaker need not be modified in any way, as the algorithm is tailored to the loudspeaker. In this chapter, two options are described whereby modifying the loudspeaker driver can also lead to enhanced bass perception. This is achieved by modifying the force factor of the driver, typically by employing either a very strong or a very weak magnet, compared to what is commonly used in typical drivers. Both these approaches also require some pre-processing of the signal before it is applied to the modified loudspeaker. Thus, this kind of BWE is a mixed approach combining mechanical and algorithmic measures. In Sec. 4.1, the influence of the force factor on the performance of the loudspeaker is reviewed, after which Secs. 4.2–4.3 discuss high force factor and low force factor drivers, respectively, and their required signal processing. Section 4.4 presents an analysis of the transient responses of these special drivers.

4.1 THE FORCE FACTOR

Direct-radiator loudspeakers typically have a very low efficiency, since the acoustic load on the diaphragm or cone is relatively low compared to the mechanical load, and in addition the driving mechanism of a voice coil is quite inefficient in converting electrical energy into mechanical motion. The drivers have a magnetic structure – which determines the force factor Bl – that is deliberately kept at an intermediate level so that the typical response is flat enough to use the device without significant equalization. It was already shown in Sec. 1.3.2.3 that the force factor Bl plays an important role in loudspeaker design; it determines among others the frequency response, the transient response, the electrical input impedance, the cost, and the weight; we will discuss various of these consequences. To show the influence on the frequency response, the sound pressure level

Audio Bandwidth Extension E. Larsen and R. M. Aarts
 ISBN 0-470-85864-8

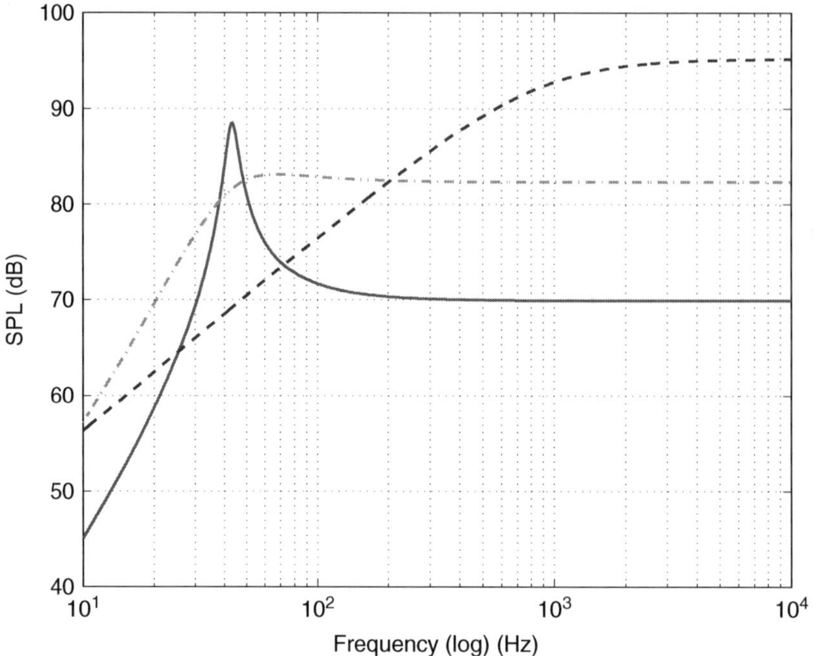

Figure 4.1 Sound pressure level (SPL) for the driver MM3c with three Bl values (low, medium, and high), while all other parameters are kept the same (1-W input power), $Bl = 1.2$ (solid), $Bl = 5$ (dash-dot), and $Bl = 22$ (dash). See Table 4.2 for the other parameters

(SPL) of a driver with three Bl values (low, medium, and high) is plotted in Fig. 4.1, while all other parameters are kept the same. It appears that the curves change drastically for varying Bl. The most prominent difference is the shape, but the difference in level at high frequencies is also apparent. While the low-Bl driver has the highest response at the resonance frequency, it has a poor response beyond resonance, which requires special treatment, as discussed in Sec. 4.3.1. The high-Bl driver has a good response at higher frequencies, but a poor response at lower frequencies, which requires special equalization as discussed in Sec. 4.2. In between, we have the well-known curve for a medium-Bl driver. To show the influence of Bl on the electrical input impedance, the magnitude is plotted for a driver with two Bl values (low and medium), while all other parameters are kept the same, see Fig. 4.2. In this plot, the curve for $Bl = 22$ is omitted; it has similar shape, but a much higher Q and a larger peak (at $R_e + Bl^2/R_t = 2206\,\Omega$). It also appears that these curves change drastically for varying Bl. The phase of the electrical input impedance is plotted in Fig. 4.3. The underlying reason for the importance of Bl is that besides determining the driving force, it also gives (electric) damping to the system. The total damping is equal to the (real part of the) radiation impedance, the mechanical damping, and the electrical damping ($(Bl)^2/R_e$), where the electrical one dominates for medium- and high-Bl loudspeakers, and is most prominent around the

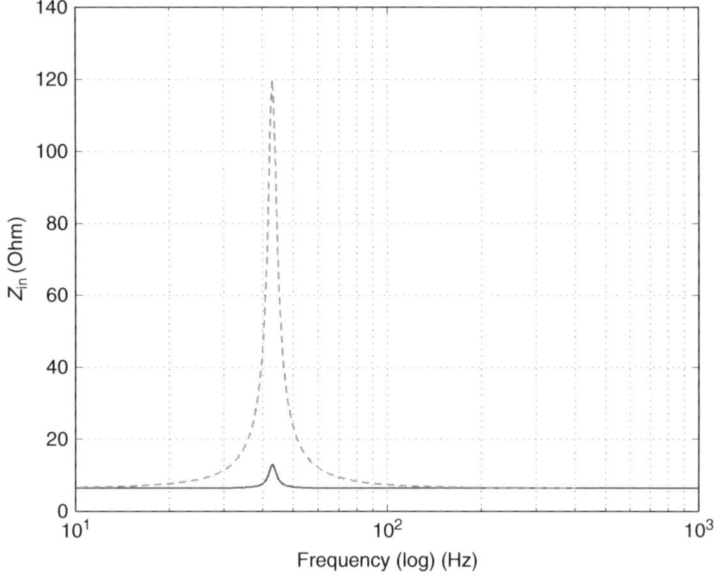

Figure 4.2 The electrical input impedance for the driver MM3c with two Bl values (low, and medium, respectively $Bl = 1.2$ (solid), and $Bl = 5$ (dash)), while all other parameters are kept the same ($L_e = 0$). See Table 4.2 for the other parameters

resonance frequency. The power efficiency given in Eqn. 1.54 can be written as

$$\eta = \frac{(Bl)^2 R_r}{R_e\{(R_m + R_r)^2 + (R_m + R_r)(Bl)^2/R_e + (m_t\omega_0\nu)^2\}}, \tag{4.1}$$

clearly showing the influence of Bl. This importance is further elucidated in the following paragraph.

A dimensionless measure of damping In Vanderkooy *et al.* [284], a dimensionless parameter was introduced to describe the relative damping due to Bl. As Bl increases, the box and suspension restoring forces become less relevant, as we shall see later, so we choose a parameter of the form

$$\frac{i\omega(Bl)^2/R_e}{-\omega^2 m_t}, \tag{4.2}$$

which is the ratio of the electrical damping force to the inertial force on the total moving mass m_t (consisting of the cone with its air load). We remove the imaginary unit and the negative sign, so the relative damping factor becomes

$$\delta = \frac{(Bl)^2}{\omega_r m R_e}. \tag{4.3}$$

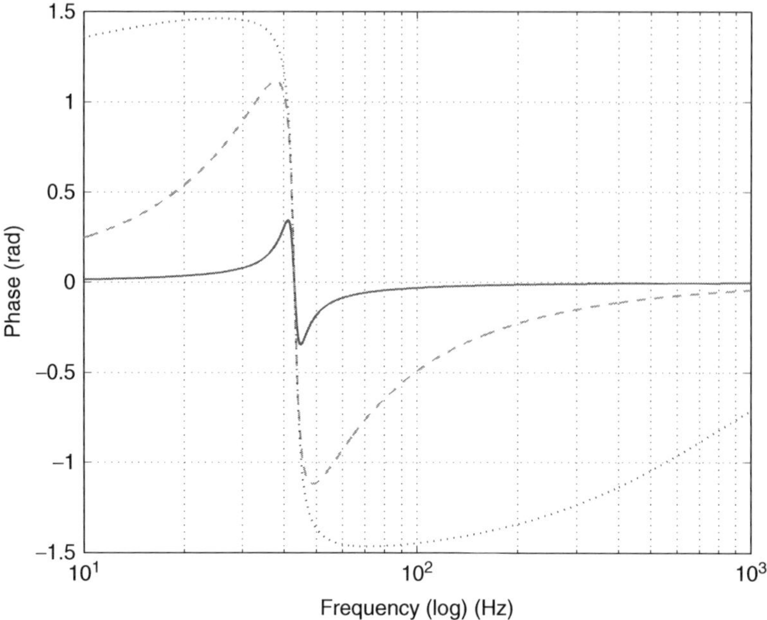

Figure 4.3 The phase [rad] of the electrical input impedance for the driver MM3c with three Bl values (low, and medium, and high, while all other parameters are kept the same (1-W input power), $Bl = 1.2$ (solid), $Bl = 5$ (dash), and $Bl = 22$ (dots). See Table 4.2 for the other parameters

The frequency ω_r can be chosen to represent the low-frequency end of the intended audio spectrum, or it could be set to a reference frequency such as 50 Hz, or the resonance frequency w_0. The reference frequency may be useful since the low-frequency cut-off of a system is significantly altered when Bl is significantly increased. Incidentally, for the usual Butterworth system aligned to frequency ω_0, δ would be $\sqrt{2}$. The drivers MM3c and HBl mentioned in Table 4.2 have $\delta = 0.059$ and $\delta = 4.43$ respectively. The common parameter Q_e, the electrical Q-factor, while similar to δ^{-1} (and at the resonance frequency there holds $Q_e = \delta^{-1}$), is predicated on a normal driver for which the resonance frequency is determined by the interaction between inertial and suspension forces. As Bl is increased, the suspension forces are less relevant, and Eqn. 4.3 is a better measure than Q_e.

4.2 HIGH FORCE FACTOR DRIVERS

In the 1990s, a new rare-earth-based material, neodymium-iron-boron (NdFeB), in sintered form, came into more common use. It has a very high flux density coupled with a high coercive force, possessing a B-H product increased by almost an order of magnitude compared to more common materials. This allows drivers to be built in with much larger total magnetic flux, thereby increasing Bl by a large factor. In Vanderkooy *et al.* [284], some features of normal sealed-box loudspeakers with greatly increased Bl were outlined.

This work focused mainly on the efficiency of the system as applied to several amplifier types, but also indicated several other avenues of interest. Figure 4.1 shows the frequency response curves (using Eqns. 1.39 and 1.40) for three Bl values: 1.2, 5.0, and 22 N/A. At the higher-Bl value, the electromagnetic damping is very high. If we ignore in this case the small mechanical and acoustic damping of the driver, the damping term is proportional to $(Bl)^2/R_e$. For a Butterworth response, the inertial term $\omega^2 m$, the damping term $\omega(Bl)^2/R_e$, and total spring constant k_t are all about the same at the bass cut-off frequency. When Bl is increased by a factor of 5, the damping is increased by a factor of 25. Thus the inertial factor, which must dominate at high frequencies, becomes equal to the damping at a frequency about 25 times higher than the original cut-off frequency. This causes the flat response of the system to have a 6-dB per octave roll-off below that frequency, as shown in the figure.

At very low frequencies, the spring-restoring force becomes important relative to the damping force at a frequency 25 times lower than the original cut-off frequency. Below this the roll-off is 12 dB/octave. Such frequencies are too low to influence audio performance, but it is clear that the system (driver and cabinet) is now no longer constraining the low-frequency performance. We could use a much smaller box without serious consequences.

How much smaller can the box be? The low-frequency cut-off has been moved down by a factor of nearly 25. The suspension stiffness k is small, and since $k_B \propto 1/V_0$, the cut-off frequency will return to the initial bass cut-off frequency when the box size is reduced by a factor of about 25: a 25-litre box could be reduced to 1 litre. Powerful electrodynamic damping has allowed the box to be reduced in volume without sacrificing the response at audio frequencies. The only penalty is that we must apply some equalization.

The equalization needed to restore the response to the original value can be deduced from Fig. 4.1, since the required equalization is the difference between $Bl = 5$ (dashed), and $Bl = 22$ (dotted) curves. Such equalization will, in virtually all cases, increase the voltage applied to the loudspeaker, since audio energy resides principally at lower frequencies. The curve levels out at just over 12 dB at low frequencies, but in actual use one might attenuate frequencies below, say, 40 Hz.

The power efficiency for very large Bl can be calculated (using Eqn. 4.1) as

$$\lim_{Bl\to\infty} \eta = \frac{R_r}{R_m + R_r}. \tag{4.4}$$

This clearly shows that the efficiency increases for decreased mechanical damping R_m.

An observation about equalization The required equalization can be calculated by the frequency response ratio $H_L(\omega)/H_H(\omega)$ using Eqn. 1.40, where the subscripts refer to the high and low values of Bl. The required equalization function for two loudspeakers with different Bl values, but identical in all other respects, can also be calculated using an alternative approach, giving new insight, which we develop now.

For the two loudspeakers to produce the same acoustic output, the shape and motion of the two pistons (or cones) must be the same. Since all other aspects of the loudspeakers are the same, this can be achieved if the total force on the pistons is the same, thus ensuring that they have the same motion. The force is derived from the electromagnetic

Lorentz force, $BlI(\omega)$. Since current $I(\omega) = V(\omega)/Z(\omega)$, where $V(\omega)$ is the loudspeaker voltage and $Z(\omega)$ is its electrical impedance, we must arrange to have $BlV(\omega)/Z(\omega)$ the same for the two conditions. Hence

$$\frac{Bl_{\mathrm{H}} V_{\mathrm{H}}(\omega)}{Z_{\mathrm{H}}(\omega)} = \frac{Bl_{\mathrm{L}} V_{\mathrm{L}}(\omega)}{Z_{\mathrm{L}}(\omega)}. \tag{4.5}$$

Note, however, that since

$$H_{\mathrm{EQ}}(\omega) = V_{\mathrm{H}}(\omega)/V_{\mathrm{L}}(\omega), \tag{4.6}$$

then

$$H_{\mathrm{EQ}}(\omega) = \frac{Bl_{\mathrm{L}}/Z_{\mathrm{L}}(\omega)}{Bl_{\mathrm{H}}/Z_{\mathrm{H}}(\omega)}, \tag{4.7}$$

a very simple relationship that indicates the importance of the electrical impedance and the force factor Bl in determining loudspeaker characteristics. We can verify the result using Eqns. 1.40 and 1.42. Note that it applies to the response at any orientation, not just on-axis, and represents a general property of acoustic transducers with magnetic drivers.

4.3 LOW FORCE FACTOR DRIVERS

The introduction of concepts such as Flat-TV and small (mobile) sound reproduction systems has led to a renewed interest in obtaining a high sound output from compact loudspeaker arrangements with a good efficiency. Compact relates here to both the volume of the cabinet into which the loudspeaker is mounted, as well as the cone area of the loudspeaker. Normally, low-frequency sound reproduction with small transducers is quite inefficient. To increase the efficiency, the low-frequency region, say 20 to 120 Hz can be mapped to a single tone, and by using a special transducer with a low-Bl value, at a very high efficiency at that particular tone. In the following section, an optimal force factor will be derived to obtain such a result.

4.3.1 OPTIMAL FORCE FACTOR

The solution to obtain a high sound output from a compact loudspeaker arrangement, with a good efficiency, consists of two steps. First, the requirement that the frequency response must be flat is relaxed. By making the magnet considerably smaller (See Fig. 4.6 at the left side), a large peak in the SPL curve (see Fig. 4.1 (solid curve)) will appear. Since the magnet can be considerably smaller than usual, the loudspeaker can be of the moving magnet type with a stationary coil (see Fig. 4.6 and Fig. 4.7), instead of vice versa. At the resonance frequency, the efficiency can be a factor of 10 higher than that of a normal loudspeaker. In this case we have, at the resonance frequency of about 40 Hz, an SPL of almost 90 dB at 1-W input power, even when using a small cabinet. Since it is operating in resonance mode only, the moving mass can be enlarged (which might be necessary owing to the small cabinet), without degrading the efficiency of the system. Owing to the large and narrow peak in the frequency response, the normal operating

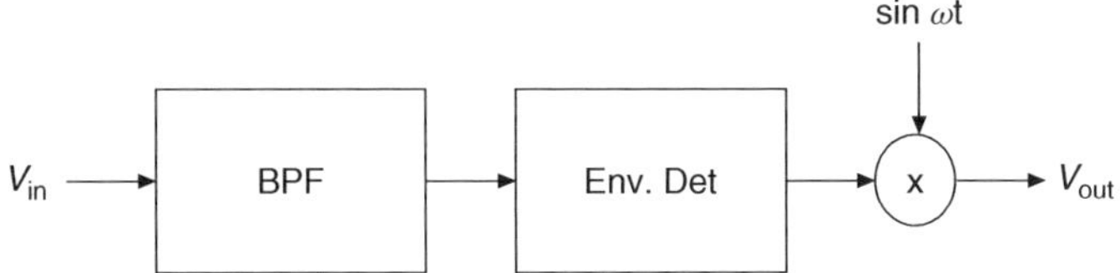

Figure 4.4 Frequency mapping scheme. The box labeled 'BPF' is a band pass filter, and 'Env. Det.' is an envelope detector, the signal V_{out} is fed (via a power amplifier) to the driver

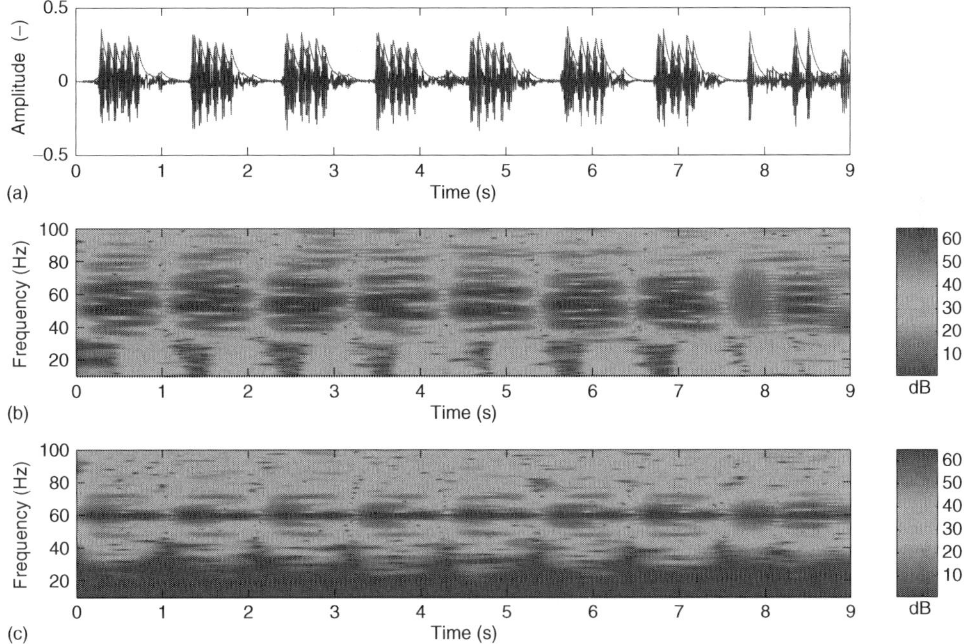

Figure 4.5 The signals before and after the frequency-mapping-processing of Fig. 4.4. (a) Shows the signal at V_{in}, and the output of the envelope detector as the thin outline along V_{in}. (b) and (c) Show the spectrogram of the input and output signals respectively

range of the driver decreases considerably, however. This makes the driver unsuitable for normal use. To overcome this, a second measure is applied. The low-frequency content of the music signal, say 20 to 120 Hz, is mapped to a slowly amplitude-modulated tone whose frequency equals the resonance frequency of the transducer. This can be done with a set-up depicted in Fig. 4.4. The modulation is chosen such that the coarse structure (the envelope) of the music signal after the mapping is the same as before the mapping, which is shown in Fig. 4.5. Part (a) shows the waveform of a rock-music excerpt (the blue curve); the red curve depicts its envelope. Parts (b) and (c) show the spectrograms of the input and output signals respectively, clearly showing that the frequency bandwidth of

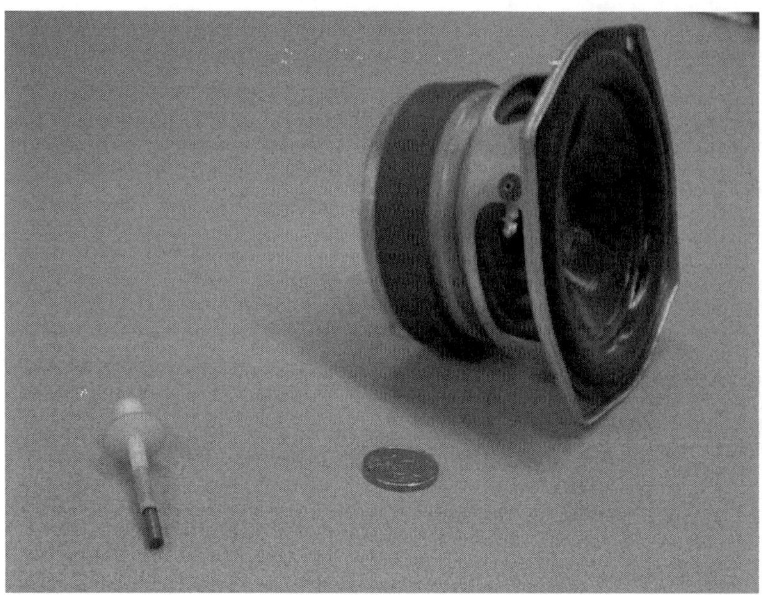

Figure 4.6 Left: Magnet system of the prototype (MM3c shown in Fig. 4.7) of the optimal low-Bl driver (only 3 g and 4.5-mm diameter) with a 50 Euro cent coin. Right: a small woofer of 13-cm diameter and about 1 kg

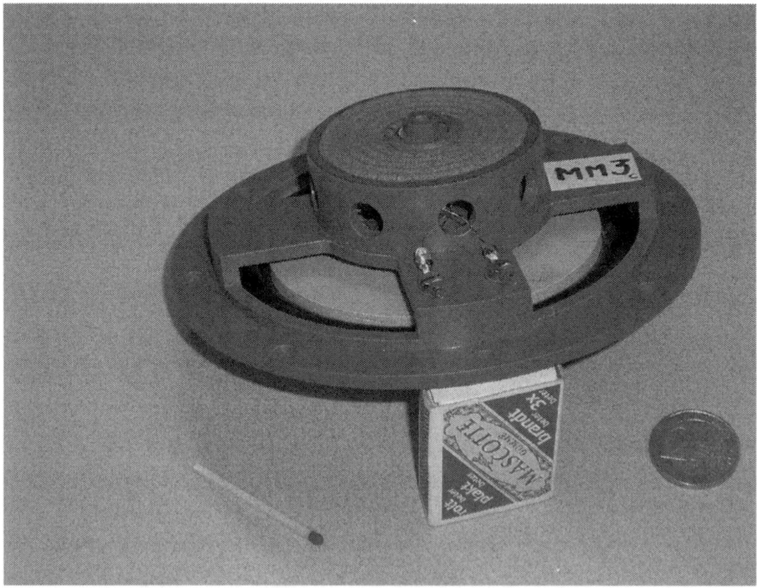

Figure 4.7 Picture of the prototype driver (MM3c) with a 2 Euro coin and ordinary matches. At the position where a normal loudspeaker has its heavy and expensive magnet, the prototype driver has an almost empty cavity; only a small moving magnet is necessary, which is visible at Fig. 4.6 (left side)

the signal around 50 Hz decreases after the mapping, yet the temporal modulations remain the same. Using Eqns. 1.37 and 1.39, the voltage sensitivity at the resonance frequency can be written as

$$H(\omega = \omega_0) = \frac{j\omega_0 SBl\rho}{2\pi r R_e(R_m + (Bl)^2/R_e)}. \tag{4.8}$$

If Eqn. 4.8 is maximized by adjusting the force factor Bl (differentiating $H(\omega = \omega_0)$ with respect to Bl and setting $\partial H/\partial(Bl) = 0$), we get

$$\frac{(Bl)^2}{R_e} = R_m. \tag{4.9}$$

Note at this point that if Eqn. 4.9 holds, we get for this particular case $Q_e = Q_m$ (see Eqn. 1.43). It appears that the maximum voltage sensitivity is reached as the electrical damping term $(Bl)^2/R_e$ is equal to the mechanical damping term R_m; in this case, we refer to the optimal force factor as $(Bl)_o$. If Eqn. 4.9 is substituted into Eqn. 4.8, this yields the optimal voltage sensitivity ratio

$$H_o(\omega = \omega_0) = \frac{j\omega\rho S}{4\pi r(Bl)_o}. \tag{4.10}$$

We find that the specific relationship between $(Bl)_o$ and both R_m and R_e (Eqn. 4.9) causes H_o to be inversely proportional to $(Bl)_o$ (which may seem counterintuitive), and thus also inversely proportional to $\sqrt{R_m}$. The power efficiency at the resonance frequency for the optimality condition obtained by substitution of Eqn. 4.9 into Eqn. 4.1 yields

$$\eta_o(\omega = \omega_0) = \frac{R_m R_r}{(R_m + R_r)^2 + (R_m + R_r)R_m}. \tag{4.11}$$

This can be approximated for $R_r \ll R_m$ as

$$\eta_o(\omega = \omega_0) \approx \frac{R_r}{2R_m}, \tag{4.12}$$

which clearly shows that for a high-power efficiency at the resonance frequency, the cone area must be large, since R_r is – according Eqn. 1.47 – proportional to the squared cone area; and that the mechanical damping must be as small as possible. This conclusion is the same as for achieving a high voltage sensitivity (given by Eqn. 4.1).

Using Eqn. 1.39, 1.40, and $v = \mathrm{d}x/\mathrm{d}t$, we get for the cone velocity

$$\frac{v(\omega = \omega_0, (Bl)^2/R_e = R_m)}{V_{in}} = \frac{1}{2(Bl)_o}, \tag{4.13}$$

which again shows the benefit of low Bl and R_m values. Further, we get, assuming the optimality condition given by Eqn. 4.9 and using Eqn. 1.45,

$$Z_{in}(\omega = \omega_0, (Bl)^2/R_e = R_m) \approx 2R_e. \tag{4.14}$$

4.4 TRANSIENT RESPONSE

In order to calculate the transient behaviour of the system, we calculate the transient response of a driver. We will determine the response to a sinusoid that is switched on at $t = 0$, and finally, the impulse response.

4.4.1 GATED SINUSOID RESPONSE

The response of a driver, with resonance frequency ω_0, to a sinusoidal signal with frequency ω_s, switched on at $t = 0$

$$v(t) = \begin{cases} 0\,, & t < 0 \\ A\sin(\omega_s t)\,, & t \geq 0 \end{cases} \tag{4.15}$$

is calculated. It is convenient to write Eqn. 4.15 in the Laplace domain

$$V(s) = \frac{A\omega_s}{s^2 + \omega_s^2}. \tag{4.16}$$

The transfer function $H(s)$ from voltage V to excursion X is

$$H(s) = \frac{1}{s^2 m_t + sR_t + k_t}, \tag{4.17}$$

where m_t is the total moving mass, R_t is the total damping (mechanical, electrical, and acoustical)

$$R_t = R_m + \frac{(Bl)^2}{R_e} + R_a, \tag{4.18}$$

and k_t the total spring constant, as described in Sec. 1.3.2.3. Hence $X(s)$ can be calculated as

$$X(s) = \frac{\dfrac{Bl}{R_e}\dfrac{A\omega_s}{s^2 + \omega_s^2}}{s^2 m_t + sR_t + k_t}. \tag{4.19}$$

Assuming that Eqn. 4.17 has complex poles (this is the case if $R_t^2 < 4m_t k_t$), they are at

$$s_{1,2} = -\beta\omega_0 \pm i\omega_n = -a \pm ib. \tag{4.20}$$

By using partial fraction expansion (see e.g. Palm [198], or Arfken and Weber [21]), we can write Eqn. 4.19 as sum of basic terms, and then the inverse Laplace transformation can be readily carried out, resulting in

$$x(t) = \frac{Bl\omega_s A}{m_t R_e}\left[\mathrm{e}^{-at}\left(k_1 \cos\omega_n t + \left(\frac{k_2 - ak_1}{\omega_n}\right)\sin\omega_n t\right) + k_3 \cos\omega_s t + \frac{k_4}{\omega_s}\sin\omega_s t\right], \tag{4.21}$$

where

$$
\begin{aligned}
a &= \frac{R_t}{2m_t},\\
b^2 &= \omega_n^2 = \frac{k_t}{m_t} - \left(\frac{R_t}{2m_t}\right)^2,\\
k_1 &= \frac{2a}{\Delta},\\
k_2 &= \frac{3a^2 - b^2 + w_s^2}{\Delta},\\
k_3 &= -k_1,\\
k_4 &= \frac{a^2 + b^2 - \omega_s^2}{\Delta},\\
\Delta &= (a^2 + b^2)^2 + 2\omega_s^2(a^2 - b^2) + \omega_s^4,
\end{aligned}
\tag{4.22}
$$

where we identify $a^2 + b^2 = \sqrt{\frac{k_t}{m_t}} = \omega_0^2$ as the driver's resonance frequency, and

$$
\begin{aligned}
\beta &= \frac{R_t}{2\sqrt{m_t k_t}},\\
Q &= \frac{\sqrt{m_t k_t}}{R_t},
\end{aligned}
\tag{4.23}
$$

as the driver's damping and quality factor Q, respectively. Further, we see that $2\beta = Q^{-1}$. Looking at Eqn. 4.21, it is clear that the time constant of the transient behaviour is determined by a^{-1}. It is proportional to the moving mass m_t, and must be small in order to get a fast response. As a special case, we consider $\omega_s = \omega_0$, that is, the driver is actuated at its resonance frequency. Furthermore, we assume that the transient part (the product with e^{-at} in Eqn. 4.21) has faded away. Then, Eqn. 4.21 reduces to

$$
x(t) = -\frac{ABl}{\omega_s R_e R_t} \cos \omega_s t. \tag{4.24}
$$

Here, we see that there is a $3\pi/4$ phase shift between the input voltage (see Eqn. 4.15) and the output excursion.

For various drivers, the lumped-element parameters are determined and listed in Table 4.2. For three of those drivers (see Table 4.1), the response to a suddenly switched sinusoid (see Eqn. 4.15) is calculated (using Eqn. 4.21). The results are shown in Fig. 4.8 for the low-*Bl* driver and the case that the driving signal is equal to the resonance frequency $f_0 = 43$ Hz, in Fig. 4.9 for a driving signal of 47 Hz, and in Fig. 4.10 for a driving signal of 86 Hz. For the two other drivers, with a medium- and high-*Bl* respectively, transient responses at resonance frequency are shown in Figs. 4.11 and 4.12 (responses at other frequencies are very similar for these two drivers). These results show that the medium- and high-*Bl* driver systems rapidly converge to their steady-state response, while this is not the case for the low-*Bl* driver system. Also, the medium- and high-*Bl* driver systems are not sensitive to deviations of the input frequency, while, again, this is not the case for the low-*Bl* driver system. This is especially obvious in Fig. 4.10, where in the first half of the time interval,

Table 4.1 Legend to the transient responses plotted in Figs. 4.8 – 4.12 of various drivers: low-Bl (MM3c), normal-Bl (AD70 652), and high-Bl (HBl) loudspeakers. The frequency f_s of the gated sinusoid was set to three different values relative to the resonance frequency f_0 of the driver: f_0, $1.1f_0$, and $2f_0$. In all cases, the amplitude was 1 V. For the entries labeled '–' no figures are included, since these are similar to the other figure referred to in the same column. The details of the lumped-element parameters are listed in Table 4.2

	MM3c	AD70652	HBl
$f_s = f_0$	Fig. 4.8	Fig. 4.11	Fig. 4.12
$f_s = 1.1f_0$	Fig. 4.9	–	–
$f_s = 2f_0$	Fig. 4.10	–	–
Bl	1.2	6.5	22
Q_e	17	0.61	0.22

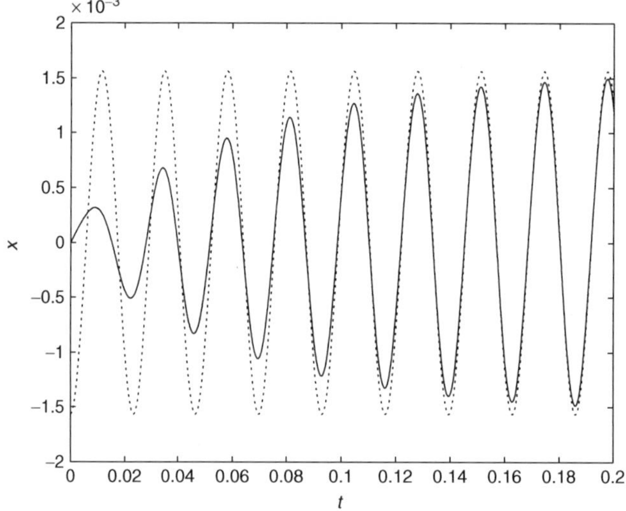

Figure 4.8 The displacement (solid line) of the low-Bl prototype driver MM3c (using Eqn. 4.21). The lumped-element parameters are listed in Table 4.2. The frequency of the driving signal f_s is equal to driver's resonance frequency $f_0 = 43$ Hz. The dotted line is the stationary value of the displacement (Eqn. 4.21 for $\lim_{t\to\infty}$)

there is significant interference between the terms with ω_n and ω_s of Eqn. 4.21, leading to severe amplitude modulation during the onset transient. Therefore, the low-Bl driver should only be used at (or near) resonance frequency.

4.4.2 IMPULSE RESPONSE

The impulse response $h(t)$ can be calculated directly (again under the assumption that $R_t^2 < 4m_tk_t$), by the inverse Laplace transform of $H(s)$ (Eqn. 4.17) as

$$h(t) = \frac{\omega_0}{\omega_n} e^{-at} \sin(\omega_n t + \phi), \tag{4.25}$$

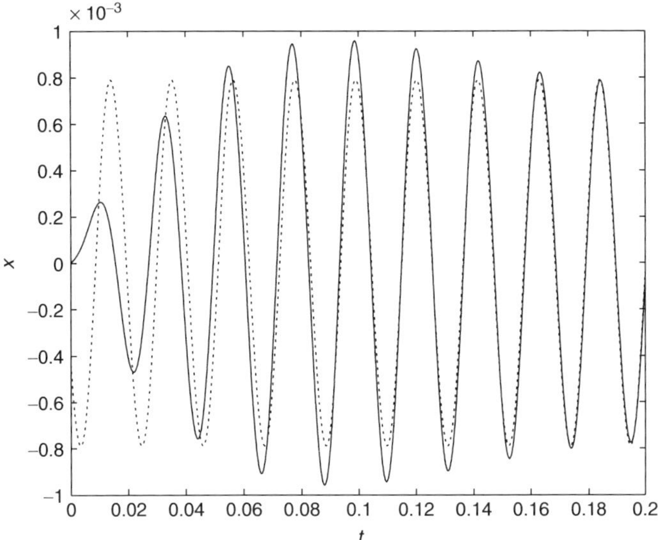

Figure 4.9 The displacement (solid line) of the low-Bl prototype driver MM3c (using Eqn. 4.21). The lumped-element parameters are listed in Table 4.2. The frequency of the driving signal f_s is equal to 47 Hz, which is 1.1 times the driver's resonance frequency $f_0 = 43$ Hz. The dotted line is the stationary value of the displacement (Eqn. 4.21 for $\lim_{t\to\infty}$)

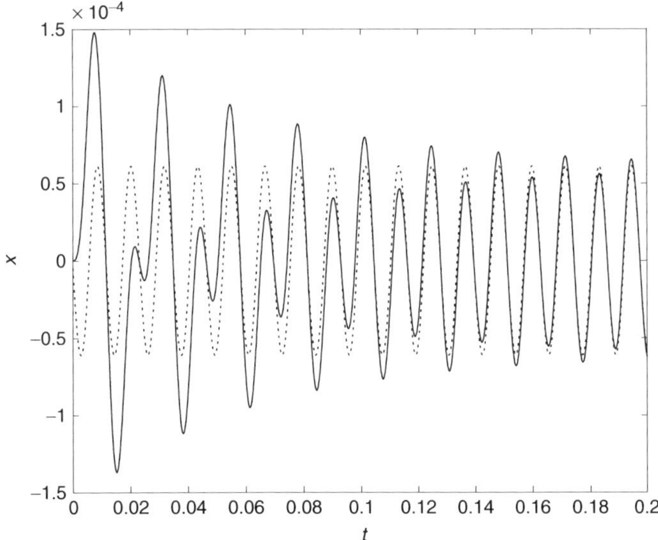

Figure 4.10 The displacement (solid line) of the low-Bl prototype driver MM3c (using Eqn. 4.21). The lumped-element parameters are listed in Table 4.2. The frequency of the driving signal f_s is equal to 86 Hz, which is twice the driver's resonance frequency $f_0 = 43$ Hz. The dotted line is the stationary value of the displacement (Eqn. 4.21 for $\lim_{t\to\infty}$)

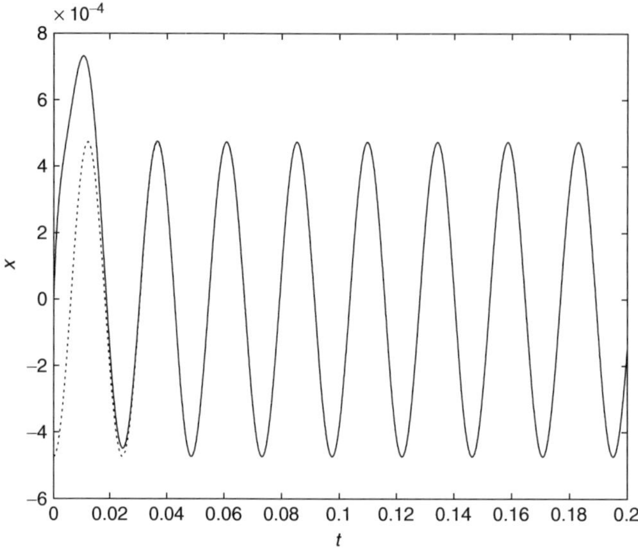

Figure 4.11 The displacement (solid line) of the medium-Bl driver (using Eqn. 4.21). The lumped-element parameters are listed in Table 4.2. The frequency of the driving signal f_s is equal to driver's resonance frequency $f_0 = 41\,\text{Hz}$. The dotted line is the stationary value of the displacement (Eqn. 4.21 for $\lim_{t\to\infty}$)

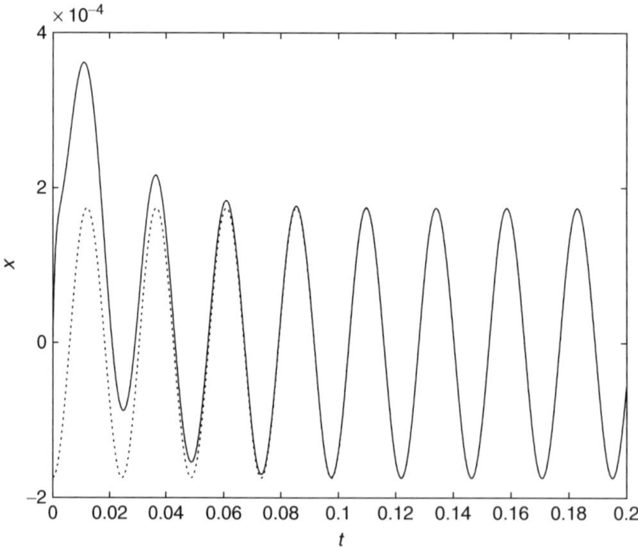

Figure 4.12 The displacement (solid line) of a high-Bl driver (HBl), (using Eqn. 4.21). The lumped-element parameters are listed in Table 4.2. The frequency of the driving signal f_s is equal to driver's resonance frequency $f_0 = 41\,\text{Hz}$. The dotted line is the stationary value of the displacement (Eqn. 4.21 for $\lim_{t\to\infty}$)

where $\phi = \arctan(b/a)$. Here, we see again that the time constant a^{-1} (the ratio $2m_t/R_t$) must be small to get a fast response (decay).

4.5 DETAILS OF LUMPED-ELEMENT PARAMETERS AND EFFICIENCY

The lumped parameters for some loudspeakers are given in Table 4.2. From these values and Eqn. 1.54, the efficiency η is calculated and plotted in Fig. 4.13, showing that the efficiency is in the range of 0.2 to 10%. As a rule of thumb, we see from Table 4.2 that for woofers k is equal to about 1 N/mm, and that $R_m \approx 1$ Ns/m, while the other parameters may differ significantly between the various drivers.

The equivalent volume of a loudspeaker is given by

$$V_{eq} = \rho c^2(\pi a^2)^2/k_t. \tag{4.26}$$

Alternatively, for a given volume of the enclosure, the corresponding k_t of the 'air-spring' can be calculated. Mounting a loudspeaker in a cabinet will increase the total spring constant by an amount given by Eqn. 4.26, and subsequently increase the bass cut-off frequency of the system. To compensate for this bass loss, the moving mass has to be increased; thus $\sqrt{k_t m}$ is increased, which changes Q_e (see Eqn. 1.43). Then Bl must be increased in order to preserve its original value. The original frequency response is then maintained, but at the cost of a more expensive magnet and a loss in efficiency. This is the designer's dilemma: high efficiency or small enclosure? To meet the demand for a certain cut-off frequency, the enclosure volume must be greater. Alternatively, the

Table 4.2 The lumped parameters for various low-frequency loudspeakers (woofers). A 4 in., and some 7, 8, 10, and 12 in. drivers, each of them with a low and high Q_e. Further, two special drivers, a (optimal) low-Bl (MM3c) and a high-Bl one (HBl). The former is an experimental driver (see Fig. 4.6 for its compact magnet system together with a more classical driver, and Fig. 4.7 for the whole driver). The high-Bl one (HBl) is discussed in Vanderkooy *et al.* [284]. See Table 1.2 for the abbreviations and the meaning of the variables

Type	R_e Ω	Bl Tm	k N/m	m_t gr.	R_m Ns/m	S cm^2	f_0 Hz	Q_m	Q_e
AD44510	6.6	3.5	839	4	0.86	54	72	2.2	1.02
AD70652	7.5	6.5	885	13.2	1.48	123	41	2.3	0.61
AD70801	6.9	2.9	1075	6.3	0.81	123	66	3.2	2.13
AD80110	6.0	9.0	971	16.5	1.38	200	39	2.9	0.29
AD80605	6.8	5.1	1205	13.4	0.84	200	48	4.8	1.05
AD10250	6.6	13.0	1124	38.5	2.74	315	27	2.4	0.25
AD10600	6.8	5.9	909	28.5	1.06	315	29	4.8	0.99
AD12250	6.6	13.0	1429	54.0	2.93	490	26	3.0	0.34
AD12600	6.9	6.0	1205	33.0	0.76	490	31	8.2	1.21
MM3c	6.4	1.2	1022	14.0	0.22	86	43	17.0	17.00
HBl	7.5	22.0	3716	56.0	0.91	490	41	16.0	0.22

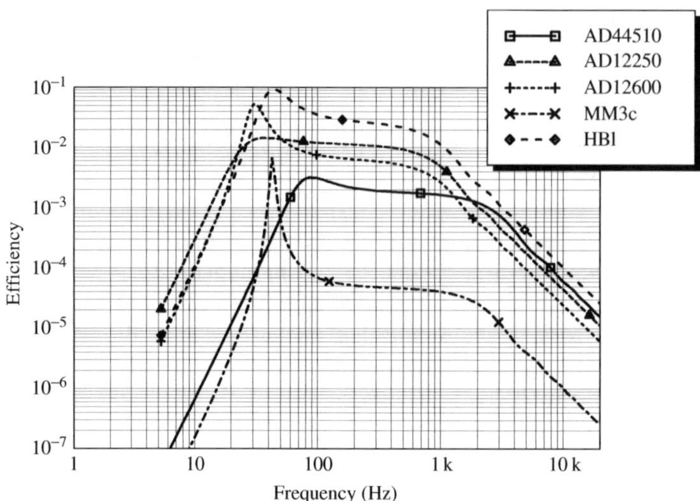

Figure 4.13 Efficiency of various loudspeakers: AD44510 (solid/square markers), AD12250 (dashed/triangle markers), AD12600 (small-dashed/'+' markers), MM3c (dot-dashed/'×' markers), and HBl (wide-dashed/'◇' markers). See Table 4.2 for the parameters. Note that not all drivers have the same cone area

efficiency for a given volume will be less than that for a system with a higher cut-off frequency. This dilemma is (partially) solved by using the low-*Bl* concept as discussed in Sec. 4.3, however, at the expense of a slightly decreased sound quality and some additional electronics to accomplish the frequency mapping. The decrease of sound quality appears to be modest, apparently because the auditory system is less sensitive at low frequencies (Sec. 1.4.5.2). Also, the other parts of the audio spectrum have a distracting influence on this mapping effect, which has been confirmed in a study by Le Goff *et al.* [158], in which detectability of mistuned fundamental frequencies were determined for a variety of realistic complex signals.

4.6 DISCUSSION

In the previous sections we have seen that the force factor B_l plays a very important role in loudspeaker design. It determines the efficiency, the impedance, the SPL response, the temporal response, the weight, and the cost. The choice concerning these parameters depends on the application. If the size of the cabinet is of less importance, a medium B_l is the simplest solution, since it does not require any other measures. On the other hand, if a small cabinet and a high efficiency are important than the low B_l system with an optimum value given by Eqn. 4.9 has the preference. This requires special electronics, however. Also, the transient response is less favorable than that of a larger magnet system and this makes it less suitable for High-Fi applications. The high B_l is in between: it has a high efficiency and good response, but it requires an expensive magnet and additional electronics. Clearly, the application dictates the parameter choice.

5

High-frequency Bandwidth Extension for Audio

5.1 INTRODUCTION

The previous chapters focused on BWE methods to extend the (perceived) low-frequency content of reproduced audio signals. The bandwidth limitation in those cases was primarily due to the transducer. Bandwidth limitation can also occur in the transmission channel (in which, for the moment, we also include signal storage). A familiar example is the telephone channel, which has a bandwidth of about 3 kHz. Speech signals transmitted through this channel are audibly bandlimited, because the bandwidth of natural speech is about 8 kHz. Methods to extend the bandwidth of speech, primarily intended for telephony applications, will be discussed in Chapter 6.

Bandwidth limitation in the transmission channel can also occur if perceptual audio coders are used at high compression ratios; for example, in recent years MPEG1 layer-3 (a.k.a. MP3) has become a tremendously popular format for audio storage and transmission. Perceptual audio coders achieve high coding efficiencies because they attempt to store signal information with a low resolution, just 'sufficiently high' for the human auditory system. For the MP3 scheme, this means, in practice, that a significant amount of distortion is introduced in the signal, but the distortion spectrum is designed such that it remains inaudible, for example, is *masked* (see Sec. 1.4.4.5). This in achieved by analysing the short-term power spectrum of the audio signal and using a masking model to compute the masked threshold, that is, a frequency-dependent curve below which distortion components will be masked by the audio signal. This is turn determines, per frequency band, how many bits are needed to code the audio signal. Now, for very high compression ratios, or equivalently, very low bit rates, the coding algorithm is not able to keep all of the distortion below the masked threshold for the full-bandwidth signal. Typically, the bandwidth is then reduced at the high-frequency end, such that the specified bit rate can be achieved for the bandwidth-limited signal, while at the same time keeping the distortion below the masked threshold. The drawback is that high frequencies are lost, resulting in a 'muffled' sound percept. Reviews of audio coding and audio signal processing can be found in, for example, Bosi and Goldberg [37] and Kahrs and Brandenburg [139]. Low-bit-rate perceptual audio coders are now being extensively used

Audio Bandwidth Extension E. Larsen and R. M. Aarts
 ISBN 0-470-85864-8

for audio storage on the Internet, distribution through Internet radio, satellite radio, and private use such as with personal computers, MP3-players, and the like.

To enhance the reproduction of high-frequency bandwidth-limited audio, BWE processing can be applied. Given that the reproduction system can reproduce high frequencies, which is usually not a problem, synthetic frequency components can be added to the signal, improving the quality thereof. Of course, the added synthetic frequency components should be derived from the available bandwidth-limited signal, and for this two approaches have been developed:

- In the first approach, the BWE algorithm is *blind*, that is, it has no information regarding the missing high-frequency components. Thus, only assumptions on the statistics of audio signals (Sec. 1.2) can be used to design such systems. The main advantages of using this approach are that such a BWE system can be applied to a wide class of signals (specifically, both to music and speech) and that there are no requirements on the signal format, because the only required information is the actual signal waveform (or spectrum). This means that such methods could also be used to enhance the quality of old (analog) recordings. It also appears that computationally efficient algorithms can be realized. The drawback is that the quality of the bandwidth-extended output signal is significantly lower than that of the original full-bandwidth signal, even though it is higher than the bandwidth-limited signal. This is due to the lack of information about the missing high frequencies. This approach is the topic of Sec. 5.4.
- In the second approach, the BWE algorithm does have a priori knowledge regarding the missing high-frequency components. This allows for a much more exact reconstruction of the original full-bandwidth signal than is possible with the blind approach, and therefore the quality of the bandwidth-extended signal can be (near) transparent, that is, indistinguishable, from the original full-bandwidth signal. The high quality of the output signal is obviously the main advantage of this approach. The drawback is that some provisions need to be taken to provide the BWE algorithm with the requisite a priori information. A successful approach is that of 'spectral band replication' (SBR), the topic of Sec. 5.5. This is a method that works in combination with an 'ordinary' audio codec[1], and stores (at a very low bit rate) some specific information about high frequencies in the coded audio stream; in this way the overall required bit rate can be significantly reduced. If the appropriate decoder is used, this information is utilized to reconstruct the missing high frequencies. A decoder that cannot use the additional information only decodes the low-frequency band, thereby insuring forward and backward compatibility. SBR has been used to enhance the coding efficiency of MP3, leading to MP3Pro, and also AAC (Advanced Audio Coding), leading to aacPlus.

A more conceptual difference between the two approaches mentioned here is that the first attempts to extend the bandwidth of a signal that is, for whatever reason, bandlimited, while the second attempt purposely limits the bandwidth of the signal, but does it in such a way that at the output a high-quality full-bandwidth signal can be recreated.

Besides creating a 'brighter', more natural sound percept, high-frequency BWE processing can potentially enhance localization of sound sources as well. Bronkhorst [40]

[1] 'Codec' is the concatenation of the words coder/decoder, referring to both the coding and decoding algorithms used for a particular coding scheme.

found that spectral cues above 7 kHz have a significant effect on localization performance – by reducing the number of front/back confusions, and enabling the listener to localize a sound source not only more accurately but also more quickly and with less head movements. See also (Chapter 8) the patent by Dempsey on p. 257.

In Sec. 5.6, a method (BWE instantaneous compression) is discussed that would properly be categorized as a blind high-frequency BWE algorithm, but it is discussed separately as it was not originally designed for BWE purposes. The original purpose of this algorithm was to enhance reproduction of centre and surround channel signals in multi-channel sound systems. Signals in these channels are often at a somewhat low level, and the algorithm was designed as a simple means to boost their level, while preventing distortion at high signal levels. The particular nature of the processing also extends the high-frequency content of the signal spectrum, and as such it is also a BWE algorithm that works well in this particular application; it is not generally applicable as are the methods of Secs. 5.4 and 5.5. Section 5.3 discusses the perceptual aspects of high-frequency BWE methods.

First, in Sec. 5.2, we briefly show that traditional methods (in particular, deconvolution) to overcome bandwidth limitations in transmission channels are not suitable for the applications as discussed in this introduction.

5.2 THE LIMITS OF DECONVOLUTION

If a wideband signal $x(t)$ is passed through a linear system $h(t)$ having, for example, a low-pass characteristic, the filtered signal $x_l(t)$ has a reduced bandwidth. We can write

$$x_l(t) = x(t) * h(t)\,, \tag{5.1}$$

where $*$ denotes convolution.

If the received signal $x_l(t)$ is used to reconstruct an estimate $\hat{x}(t)$ of the original $x(t)$, we need to find a filter $g(t)$ such that $x(t) * g(t) = \delta(t - \tau)$, with $\tau > 0$. In that case, we would have

$$\hat{x}(t) = x_l(t) * g(t) = x(t) * f(t) * g(t) = x(t) * \delta(t - \tau) = x(t - \tau)\,, \tag{5.2}$$

a perfect reconstruction, up to a finite time delay. The only condition on $g(t)$ is that it must be stable, which means that all its poles must lie within the unit circle (Sec. 1.1). Now $g(t)$ is simply the inverse of $f(t)$ (up to the time delay τ), which means that all of $f(t)$'s zeros must lie within the unit circle, implying that $f(t)$ be minimum phase. Because $f(t)$ must also be stable, its poles will also lie inside the unit circle, and by the same argument as before, $g(t)$ must then also be minimum phase. So we find that the inverse of $f(t)$ can only be obtained if it is minimum phase; the inverse filter $g(t)$ will then also be minimum phase. If $f(t)$ is not minimum phase, a stable inverse filter does not exist. See, for example, Neely and Allen [184] for a more elaborate discussion, in the context of inverting room impulse responses. The process of obtaining $\hat{x}(t)$ from $x_l(t)$ is called deconvolution, or inversion. In practice, there is a complicating factor in that the received signal $x_l(t)$ will be corrupted by additive noise, wherefore Eqn. 5.1 becomes

$$x_l(t) = x(t) * f(t) + \epsilon(t), \tag{5.3}$$

where $\epsilon(t)$ is the noise. The optimal filter, in the sense that $\hat{x}(t) - x(t)$ is minimized in least-squares sense, is then given in the frequency domain as (e.g. Berkhout *et al.* [29], [30])

$$G(f) = \frac{F^*(f)}{|F(f)|^2 + \sigma_\epsilon^2(f)}, \tag{5.4}$$

with $\sigma_\epsilon^2(f)$ the frequency-dependent variance (power) of $\epsilon(t)$, and $F^*(f)$ the complex conjugate of the Fourier transform of $f(t)$. This optimal filter is also called a Wiener filter, although the solution can be improved if the short-term spectrum $\Xi(f,k)$ of $x(t)$ is known (signal frame k), in which case the time-varying Wiener filter becomes

$$G(f,\, k) = \frac{\Xi(f,k)F^*(f)}{\Xi(f,k)|F(f)|^2 + \sigma_\epsilon^2(f)}. \tag{5.5}$$

Of course, $\Xi(f,k)$ is not known, but a long-term average spectrum $\overline{X}(f)$ might be known or estimable, and could be used as well. Then the estimated spectrum $\hat{X}(f)$ becomes

$$\hat{X}(f) = \frac{\{\overline{X}(f)\}^2|F(f)|^2}{\overline{X}(f)|F(f)|^2 + \sigma_\epsilon^2(f)}. \tag{5.6}$$

For high signal-to-noise ratio (SNR), the Wiener filter can be approximated as

$$G(f) \approx F^{-1}(f), \quad \sigma_\epsilon^2(f) \ll |F(f)|^2, \tag{5.7}$$

such that $g(t)$ is simply the inverse of $f(t)$ as we found previously, and does not depend on the signal spectrum. For low SNR this becomes, for the cases of Eqns. 5.4 and 5.5, respectively

$$\left.\begin{array}{rcl} G(f) & \approx & F^*(f)/\sigma_\epsilon^2(f) \\ G(f,\, k) & \approx & \Xi(f,k)F^*(f)/\sigma_\epsilon^2(f) \end{array}\right\} \quad \sigma_\epsilon^2(f) \gg |F(f)|^2, \tag{5.8}$$

which is the matched filter, as also known from signal detection theory, and does depend on the signal spectrum. For a given situation, the approximations in Eqns. 5.7–5.8 can be valid in different frequency bands. So those frequency bands having a low SNR will be strongly attenuated, and the effect of $f(t)$ can only be inverted if the SNR is high or intermediate. The conclusion is that for bandlimiting operations, deconvolution is not very effective, because the bandlimited frequency regions will usually have a poor SNR, and the signal in those bands can therefore not be retrieved. This is true for a telephone network, but, in particular, also for perceptually coded audio in which high frequencies have been eliminated to reduce the required bit rate. In those cases, the high-frequency band does not contain any useful signal any more, and other (non-linear) methods must be used to restore (some of) the original signal parts. Because in audio applications there is in general very little a priori information about the nature of the signals, more specialized deconvolution methods are not practical.

As an example, we use the situation of speech transmission through the telephone network. Although a full treatment of speech enhancement processing is deferred until Chapter 6, we use this example because (1) the bandlimitation (being the telephone network) is very well defined, and (2) the long-term average spectrum of speech is known. For general audio applications, both the bandlimitation as well as the signal spectrum are not well defined (e.g. have a large variability). It should however be understood that the forthcoming arguments regarding the limitation of deconvolution apply equally well to speech transmission through the telephone network as to more general situations. Thus, consider Fig. 5.1, which shows the approximate filtering characteristic of the telephone channel $F(f)$ (see also Chapter 6 and Fig. 6.1) in (a) (solid line), together with an assumed white noise spectrum $\sigma_\epsilon(f)$ at -30 dB (dashed line). Application of Eqn. 5.4 leads to a deconvolution filter $G_1(f)$, the magnitude of which is shown in (b) (solid line). Taking into account the long-term average spectrum of speech $S(f)$, shown in (c) (solid line, calculated using Eqn. 1.24 and assuming an average male pitch of 125 Hz), we find using Eqn. 5.5 a different deconvolution filter $G_2(f)$, also shown in Fig. 5.1 (b) (dashed line). We see that $G_2(f)$ has a broad peak around 150 Hz and a narrower peak at about 4 kHz. The final received speech signal $S_r(f) = S(f)F(f)G_2(f)$ has a long-term average spectrum as shown in (c) (dashed line). In contrast to the speech signal, $S_0(f)$ would be received without any deconvolution filtering, shown as the dash-dotted line in (c), the effect of $G_2(f)$ is to accurately invert the telephone channel filtering down to about 200 Hz and up to about 4 kHz. Although this is an improvement, a lot of energy in the speech signal is still not recovered, as is clear from Fig. 5.1. In particular high frequencies, above 4 kHz (as contained mostly in fricatives such as /s/ and /f/), are not well reproduced. Note that the Wiener filter is optimal from a signal's point of view, and does not necessarily yield that linear filter that gives the best perceptual result.

We do not give examples for the case of the music signals, as music spectra are highly variable, and the bandlimiting is highly dependent on the coding algorithm and bit rate used (assuming the bandlimitation occurs through perceptual coding). The reader will understand that similar arguments as those made above for the example of speech through the telephone channel lead to similar conclusions in that deconvolution is not a practical method to restore highly bandlimited music signals.

5.3 PERCEPTUAL CONSIDERATIONS

In this section, the characteristics the synthesized high-frequency components should have to properly enhance audio reproduction, by considering pitch, timbre, and loudness are discussed. These considerations are useful to know what aspects of the available narrowband signal should be reflected in the high band; for the high-frequency BWE codec of Sec. 5.5 it also useful as it helps to realize what information the encoder should store (and what is irrelevant), to be used as a priori information for the decoder.

5.3.1 PITCH (HARMONIC STRUCTURE)

The pitch of a complex tone is determined by the frequencies of the constituent partials (harmonics), see Sec. 1.4.5. The strongest pitch percepts are obtained when low-order (resolved) harmonics are present, but complex tones with only high-order (unresolved)

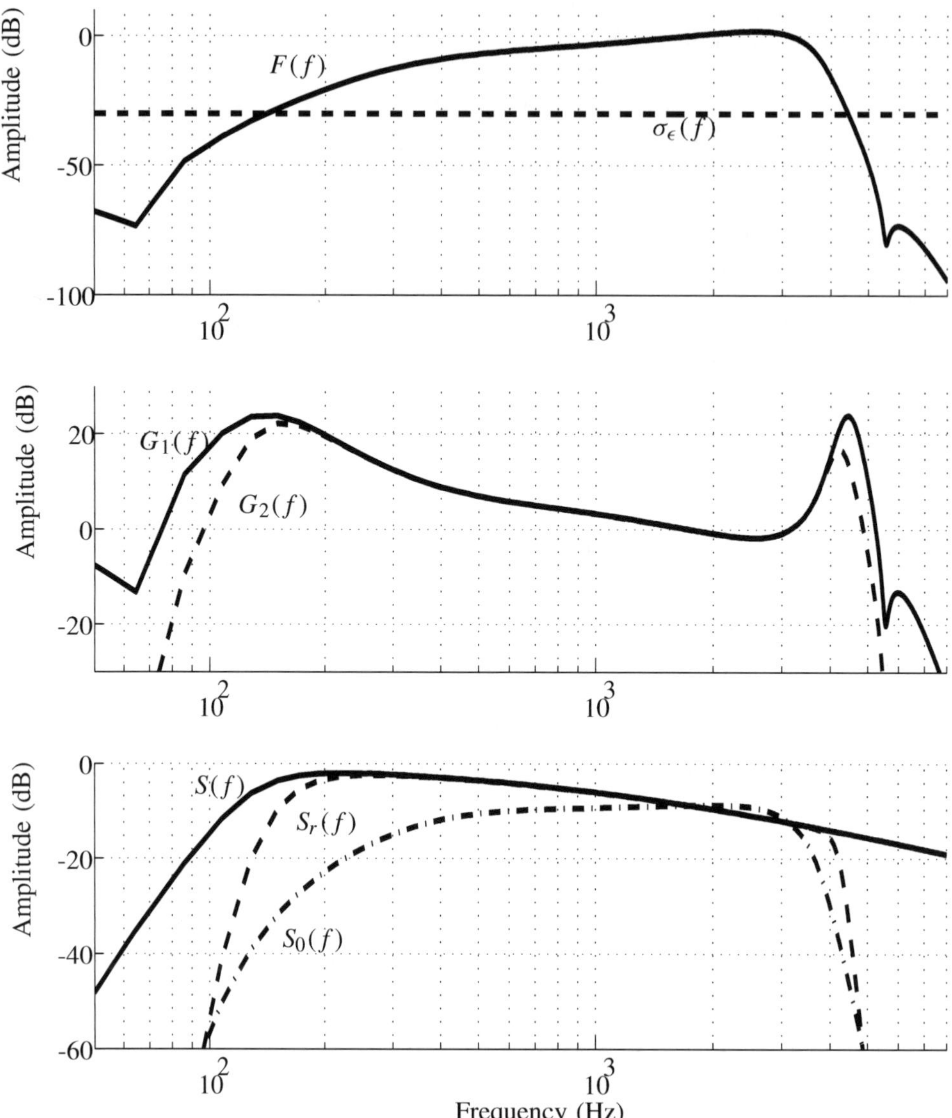

Figure 5.1 (a) Shows the approximate filter characteristic $F(f)$ of the telephone network (solid line), and an assumed white noise spectrum $\sigma_\epsilon(f)$ at -30 dB. (b) Shows the optimal filter $G_1(f)$ to invert this channel effect according to Eqn. 5.4 (solid line); taking into account the long-term average spectrum of speech $S(f)$ (shown in (c), solid line), the optimal filter $G_2(f)$ becomes as is shown by the dashed line (b), using Eqn. 5.5. Finally, (c) shows the long-term average spectrum of natural speech $S(f)$ and the received speech $S_0(f)$ without deconvolution filtering (dash-dotted line), and also for the received speech $S_r(f)$ signal using the Wiener filter of Eqn. 5.5. It can be seen that the Wiener filter improves reproduction down to 200 Hz and up to about 4 kHz, but that a significant portion of speech energy (mainly that above 4 kHz) is not recovered

harmonics also yield pitch percepts, although weaker. Ritsma [227] determined that harmonics 3–5 are dominant in the perception of pitch. Such low-order harmonics would in most cases fall outside the frequency range in which synthetic frequency components are generated by high-frequency BWE algorithms, as the lower limit of this range is typically 4 kHz at least, but may be up to over 10 kHz. Therefore, frequency components added by high-frequency BWE would typically be unresolved harmonics.

We assume that the input signal $x(t)$ is a complex tone with fundamental f_0. If $x(t)$ is bandlimited, only a finite number N_b of harmonics will be present at integer multiples of f_0. The high-frequency BWE algorithm should add additional frequency components at kf_0, with $k = N_b + 1, N_b + 2, N_b + 3, \ldots$. Of course, without a priori information, the correct amplitudes of these harmonics are unknown, but in most cases a gradually decaying amplitude spectrum is suitable. By using a proper harmonics generator (non-linear device, see Sec. 5.4.1), it can be ensured that the harmonic amplitudes do indeed gradually decay.

If the generated frequency components do not fall onto the regular $k = N_b + 1, N_b + 2, N_b + 3, \ldots$ pattern, a variety of effects could occur. If the spacing between the partials is incorrect, or if the partials are shifted by an amount not equal to a multiple of f_0, the added harmonics will elicit a pitch at a different frequency. The signal comprising these high harmonics would then be heard separately from the original complex tone, that is, the signal bands segregate (see Sec. 1.4.7). Such a translation of the harmonic 'grid' could occur, for example, if the higher harmonics are generated through spectral folding (Sec. 6.3.3.1).

Not all musical signals contain harmonically related frequency components however, and particularly at higher frequencies noise-like signals can occur, for example, percussion. In such cases, the extended frequency spectrum should also be noise-like.

5.3.2 TIMBRE (SPECTRAL ENVELOPE)

The explicit goal of high-frequency BWE is to extend the spectral envelope to high frequencies, thereby modifying the sound's timbre; the loss of high frequencies is the reason that music and speech sounds muffled. As discussed in Sec. 1.4.6, timbre depends on a number of variables, including amplitude spectrum, phase spectrum, and temporal envelope (in particular, attack and decay times).

- The temporal envelope in a high-frequency band should be broadly similar to that in a low-frequency band for most typical audio signals, so it would suffice if the BWE algorithm ensures a more or less linear relationship between the two. This will be the case if the non-linear device (NLD) is a homogenous system (Sec. 1.1.1).
- The phase spectrum is considered to be the most unimportant aspect in high-frequency BWE applications. The lower limit of generated frequency components is typically 4 kHz (but possibly much higher), and at these frequencies the auditory system is fairly insensitive to phase. It is possible that phase changes of adjacent frequency components lead to modifications of the interference pattern produced at particular locations on the basilar membrane (BM). Because auditory filters broaden with increasing frequency (e.g. being 672 Hz wide at 4 kHz and 888 Hz wide at 8 kHz, according to Eqn. 1.88), the likelihood of such interactions increases at high frequencies. It is conceivable that certain phase changes could significantly change the overall amplitude of vibration

at particular locations of the BM, and this would cause a change in neural response (because neural response is directly linked to the amplitude of the BM vibration). In this way, the amplitude spectrum as sensed by the auditory system could be modified by phase changes of the physical signal spectrum. These effects are difficult to predict, in particular because the signals arriving at the ears have a more or less random phase (in the sense of a deterministically chaotic system) owing to the variations of the impulse responses between loudspeaker and ears. Also, a study by Plomp and Steeneken [210] found the effect of phase spectrum on timbre to be small compared to the effect of amplitude spectrum, although in that study complex tones containing low-order (<10) harmonics were used. For all these reasons, it is considered impractical and not necessary to control for the phase of the synthetic high-frequency components.

- The amplitude spectrum, directly determined by the amount of high frequencies added through the high-frequency BWE algorithm, is known to be important in determining timbre. As before in Secs. 2.2.2 and 3.2.2, we model only the brightness aspect of timbre that is closely linked to amplitude spectrum. Brightness is modelled by the spectral centroid C_S (Eqn. 1.95 in Sec. 1.4.6), with higher values for C_S implying a brighter sound percept.
  If the original full-bandwidth signal is known, the BWE algorithm should obviously be designed such that the reconstructed high frequencies closely match the original high frequencies in amplitude spectrum. For blind BWE algorithms, the original high-frequency spectrum is unknown, and the best approach is to smoothly 'extrapolate' the signal spectrum to high frequencies. Typically, audio signals have gradually decaying spectra (although resonances do occur). Such a 'smooth' extrapolation can be ensured by a proper choice of the NLD, as is discussed in Sec. 5.4.1

5.3.3 LOUDNESS (AMPLITUDE)

The loudness of the harmonics signal is directly related to its amplitude. However, if properly generated, the added harmonics will not be perceived separately, but as integral part of the original narrowband signal (e.g. grouping will occur). This would also be the case if the extended signal does not consist of regularly spaced harmonics, but is noise-like. Therefore, we should consider the effect on the loudness of the, originally narrowband, tone when adding higher harmonics.

Standardized loudness models such as ISO532A and ISO532B compute loudness on the basis of the long-term amplitude spectrum of the signal. The amplitude spectrum is specified in narrowbands (e.g. one-third octave bands for ISO532A and 0.1 Bark bands for ISO532B) followed by an integration over frequency, also allowing for masking effects. The details of each procedure differ and are explained in Sec. 1.4.4.2. The main conclusion is that accepted models of loudness perception only take into account the amplitude spectrum to compute loudness. Even Glasberg and Moore's [90] more recent loudness model that can be used for time-varying signals only takes the short-term amplitude spectrum into account. So if a high-frequency BWE algorithm exactly reconstructs the amplitude spectrum of the high-frequency band, the reconstructed signal should have the same loudness as the original full-bandwidth signal. This is of course only possible for BWE algorithms that employ a priori information. For blind algorithms, the reconstructed high-frequency band will deviate from the original high-frequency band. Depending on the pattern and magnitude of these deviations, loudness of the reconstructed signal will not be identical to the original signal's loudness. However, some of these deviations might not be perceptible, as masking effects can reduce or eliminate the contributions of some frequency bands to the total loudness. Also, the largest contributions to loudness will derive from intermediate frequency bands, around 1–4 kHz, where absolute thresholds are lowest (ear is most sensitive). For typical high-frequency BWE applications, the synthesized high frequencies will have a lower limit of at least 4 kHz, and possibly much higher, so the entire contribution of the synthesized high frequencies is probably fairly small anyway.

5.3.4 EFFECTS OF HEARING LOSS

Figure 1.19 shows hearing loss for a group of otologically normal males of various ages (20–70 years) in terms of the 50th percentile points, as a function of frequency. For frequencies below about 1 kHz, the loss remains below 12 dB, but above 1 kHz the amount of loss increases rapidly. At 4 kHz, the amount of hearing loss for a 70-year-old male is, on average, about 42 dB, and at 8 kHz (the highest frequency that was included) this is as much as 60 dB. The same trend is observed for females, although somewhat smaller values for hearing loss are typical. The implication is that older persons, on average, will not perceive high frequencies contained in speech and music signals. The situation is probably more aggravated for the latter category, as music signals contain more energy at higher frequencies than does speech. For these persons, high-frequency bandwidth limitation might not be perceivable at all, and conversely, they might not detect any enhancement of the high-frequency spectrum obtained through high-frequency BWE processing. This is confirmed by the fact that for A/B tests of high-frequency BWE systems younger listeners often perceive clear enhancement, but no or little difference is detectable for older listeners (who did not use hearing aids). From the limited experience gained through informal tests of typical implementations of high-frequency BWE systems, this seems to be the case for persons in the age group of approximately 40 to 60 years old (and presumably older persons as well, although no listeners in that age group had been tested).

For individual listeners, this problem could be (at least partly) overcome by a linear filter that emphasizes signal energy in those regions where hearing loss is severe. But in almost all practical applications such flexibility is not implementable, and probably not even desirable, as reproduced signals can be intended for a group of listeners. In the latter case, a design must be sought that is the best 'on average', and definitely not annoying for any single listener. In practice, this probably means that a high-frequency BWE system would be designed to sound as good as possible for persons with no, or little, hearing loss (typically younger listeners); persons with high-frequency hearing loss (typically older listeners) will therefore, on average, benefit less or not at all, from high-frequency BWE processing (unless the hearing loss is properly compensated for by a hearing aid).

5.3.5 CONCLUSIONS

For a high-frequency BWE algorithm to resynthesize a signal with correct timbre and loudness, it suffices to match the spectral envelope of the original full-bandwidth signal. Correct reproduction of the spectral fine structure is essential for a proper grouping of the synthesized high frequencies with the low frequencies. Persons with significant high-frequency hearing loss will not benefit from high-frequency BWE methods.

5.4 HIGH-FREQUENCY BANDWIDTH EXTENSION FOR AUDIO

Although this whole chapter is devoted to high-frequency BWE for audio applications, the title of this section reflects that here we will discuss implementations analogous to the general structures discussed in Chapters 2 to 3. Other sections in this chapter present alternate structures for high-frequency BWE algorithms. The general structure presented here is shown in Fig. 5.2. Note the correspondence with BWE structures for low-frequency psychoacoustic BWE (Fig. 2.4) and low-frequency physical BWE (Fig. 3.1). Again, there are two branches, the lower of which simply delays the input signal $x(t)$ such that it is later added exactly in phase with the processed signal from the upper branch. The processing consists of two filters and a non-linear device (NLD). The first filter, FIL1, extracts the highest octave present in $x(t)$, which is then the input for the NLD. The non-linear processing generates a harmonics signal, which is filtered by FIL2 to obtain a suitable spectrum. After scaling, the resulting signal is added back to $x(t)$ to yield the bandwidth-extended output $y(t)$. In the remainder of this section, we will explore the various processing steps in more detail.

Note that the signal $x(t)$ must have enough 'empty' bandwidth at the high-frequency end to synthesize the higher harmonics. At a sample rate f_s, the highest frequency present in the signal is maximally equal to the Nyquist frequency, $f_N = f_s/2$. If $x(t)$ contains energy at frequencies higher than $f_N/2 = f_s/4$, then $x(t)$ first needs to be upsampled. In all cases, there must be at least one additional octave above the highest frequency of $x(t)$.

5.4.1 NON-LINEAR DEVICE

To ensure that the synthetic high-frequency band covaries in amplitude with the bandlimited input signal, it is necessary to use a harmonics generator that is homogenous, i.e. scales the output proportional to the input (Sec. 1.1.1). This also has the beneficial property that the relative amount of harmonics generated is independent of the input level. We intend to extend the bandwidth of $x(t)$ by one octave; this yields a significantly

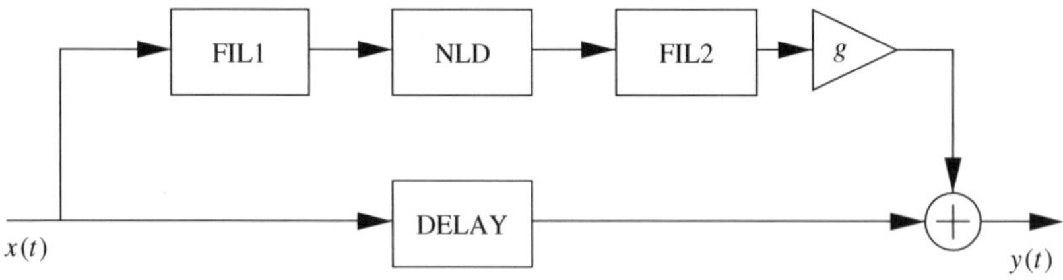

Figure 5.2 High-frequency BWE system. FIL1 extracts the highest octave present in $x(t)$, and harmonics of this signal are generated by NLD. The harmonics spectrum is shaped by FIL2. The harmonics signal is then scaled and added to the delayed input signal to form the bandwidth-extended signal $y(t)$

brighter percept, yet does not suffer from artefacts that have sometimes been observed when extending the bandwidth even further.

As FIL1 extracts the highest octave of $x(t)$, which we denote as $x_0(t)$, the NLD must therefore double the frequencies present in $x_0(t)$, leading to the harmonics signal $x_h(t)$. Frequency doubling can be very efficiently done through rectification, as the spectrum of a rectified pure tone consists mainly of its double frequency; the other components are higher even harmonics, but these decay by 12 dB per octave (Eqn. 2.19 in Sec. 2.3.2.2) and are thus quite weak. A rectifier is also a homogenous system. The intermodulation distortion of a rectifier, given a two-tone input signal, was analysed in Sec. 2.3.2.2, and displayed in Fig. 2.9. It was shown that the relative amount of intermodulation distortion could be fairly high if multiple components of comparable amplitude are present in the input signal. The results of this analysis cannot be directly used for high-frequency BWE though, as in all cases the used expression was for continuous-time implementations of the rectifier. As those sections dealt with low-frequency psychoacoustic BWE applications, that approach was valid, because the frequencies of interest were at least two orders of magnitude smaller than the sample rate. It can be shown that in the limit of very high sample rates, expressions for NLD output spectra in continuous and discrete time are equal. In the present case however, the frequencies of interest are in the same order of magnitude as the sample rate. Specifically, if the spectrum of $x_0(t)$ lies in the range $[f_s/8, f_s/4]$, which would be fairly typical, high harmonics could only be added up to $f_s/2$ maximally. Thus, harmonics higher than the second harmonic cannot even exist in such cases. This obviously alters the expressions for the output spectrum of a rectifier, and also necessitates a re-evaluation of the relative amount of intermodulation distortion, given a multiple-component input. The latter also differs from low-frequency psychoacoustic BWE applications in another aspect, namely, for low-frequency psychoacoustic BWE it is probably reasonable to assume that no more than two frequency components will be present at the input to the NLD, as FIL1 in that case is typically a band-pass filter with a bandwidth of about 50 or 100 Hz. For high-frequency BWE applications, FIL1 is a band-pass filter with a bandwidth of several thousand hertz, and will in most cases contain many frequency components. The correct expression for computing the output spectrum of a rectifier in discrete time, given an arbitrary periodic input signal, is Eqn. 2.116. The expression is fairly complex, and with the added variability of input signals (in terms of number of components, and their frequencies), we have not derived expressions to evaluate the amount of intermodulation distortion energy relative to the amount of harmonic energy, as was done for the simpler case in Sec. 2.3.2.2. Rather, the quality of the bandwidth-extended signal has been judged perceptually, using a variety of repertoire, and the performance is generally considered to be good.

5.4.2 FILTERING

As with low-frequency psychoacoustic BWE and low-frequency physical BWE systems, for high-frequency BWE the signal applied to the NLD needs to be a specific frequency band, and the output of the NLD has to be shaped properly to yield a proper timbre. Thus, it is necessary to use filters before and after the NLD, as in Fig. 5.2.

For both reasons mentioned in Sec. 2.3.3.3, it is beneficial to use linear-phase filters. The first reason was that non-linear-phase filters can lead to interference between the processed (harmonics) signal and the original bandlimited signal, in the limited frequency band where FIL1 and FIL2 overlap. As this is only a small frequency region, this might not be as important as with low-frequency psychoacoustic BWE systems. The other reason is that non-linear-phase filters can give rise to large variations in group delay, which might lead the synthetic high-frequency signal to group poorly with the lower-frequency bandlimited input signal. If FIL1 and FIL2 are both linear phase, their processing delay can be exactly compensated for by a delay of the input signal, such that both harmonics signal and input signal can be added in phase to form the bandwidth-extended output signal. Because the bandwidths of FIL1 and FIL2 are fairly large compared to the sample rate, it is feasible to implement these using FIR filters. Alternatively, IIR filters can be used in the method as described in Sec. 2.3.3.3.

Because a typical application for a blind high-frequency BWE system would be to enhance bandlimited signals as received from, for example, Internet radio, the bandwidth of the incoming signal is not known a priori. Therefore, the passbands of FIL1 and FIL2 need to be adjustable to be able to adapt to whatever the momentary signal bandwidth is. Two methods could be used to implement high-frequency BWE, depending on the bandwidth of the input signal. The first method simply assumes that the signal bandwidth is equal to the Nyquist frequency, that is, half the sample rate. Therefore, the input signal is first up-sampled by a factor of 2, after which the additional octave is 'filled' with the synthesized higher frequencies. Although this method is not guaranteed to work because of the simple assumptions, in practice, for Internet Radio applications it has demonstrated to work quite well. A second, in principle more reliable, method is to analyse the energy content of the signal in various frequency bands, for example, through a number of broad band-pass filters. In most cases, high-frequency BWE will be applied to perceptually coded audio, and in those cases the bandwidth of the signal can be detected by analysing the coefficients of the encoded audio stream directly.

5.4.2.1 Filter 1

Assume that the bandwidth of the input signal is known (or estimated), and the highest frequency component present is f_{h}. Further, assume that the sample rate $f_{\mathrm{s}} \geq 4f_{\mathrm{h}}$, possibly through upsampling prior to BWE processing. As the NLD, being a rectifier, generates second harmonics of the input signal, FIL1 should be band pass between $f_{\mathrm{h}}/2-f_{\mathrm{h}}$. The high-pass flank of FIL1 can be designed as a second-order filter, while for the low-pass flank a somewhat higher order, say fourth order, is better. This prevents frequencies $f' > f_{\mathrm{s}}/4$ from entering the NLD (only if $f_{\mathrm{h}} \approx f_{\mathrm{s}}/4$); if such frequencies did enter the NLD, they would end up as aliased components at low frequencies, because $2f' > f_{\mathrm{s}}/2$. Although any low-frequency component generated by the NLD would be filtered out by FIL2, it is generally beneficial to keep the number of frequency components entering the NLD as small as possible, to minimize intermodulation distortion.

FIL1 could, in principle, be implemented as a filterbank, with each output driving a separate NLD, the aim of which would be to minimize intermodulation distortion. Some informal testing revealed that this strategy does not seem to lead to a significantly better-quality signal, however.

5.4.2.2 Filter 2

The input signal for FIL2 is the harmonics signal as processed by the NLD. Because the input of the NLD is a frequency band $f_h/2-f_h$ (the highest octave present in the bandlimited input signal), the NLD output consists primarily of the second harmonics of these components, that is, the frequency band f_h-2f_h. However, there will be intermodulation distortion components at frequencies below f_h, which have to be eliminated. Therefore, FIL2 has a high-pass flank of at least fourth order. This ensures that the synthesized frequency components are only added at frequencies higher than those contained in the input signal. Depending on the sample rate f_s, a low-pass flank may or may not be required. If $f_s = 4f_h$, then the harmonics signal extends maximally up to $2f_h = f_s/2$, and a low-pass flank is not required. If the sample rate is higher, a low-pass flank can be implemented at a cut-off frequency of $2f_h$. The order of this low-pass flank can be quite low, as the harmonics signal generated by the NLD (rectifier) decays rapidly.

Figure 5.3 shows an example implementation of both FIL1 and FIL2. The input signal has a bandwidth $f_h = 4$ kHz, and the sample rate has been converted to 16 kHz. FIL1 is a Butterworth band-pass filter from 2 to 4 kHz, with a second-order high-pass flank and an eighth-order low-pass flank. FIL2 is a high-pass filter at 4 kHz, with an eighth-order flank. It is not necessary to implement a low-pass flank for FIL2, as frequencies higher than 8 kHz do not exist (as the sample rate is 16 kHz).

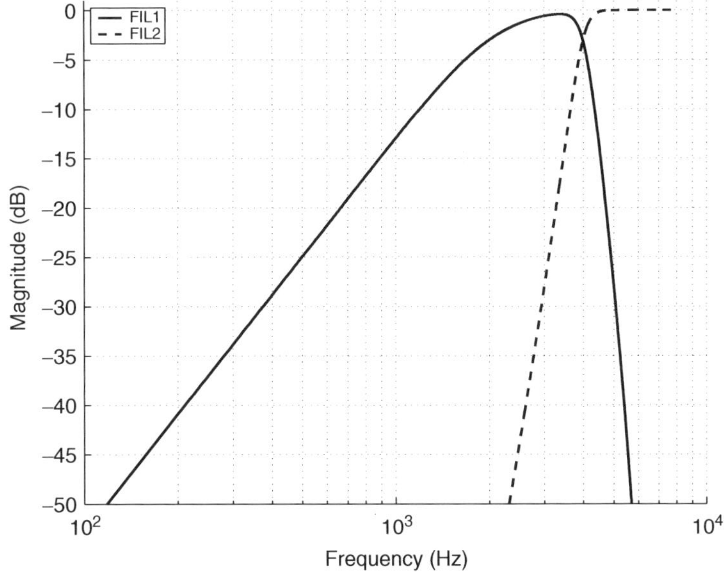

Figure 5.3 Example implementations for FIL1 and FIL2 of a high-frequency BWE system. It is assumed that the highest frequency present in the input signal is 4 kHz, such that FIL1 extracts the highest octave therein. The NLD predominantly generates second harmonics of these components, after which FIL2 ensures that any intermodulation distortion component below 4 kHz is removed

5.4.3 GAIN OF HARMONICS SIGNAL

As the high-frequency spectrum is not known a priori in a blind high-frequency BWE system, the gain of the high-frequency spectrum relative to the low-frequency spectrum is unknown. In practice, a gain value must be chosen that sounds well 'on average'. It is thus inevitable that on many occasions the high-frequency spectrum is either too strong or too weak, compared to the actual high-frequency spectrum. However, these deviations are not excessive, and in nearly all cases the bandwidth-extended signal is judged as more natural compared to the bandlimited signal.

A conceptually simple improvement would be to use an adaptive gain. As there is no a priori information, the control signal for the gain variations would have to be derived from the bandlimited input signal. Some preliminary experiments indicated that this could lead to a more accurate high-frequency percept. Specifically, the gain control signal was derived by matching the energy of the artificially generated high-frequency band to the energy of the actual high-frequency band, short (~20 ms) time frames. This gain control signal was then used to scale the synthesized frequency components. Obviously, this is not possible in an actual application, but it demonstrated that it is possible to improve the quality of the described blind high-frequency BWE algorithm by relatively simple means. Figure 5.4 shows an example, in which a 10-s fragment of pop music, bandlimited to 11 kHz, was processed by the described high-frequency BWE algorithm (FIL1: 5.5–11 kHz, FIL2: high pass at 11 kHz, NLD: rectifier); signal energy in 20-ms frames was computed for the synthetic high frequencies. This was compared to the actual signal energy above 11 kHz (the full-bandwidth signal was also available), the result of which is shown in the figure.

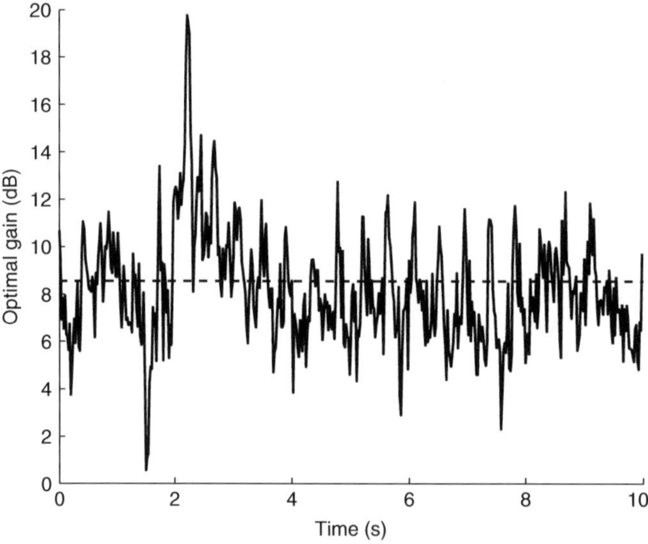

Figure 5.4 'Optimal gain' calculated for a 10-s pop music signal. The gain was derived by comparing the actual signal energy above 11 kHz with the energy of the synthetic high-frequency signal as generated by the high-frequency BWE algorithm. The dashed line indicates the mean (~8.2 dB)

This 'optimal gain' (in the sense that it matches the high-frequency energy) varies around a fairly stable mean value of ~8.2 dB, although occasional large deviations occur (e.g. around 2.5 s). Although the 'mean optimal gain value' varies per repertoire, a 'grand mean value' can be chosen such that fairly good results are obtained for most repertoire. Note that appreciation of individual listeners will vary, which may in part be due to differences in hearing loss at high frequencies (Sec. 5.3.4).

The problem in deriving a practical gain control signal using only information from the bandlimited input signal might conceptually be solved in similar fashion as is done for speech BWE algorithms. Section 6.5 describes what features of the narrowband speech signal are thought to carry some information regarding the high-frequency spectral envelope. For speech, this approach works reasonably well, but it might be much more difficult for music, as music signals have a much larger range of variability than speech signals. Also, the amount of speech bandwidth limitation is well defined through the telephone channel (being about 300–3400 Hz), but this is not the case for the more general situation where bandwidth limitation occurs through perceptual coding. Depending on the bit rate and the coder implementation, the bandwidth can vary from less than 4 kHz to full bandwidth (~22 kHz). For each degree of bandlimitation, another set of parameters would have to be defined to translate narrowband signal features to high-band spectral envelope. Therefore, it remains to be seen if such an approach could work, while remaining practically feasible, for general audio applications.

5.5 SPECTRAL BAND REPLICATION (SBR)

Spectral Band Replication (SBR) is a technique to enhance the efficiency of perceptual audio codecs (Ekstrand [111], Kunz [153], Schug *et al.* [241]). High-frequency components of an audio signal are reconstructed from low-frequency components by the decoder, such that the encoder need only encode the low-frequency part. In this fashion, a bit-rate reduction can be achieved while maintaining subjective audio quality. The basic idea of SBR is based on the observation that characteristics of high-band signals typically exhibit quite a high correlation with those of the lowband signals. Therefore, it is often possible to replace the high band with a transposed version of the lowband, avoiding the need to transmit the high-band signal at all. This can obviously reduce the required bit rate. SBR encodes a bandlimited version of the audio signal using conventional means, and then recreates the high band in the decoder. The difference with blind methods, such as those discussed in Sec. 5.4, is that the encoder provides a very small amount of additional control information (5–10% of the total), which the decoder uses to shape the high-band spectrum. This process is illustrated in Fig. 5.5. The control information is multiplexed with the encoded data into a single bitstream; the decoder first de-multiplexes the bitstream, decodes the lowband signal, and uses a high-frequency BWE algorithm to recreate the high-band signal, thereby using the control data to optimize the BWE processing.

The most important part of the SBR data is the information describing the spectral envelope of the original high-band signal (Dietz *et al.*[61]). Its main design goal is to use it as an equalizer without introducing annoying aliasing artefacts, and to provide good spectral and time resolution. The core algorithm of SBR consists of a 64-band, complex-valued polyphase filterbank (QMF). At the encoder side, an analysis QMF is

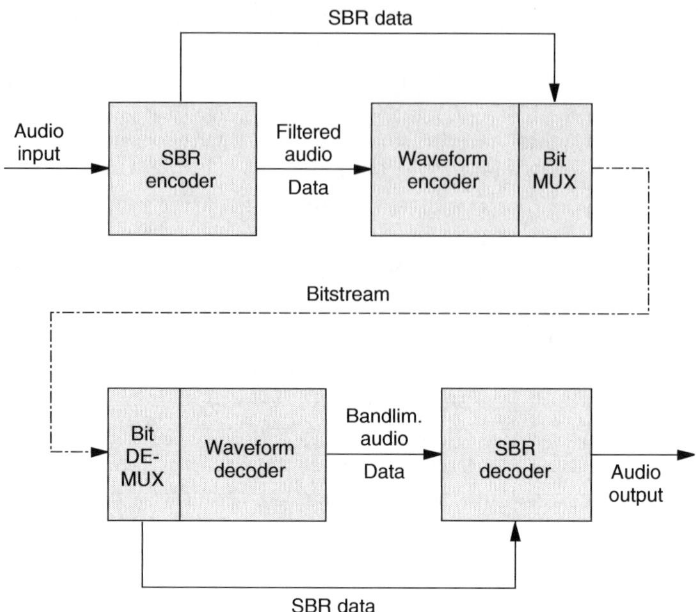

Figure 5.5 Spectral band replication (SBR) acts as pre-processor at the encoder and as post-processor at the decoder. A small amount of control data is provided along with the encoded lowband signal, which the decoder uses to optimize the high-frequency BWE algorithm

Table 5.1 SBR data rates for a number of example configurations. The total bit rate (audio coding and SBR data) is shown in the left-hand column, the number of used frequency bands in the middle column, and the SBR data rate in the right-hand column

Bit rate mono [kb/s]	SBR freq. range # QMF bands	SBR data rate [kb/s]
16	21	1.2
24	24	2.0
32	29	2.5
48	32	3.5

used to obtain energy samples of the original input signal's high band, which are used as reference values for the envelope adjustment at the decoder side. In order to keep the overhead low, the bitstream format of aacPlus[2] allows to group the QMF bands into scalefactor bands. By using a Bark-scale-oriented approach, grouping frequency bands may result in wider scalefactor bands the higher the frequency gets. Table 5.1, from

[2] The combination of AAC with SBR is named aacPlus, which is a registered trademark of Coding Technologies.

Dietz *et al.* [61], shows typical SBR data rates for a number of example configurations. The SBR method is obviously non-blind, as control parameters are used to create the high-frequency signal; however, a blind mode is possible as well, as explained in the patent discussed in Chapter 8.

In some cases, subjectively unsatisfactory results are produced when the low- and high-frequency bands are weakly correlated. This can occur with signals that are predominantly harmonic in the low-frequency range, but more noise-like in the high-frequency range (or vice versa), for example, having tonal instruments at low frequencies together with a hi-hat or cymbals at high frequencies. In such cases, additional information is encoded to indicate the need for synthesizing additional noise or additional tonal components at the decoder, such that the reconstructed high band will be similar to the original.

The combination of SBR technology with the conventional waveform audio coder standardized in MPEG, Advanced Audio Coding (AAC), is discussed in Ehret *et al.* [62]. With this enhanced audio coding scheme, called aacPlus, it is possible to achieve high-quality stereo audio at bit rates as low as 40 kb/s. The structure of the aacPlus decoder is shown in Fig. 5.6. After demultiplexing the aacPlus bitstream, the standard AAC bitstream is converted into a bandlimited audio signal. Then the SBR decoder generates high frequencies from the QMF-filtered bandlimited audio, ensuring a proper spectral envelope by using the SBR data. The high- and low-band QMF signals are then synthesized into a full-bandwidth output signal.

SBR technology is especially interesting in applications in which very high compression efficiency is desired, usually motivated by cost or physical limitations. Examples of such application areas are digital broadcasting and mobile applications. An overview of the latest developments with respect to the standardization process of aacPlus within MPEG-4 and subjective verification results are given in Ehret *et al.* [62], while implementations are described in Homm *et al.* [112].

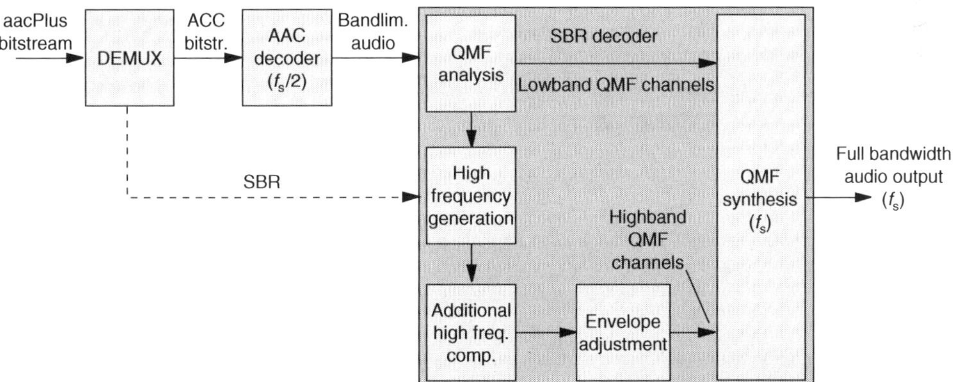

Figure 5.6 Block diagram of the aacPlus decoder. After demultiplexing the aacPlus bitstream, the standard AAC bitstream is converted into a bandlimited audio signal. Then the SBR decoder generates high frequencies from the QMF-filtered bandlimited audio, ensuring a proper spectral envelope by using the SBR data. The high- and low-band QMF signals are then synthesized into a full-bandwidth output signal

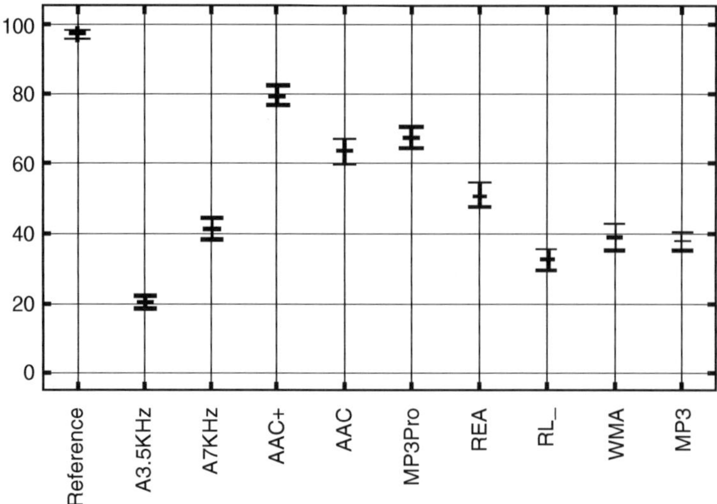

Figure 5.7 MUSHRA test results for various codecs at a stereo bit rate of 48 kb/s. The best result was obtained by aacPlus, followed by MP3Pro, which both use SBR technology

aaCPlus has been subjectively evaluated by several independent listening test sites. The results of all these tests have shown that aacPlus is a very good codec. For example, Fig. 5.7 shows the results of a MUSHRA test[3], carried out in the course of the EBU Internet audio evaluation, in which several audio-coding schemes were compared (cited by Dietz *et al.* [61]). Eight codecs were tested, and their results can be compared to two standards, namely the original, or reference signal (score nearly 100), and a 3.5-kHz low-pass filtered signal (score 20). The figure shows the result of the test for a stereo bit rate of 48 kb/s. The aacPlus decoder was judged as yielding the highest quality signal, with a score of 80 (on the border between 'good' and 'excellent'). The average score of the other codecs was about 50 ('fair'). The score of the core AAC codec was about 65. Another codec that has been integrated with SBR is MP3, called MP3Pro (Gröschel *et al.* [101]). Figure 5.7 shows that MP3Pro has the second-highest test score, nearly 70. The core MP3 coder received a score of just below 40.

5.6 HIGH-FREQUENCY BANDWIDTH EXTENSION BY INSTANTANEOUS COMPRESSION

5.6.1 INTRODUCTION AND ALGORITHM

A special form of BWE can be achieved by audio compression. This approach is especially suitable for multi-channel sound reproduction, in which the processed signals are predominantly speech or special effects. One of the disadvantages of multi-channel material is

[3]The MUSHRA scale range is 0–100, where 0–20 means 'bad', 20–40, 'poor', 40–60, 'fair', 60–80, 'good' and 80–100, 'excellent audio quality'.

that the surround-sound signal is often at a very low level. If the surround signal is simply linearly increased, it can become too dominant, or even lead to audible distortion in either the amplifier or loudspeaker. The same is true for the centre signal, which is often used for dialogues. Here we develop and analyse in instantaneous compression, algorithm that can enhance signals for centre and surround channels. This method is not generally applicable, because for music it does not yield good results; therefore the algorithm is not applied to the left/right loudspeakers of the multi-channel system.

Whereas the initial goal of the described compression algorithm was to overcome the problems of low signal level in centre and surround channels, it was also realized that it is a special kind of a BWE system. At high signal levels, where compression is most active, harmonic frequencies are generated and add some 'brilliance' to the sound. In contrast to what is normally desired of a BWE system, this 'BWE compressor' is not a homogeneous system (i.e. it does not scale its output proportionally to its input, see Sec. 1.1.1), and it is most effective at high signal levels (while e.g. low-frequency psychoacoustic BWE should be more effective at low signal levels, see Sec. 2.3.4). It is also different from other BWE algorithms as it uses the entire bandwidth of the input signal to generate harmonics, that is, the BWE compressor consists only of a non-linear device (NLD), without pre- or post-processing.

The BWE compressor uses a function that has a gain at low and moderate signal levels, but an attenuation at high signal levels. It is different from more usual compressors in that it is memoryless, that is, it is an instantaneous compressor. Any anti-symmetric monotonous function with a positive but decreasing derivative can be used in principle. During experiments, it appeared that the function

$$y(x) = c_1 \tanh(c_2 x), \tag{5.9}$$

plotted in Fig. 5.8 (for $c_1 = c_2 = 1$) is a suitable choice. The constant c_1 determines the maximum output level and c_2 determines the gain at low signal levels. During experiments, it appeared that for $|x| \leq 1$ suitable values for these constants are $c_1 = 0.763$, and $c_2 = 4.19$. For these values, the instantaneous input–output function is shown in Fig. 5.9. Using the Taylor series expansion,

$$\tanh(x) = x - \frac{x^3}{3} + \frac{2x^5}{15} - \frac{17x^7}{315} + \cdots \quad \text{for } |x| < \pi/2, \tag{5.10}$$

we get for small x that $y/x = c_1 c_2 \approx 10\,\text{dB}$ for the given values of c_1, c_2.

5.6.2 ANALYSIS OF HARMONICS GENERATION

In order to study the bandwidth extension of the NLD given by Eqn. 5.9, we assume an input signal $x(t) = A\sin(2\pi t)$, and calculate the coefficients b_n of the Fourier series of $y(x)$

$$\tanh(A \sin 2\pi t) = \sum_{n=0}^{\infty} b_n \sin 2\pi(2n+1)t, \tag{5.11}$$

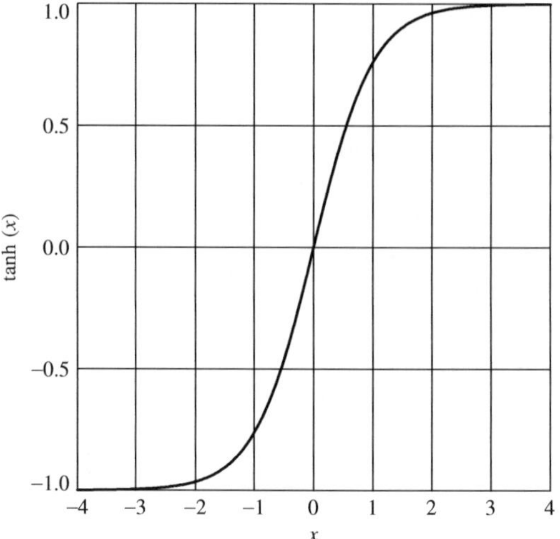

Figure 5.8 The function tanh(x), used as BWE compressor (Eqn. 5.9)

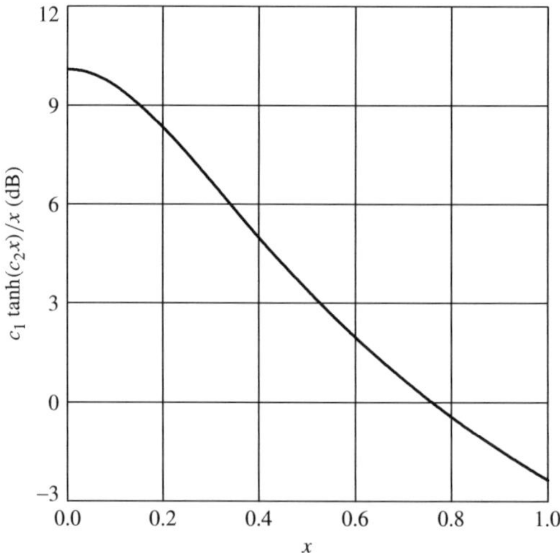

Figure 5.9 The instantaneous input–output 'transfer function' of the BWE compressor: $c_1 \tanh(c_2 x)/x$, for $c_1 = 0.763$ and $c_2 = 4.19$. These values are suitable for signals that are ± 1 at full scale

After some calculations[4] involving the calculus of residues, we find

$$b_n = \frac{8}{A} \sum_{k=0}^{\infty} \frac{1}{u_k^{2n}(1 + u_k^2)}, \tag{5.12}$$

[4] Private communication with A.J.E.M. Janssen, Dec. 2002.

and

$$u_k = \frac{\pi(k+1/2)}{A} + \left(1 + \left(\frac{\pi(k+1/2)}{A}\right)^2\right)^{\frac{1}{2}}, \quad k \in \mathbb{N}_0. \tag{5.13}$$

For large A, we can approximate the b_n by ignoring terms containing powers of $1/A$ higher than five, which yields

$$b_n = \frac{4}{\pi}\frac{1}{2n+1} - \frac{(2n+1)\pi}{6A^2} - \frac{7(2n+1)\pi^3}{60A^4}(2 - \frac{1}{6}n(n+1)), \quad n \in \mathbb{N}_0. \tag{5.14}$$

For very large A, the output signal will tend to a square wave, which can be made explicit by taking $\lim_{A\to\infty}$ and showing

$$\lim_{A\to\infty} b_n = \frac{8}{A}\sum_{k=0}^{\infty}\frac{1}{u_k^{2n}}\frac{1}{(1+u_k^2)} = \frac{4}{\pi}\frac{1}{2n+1}, \tag{5.15}$$

so that we get, as we should, the familiar Fourier series coefficients of a square wave. On the other hand, for very small A, the output and input signals are proportional, because we get $b_0 = A$ and $b_k = 0$ for $k \in \mathbb{N}\backslash\{0\}$. This was also directly obvious from Eqn. 5.10.

5.6.3 IMPLEMENTATION

With analog components, Eqn. 5.9 can be easily implemented, using a long-tail pair with two transistors. On a digital platform, there are several possibilities. If the platform used is capable of directly implementing Eqn. 5.9, this would be the easiest way. If this is not the case, Eqn. 5.9 can be approximated by a power series. The Taylor series expansion of Eqn. 5.10 is not suitable, since this is only accurate for small $|x|$, while we are interested in the range $|c_2x| \le 1$ (where $|x| \le 1$). Therefore, we use a power series with an ℓ_∞−norm using a NAG [182] routine, based on a Chebyshev approximation (Barrodale and Phillips [25]). It appears impractical to use only one polynomial for the whole range, and therefore we use (for the case that $c_2 = 4.2$) two ranges, namely $|c_2x| \le 1$ and $|c_2x| > 1$. This yields the following result

$$\tanh(x) \approx \hat{y}(x) = x\sum_{k=0}^{3} a_k x^{2k} \qquad \text{for} \quad |x| \le 1 \tag{5.16}$$

and

$$\tanh(x) \approx \hat{y}(x) = \text{sign}(x)\sum_{k=0}^{7} b_k |x|^k \qquad \text{for} \quad 1 < |x| \le 4.2 \tag{5.17}$$

where $z = c_2x$.

The order of the approximation is chosen such that the maximum error is equal to about 2^{-15}, which is suitable for 16-bit systems. If a lower or higher degree of approximation

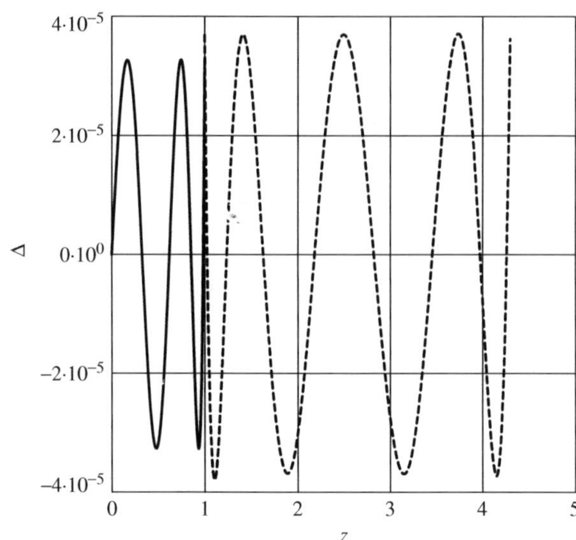

Figure 5.10 The error $\Delta = \tanh(z) - \hat{y}(z)$. The solid line shows the approximation error of Eqn. 5.16; the dotted line shows the approximation error of Eqn. 5.17. The coefficients a_k and b_k are as given in Sec. 5.6.5

is required, all coefficients a_k and b_k have to be recomputed again, since, as opposed to a Taylor series, the coefficients of a Chebyshev approximation depend on the order of the approximation. The approximation error, using Eqns. 5.16 and 5.17, and the coefficients a_k and b_k as given in Sec. 5.6.5, is plotted in Fig. 5.10.

5.6.4 EXAMPLES

Here we present some example signals and their processed versions, to illustrate the effects of BWE compression processing. Figure 5.11 shows four histograms displaying the amplitude distribution of two different input and output signals. All signal values were contained in $[-1,\ 1]$, and the processing used coefficients $c_1 = 0.763$ and $c_2 = 4.19$. The histograms have 50 equally spaced bins, and the value displayed for each bin is the log (base 10) of the number of occurrences that the signal value was in the bin range. Figure 5.11 (a) shows the amplitude distribution for a 60-s fragment of a pop music signal, which is nearly full scale, assuming that the transducer limits are ± 1. A linear amplification of this signal would lead to clipping distortion. Part (c) shows the signal distribution after BWE compression. It is obvious that the maximum values of the signal have been reduced to $c_1 = 0.763$, and that the overall distribution has become flatter (as low-valued samples of the signal have been amplified). Part (b) displays the amplitude distribution for the same signal as in part (a), but scaled down by a factor of 10. BWE compression leads to an amplitude distribution as shown in part (d). Because of the small signal values, the compressor operates in its linear region, and as

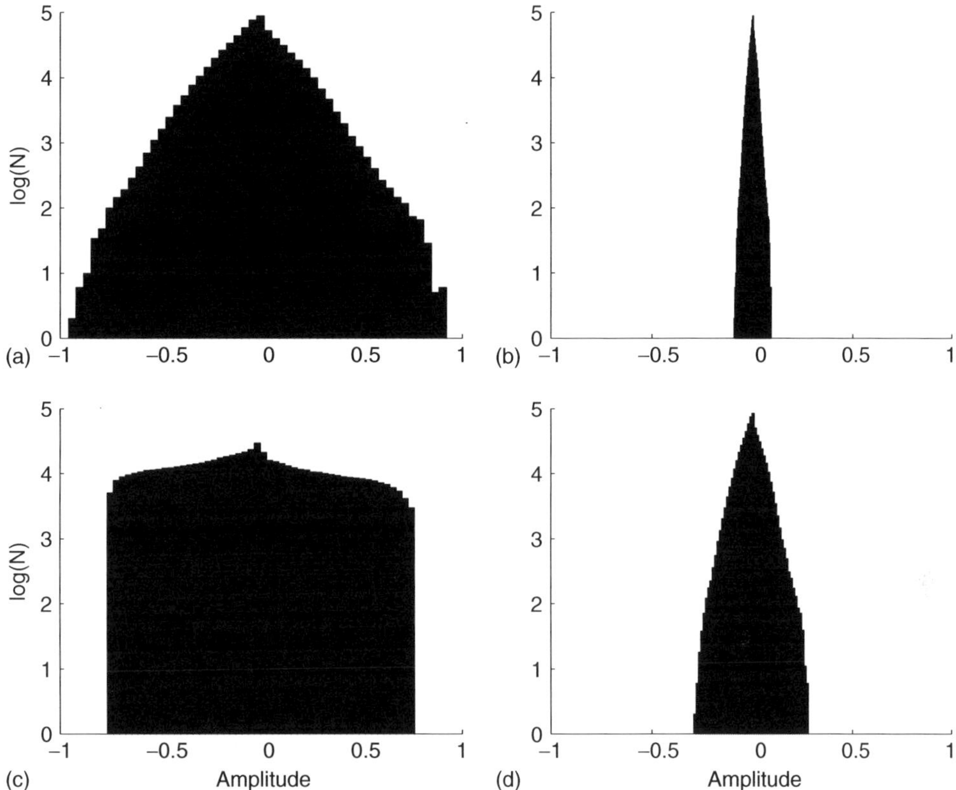

Figure 5.11 Histograms of a 60-s excerpt of music. Histogram values indicate number of occurrences per bin (log–base 10) of the signal value. Part (a) shows a signal that is close to full scale, and part (c) shows the BWE-compressed version. Note that the maximum signal value has decreased and the distribution has become flatter. Part (b) shows the signal of part (a), but scaled by a factor of 0.1. Part (d) is its BWE-compressed version, which shows only a linear scaling and no change in shape of the distribution (no flattening)

a result the signal distribution does not change shape (it has not become flatter as in part (b)).

It is also instructive to visualize the modifications generated by BWE compression in the time–frequency domain. Figure 5.12 (a) shows the spectrogram of the first 10 s of the input signal (the amplitude distribution of which is shown in Fig. 5.11) (a); in all spectrograms, black indicates high energy, and white indicates low energy (dB scale). Note that there is a gradual roll-off above 3 kHz and an abrupt high-frequency limit at about 5 kHz. The spectrogram of the BWE-compressed signal, shown in part (b), displays much more energy in the high-frequency region, up to the Nyquist frequency. The transients (recognizable as dark vertical lines) are clearly enhanced. Also, some of the complex tones have enhanced harmonics, for example, the harmonics below 1000 Hz, just before

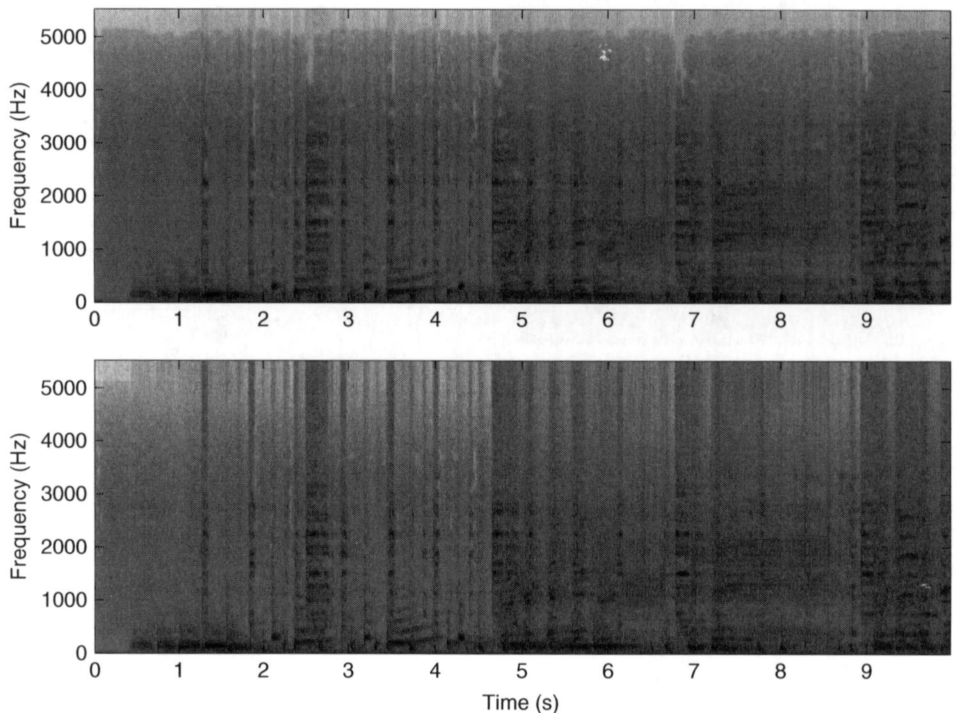

Figure 5.12 Spectrograms of a 10-s excerpt of music (the amplitude distribution of which is shown in Fig. 5.11) (a), and its BWE-compressed output in part (b). The input signal is somewhat bandlimited, whereas the output shows enhanced transients and more high-frequency content

4 s; although difficult to see in this fashion, it can be clearly observed when switching the spectrograms on a computer screen.

Figure 5.13 shows two spectrograms, the upper one displaying the time–frequency energy distribution of the same input signal as previously, but scaled down by a factor of 10 (as in Fig. 5.11 (b)). Note that all signals have been normalized before time–frequency analysis, such that any changes in the spectrograms are not due to overall level effects, but indicate relative changes in energy distribution in the time–frequency domain. Figure 5.13 (b) shows the BWE-compressed output, and exhibits only a modest enhancement of high frequencies, as we would expect, given that for low levels the compressor operates in its (near) linear regime. Also, the enhancement of low harmonics is not as pronounced as for the higher-level signal of Fig. 5.12.

5.6.5 APPROXIMATION OF THE FUNCTION tanh(**Z**)

In order to derive an approximation of the function $\tanh(x)$, a power series with an ℓ_∞–norm and a NAG [182] routine, based on a Chebyshev approximation (Barrodale

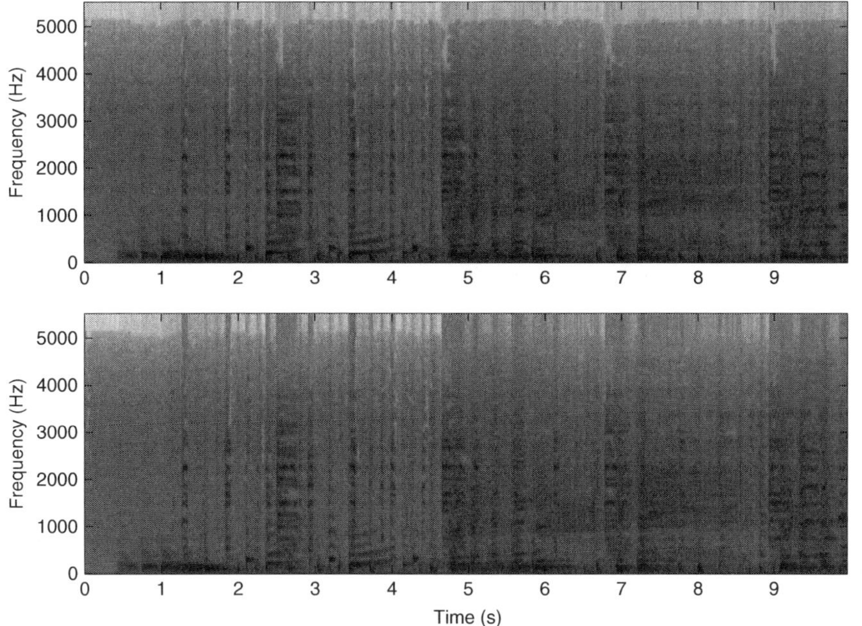

Figure 5.13 Spectrograms of a 10-s excerpt of music, the same as in Fig. 5.12 (a), but scaled down in amplitude by a factor of 10. Its BWE-compressed output is shown in part (b). Because of the low signal level of the input, the compressor operates in its (near) linear regime, and there is only a modest enhancement of high frequencies; overall the two spectrograms are very similar (much more so than the spectrograms of Fig. 5.12)

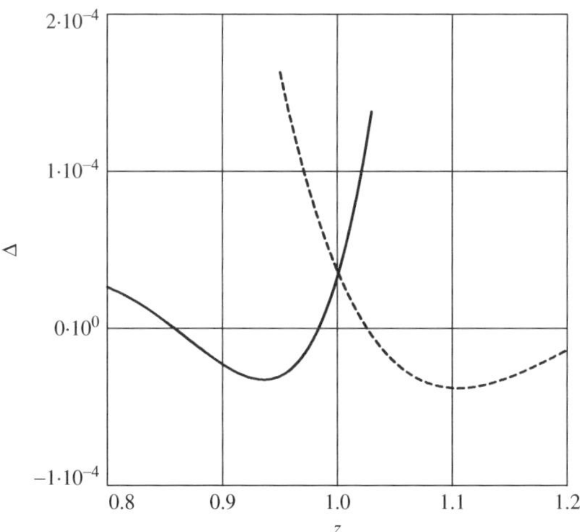

Figure 5.14 The error $\Delta = \tanh(z) - \hat{y}$; same as Fig. 5.10, but zoomed in around $z \approx 1$

and Phillips [25]), was used. Using Eqn. 5.16 and 5.17, we get the following algorithm:

```
x =~c_2*x

IF |x| <= 1 THEN

x2 = z^2
y=c_1*x*(0.9997 + x2*(-0.3289 + x2*(0.1154 -~x2*0.02465)))

ELSE

xa = |x|
y = c_1*sign(x)*(-0.1694 + xa*(1.6489 + xa*(-0.9587 + xa*(0.2713
    + xa*(-0.02786 + xa*(-0.003742 + ...
xa*(0.001199 + xa*(-0.00008518)))))))

END
```

This yields an approximation error as plotted in Fig. 5.14. To avoid a discontinuity in the transition area between both approximations ($x \approx 1$), the coefficients are chosen such that the sign of the errors for both approximations are the same, and the magnitudes are about equal.

6

Bandwidth Extension for Speech

Peter Jax

Institut für Nachrichtengeräte und Datenverarbeitung (IND);
Rheinisch-Westfälische Technische Hochschule (RWTH) Aachen

In this chapter, the problem of speech enhancement by artificial bandwidth extension is addressed. Whereas in the preceding chapters the signal processing was mostly based on properties of the human auditory system, that is, of the signal sink, the bandwidth extension of speech signals uses properties of the signal source. Hence, here we restrict our view to those bandwidth extension approaches that perform adaptive signal processing according to the well-known time-varying source-filter model of speech production. Note that any of the methods described in the other Chapters 2, 3 and 5 in general can as well be applied to speech signals yet often with lower quality than specialized algorithms due to the lower amount of a priori information that is utilized.

The typical application of bandwidth extension for speech is due to the basic design of speech transmission systems: in current digital public telephone systems the acoustic bandwidth of the transmitted speech signal is usually still limited to the frequency range of the old analogue telephone system, that is, to about 300 Hz to 3.4 kHz. This bandwidth limitation causes the characteristic sound of *telephone speech*.

The minimum requirements on the bandwidth of analogue speech communication systems was specified in the CCITT Red Book from 1961 (see, e.g. Schmidt and Brosze [237]): at the cut-off frequencies of 300 Hz and 3.4 kHz the transmission level may be attenuated by no more than 10 dB with regard to the level at the reference frequency of 800 Hz (ITU-T Rec. G.132 [121], Rec. G.151 [122]). The reasons for the bandwidth limitation at that time were the use of analogue frequency-division multiplex transmission with a frequency grid of 4 kHz, and the optional use of sub-audio telegraphy for out-of-band signalling. The minimum bandwidth of 300 Hz to 3.4 kHz was specified to guarantee an intelligibility of sentences of about 99% from clean telephone speech.

Nowadays, the public telephone system has almost completely been converted to digital transmission techniques. According to the international standard ITU-T Rec. G.711 [123], the speech signals are sampled at a sampling frequency of 8 kHz, and the samples are quantized using the A-law respectively μ-law PCM-encoding laws, yielding a bit rate of 64 kb/s. A strict upper limit of 4 kHz on the transmitted frequency range is enforced by

Audio Bandwidth Extension E. Larsen and R. M. Aarts
 ISBN 0-470-85864-8

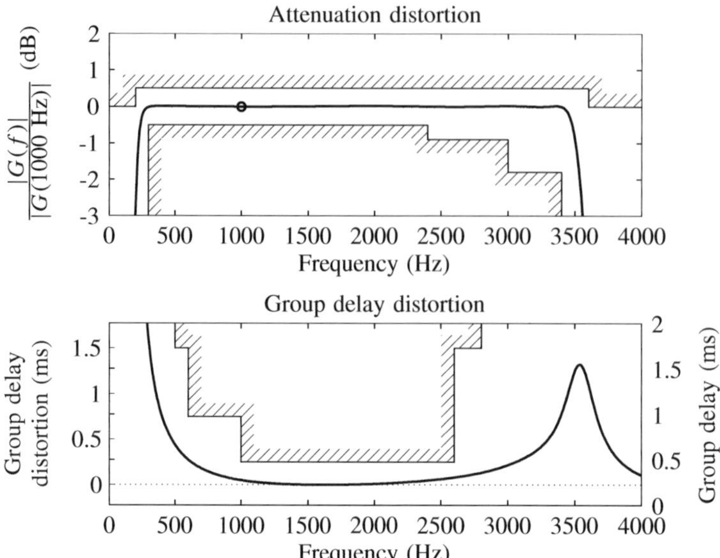

Figure 6.1 Design constraints from ITU-T Rec. G.712 [124, Sec. 1, 2] for PCM speech transmission. The reference value (0 dB at 1000 Hz) for the attenuation distortion is marked by the filled circle in the upper diagram. The minimum value of the group delay (here: 224 µs at 1621 Hz) is taken as the reference for the group delay distortion. The solid curve gives an example of admissible filter characteristics

the sampling frequency of 8 kHz. Because the implementation of digital circuits in existing networks was performed by successive replacements of analogue circuits, the constraints of the old analogue system applied. The required performance characteristics of PCM transmission channels are specified in detail in the standard ITU-T Rec. G.712 [124]. The design constraints with respect to attenuation and group delay are illustrated in Fig. 6.1.

For mobile radio telephony systems, a further limitation of the frequency range is specified to reduce the amount of disturbing low-pass background noise. In GSM, for example, both the sending and receiving sensitivity of headset or handset mobile terminals shall provide an attenuation of at least 12 dB for low frequencies below 100 Hz (ETSI Rec. GSM 03.50 [68]).

Compared to natural speech, telephone speech has a significantly degraded quality: the removal of low frequencies below about 300 Hz leads to a reduction of the loudness of the speech, leading to a 'thin' voice. In spite of this absence of the fundamental harmonic in the bandlimited speech, a human listener can still perceive the virtual pitch from the harmonic structure of the remaining overtones (Zwicker and Fastl [309], Terhardt [267]), see Sec. 1.4. The elimination of high-frequency components beyond 3.4 kHz, on the other hand, leads to a reduction of the transparency and articulateness of the speech. The bandlimited telephone speech sounds somewhat 'muffled'.

Because both the high- and low-frequency speech components contain some speaker-dependent characteristics, their absence in the bandlimited speech makes it sometimes difficult for a human listener to identify the conversational partner.

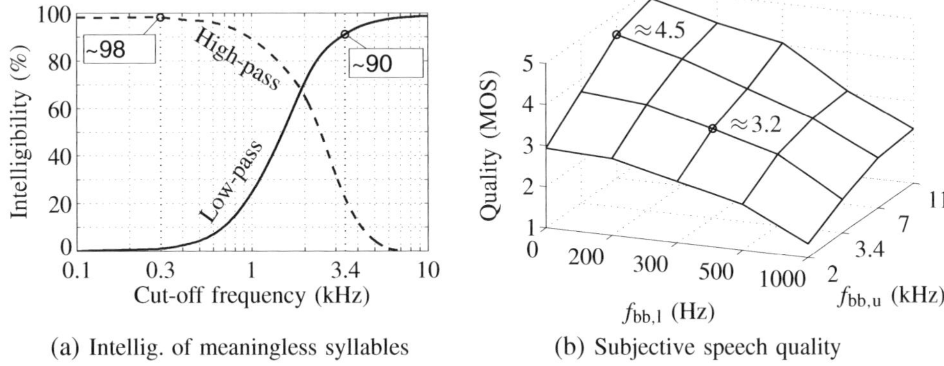

(a) Intellig. of meaningless syllables

(b) Subjective speech quality

Figure 6.2 Impacts of a bandwidth limitation on speech intelligibility and subjective quality. In part (a), the intelligibility of meaningless syllables in low-pass respectively high-pass filtered speech is illustrated (data from Terhardt [267]). (b) This compares the speech quality, measured in terms of the subjective *mean opinion score* (MOS), of band-pass-filtered speech with different lower ($f_{\mathrm{bb,l}}$) and upper ($f_{\mathrm{bb,u}}$) cut-off frequencies (data from Krebber [148])

Speech Intelligibility

The relevance of high- and low-frequency speech components for the speech intelligibility is pointed out in Fig. 6.2 (a). The diagram shows the intelligibility of (individually) low-pass or high-pass filtered meaningless syllables (French and Steinberg [76], Terhardt [267]). It can be observed that the intelligibility is quite high for the band limits of the telephone band-pass: low-pass filtering with a cut-off frequency of 3.4 kHz yields intelligibilities around 91%, while high-pass filtering at 300 Hz leads to an intelligibility of about 98%.

The intelligibility of meaningless syllables from telephone speech is about 90%, thus making it sometimes necessary to use the spelling alphabet to communicate words that cannot be understood from the context, for example unknown names. The intelligibility of whole *sentences* from clean telephone speech, however, is around 99% (Brosze *et al.* [41], Schmidt and Brosze [237]). Thus, potential benefits of bandwidth extension in terms of the intelligibility of sentences seem to be quite small. Nevertheless, an improvement of the intelligibility of syllables would make the communication more comfortable and less strenuous in many cases, that is, the *listening effort* would be reduced.

Subjective Speech Quality

Listening experiments have shown that the acoustic bandwidth of speech signals contributes significantly to the perceived speech quality (Krebber [148], Voran [291]). This fact is illustrated in the right diagram of Fig. 6.2 (b), which shows the results of evaluations of subjective speech qualities for clean band-pass-filtered speech. The speech quality is expressed in terms of the *mean opinion score* (MOS), which reflects the subjective rating by human listeners on a scale between one (unacceptable quality) and five (excellent quality). The two points in Fig. 6.2 (b) that are marked by circles indicate the scores for telephone speech (3.2 MOS points) and 'wideband' speech (4.5 MOS points), respectively.

Starting from the bandwidth of telephone speech (300 Hz to 3.4 kHz), the expansion of the bandwidth both towards low and high frequencies leads to significant gains of the achieved MOS scores. The best scores are obtained by a symmetric expansion towards low *and* high frequencies. In comparison to telephone speech, typical *wideband speech* with a bandwidth of 50 Hz to 7 kHz yields a considerable maximum gain of about 1.3 MOS points.

6.1 APPLICATIONS

Owing to the importance of the acoustic bandwidth for speech intelligibility, and especially for the subjective quality, it seems to be worthwhile to aim at an expansion of the transmitted acoustic speech bandwidth. Particularly, in digital communications and hands-free telephony, there is a demand for enhancing the subjective speech quality. True wideband speech communication requires a modification of the transmission link – enhanced speech codecs have to be employed on both sides of the link. Accordingly, several wideband speech-coding schemes have been investigated in the past, aiming at the increase of the acoustic bandwidth to 50 Hz to 7 kHz. In the 1980s, the G.722 codec was standardized by ITU, with bit rates of 64, 56, and 48 kb/s mainly targeting the applications of teleconferencing and ISDN telephony (ITU-T Rec. G.722 [125], Maitre [165]). Later the G.722.1 codec [126] was added with bit rates of 32 and 24 kb/s. Recently, the *adaptive multi-rate wideband* (AMR-WB) codec algorithm (several modes with bit rates from 23.85 down to 6.6 kb/s) was developed and standardized by 3GPP and ETSI [2], Bessette *et al.* [31]. This codec family has also been adopted by the ITU [127]. The implementation of the AMR-WB codec is projected for GSM and 3GPP WCDMA networks.

However, for economical reasons, the bandwidth limitation is not likely to change *on a broad scale* in the near future. It is very likely that, at least for some transitional period, the telephony network will be a mixed network, comprising both narrowband- and wideband-capable terminals.

An alternative approach towards an enhanced acoustic bandwidth of the received speech signal is artificial *bandwidth extension* (BWE) of speech. The challenge of BWE in speech transmission is illustrated in Fig. 6.3: the wideband microphone signal s_{wb} is band-pass filtered prior to analogue-to-digital conversion and transmitted across the telephone network. At the receiving terminal, only the narrowband signal s_{nb} is available. This bandlimited speech signal is analysed by the bandwidth extension system. The missing low- and/or high-frequency signal components are estimated and added to the received base-band components. By this, the algorithm determines an estimate $\tilde{s}_{\mathrm{wb}}$ of the wideband speech that is passed on to the loudspeaker.

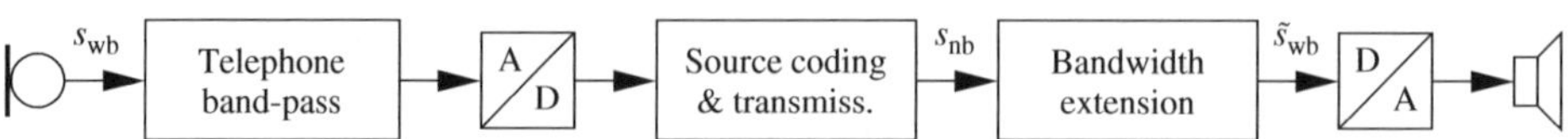

Figure 6.3 Artificial bandwidth extension in digital speech transmission

The application of bandwidth extension is, in principle, independent of the sending side of the transmission link and of source coding and transmission methods. Hence, the bandwidth extension approach is fully compatible with the existing speech communication infrastructure. It must be emphasized that the concept of artificial bandwidth extension should not be considered to be antagonistic to true wideband coding – on the contrary, it constitutes a harmonious extension to wideband speech services, because it can help reduce the quality variations between the different speech signals in a mixed mode network. Possible fields of application for artificial bandwidth extension systems include the following ones:

- Artificial bandwidth extension can be implemented in a (receiving) terminal equipment as depicted in Fig. 6.3. Then, the user of the terminal gets an improved speech quality, albeit the sending terminal is only capable of narrowband speech transmission. The implementation of bandwidth extension is attractive for manufacturers with respect to the competition on the terminal market.
  It must be noted that there are certain physical constraints caused by the rather small size of modern mobile handsets, particularly for playing low-frequency signals via small loudspeakers (compare Chap. 2). Some loudspeakers have lower cut-off frequencies of up to 1000 Hz, particularly if the small loudspeaker of a mobile phone is operated in hands-free mode. With handsets, the transfer function from the loudspeaker to the ear strongly depends on the positioning of the handset at the ear. If the auricle is not tightly sealed, an acoustic leakage occurs, which impairs the transfer function particularly at low frequencies (Krebber [148]). In many cases, with the aforementioned physical constraints, physical speech bandwidth extension towards *low* frequencies does not make much sense since the extended signal components cannot be provided to the listener.
  For the design of the bandwidth extension algorithm, it should be regarded that, in general, source coding has been applied within transmission. For example, in ISDN the A-/μ-law, PCM-encoding rules from ITU-T Rec. G.711 [123] are used, or in GSM one of the speech codecs specified in ETSI is utilized. It can be observed, however, that coding distortions do not have a major detrimental effect on bandwidth extension, but on the other hand the extension algorithm can benefit from adopting dequantized parameters from the speech decoder.
- In a mixed mode speech communication network, comprising both narrowband- and wideband-capable terminals, artificial bandwidth extension can be implemented within network nodes for transcoding from narrowband codecs to wideband codecs. This is especially beneficial if switchings between narrowband and wideband transmission modes occur, for example, due to handovers in mobile radio access networks [1, Sec. 27].
- If so-called wideband speech (typical frequency range: 50 Hz to 7 kHz) is already available, it is possible to perform bandwidth extension towards 'super-wideband' speech, that is, with a target frequency range of up to 16 kHz. For example, in low bit-rate MPEG coding, a special *speech mode* without need to send extra side information as in spectral band replication (SBR) audio coding is possible, or the speech quality of wideband speech codecs can be improved further. This application is even more promising than the extension of telephone speech because the uncertainty of 'super

high frequency' speech components (e.g. between 7 and 16 kHz) is lower and more information can be gained from the available wideband speech signal.

- Bandwidth extension techniques are commonly used within wideband speech codecs, for example, in Dietrich[60], Taori *et al.* [265], in the split-band CELP (SB-CELP) family of speech codecs, such as Paulus [206], Schnitzler [238], Erdmann *et al.* [67]), and in the AMR-WB codec [3]. However, in these approaches mostly the bandwidth extension is applied to a quite narrow frequency band at very high frequencies, for example, only to the signal components between 6 and 7 kHz. Furthermore, the extension can be supported by transmitting side information (compare the *spectral band replication* (SBR) techniques for audio coding in Sec. 5.5).
- One of the first investigated applications of artificial bandwidth extension aimed at the improvement of the quality of telephone contributions in broadcast programmes Croll [54]. If telephone speech is interposed between passages of studio speech, it can become distracting for the listener, because understanding the two different types of speech requires different levels of concentration. By bandwidth extension, the quality of the enhanced speech comes closer to that of studio speech. If the telephone contribution is from a professional correspondent, pre-collected a priori knowledge about the characteristics of the original voice can be made available to the extension algorithm.
- Artificial bandwidth extension can be applied to enhance the acoustical quality of historical recordings of speech. In this application, no real-time processing is required, and the parameters of the algorithm can be tuned manually. If additional wideband recordings of the speaker are available, they can be used to determine the particular voice characteristics.

6.2 FROM A SPEECH PRODUCTION MODEL TO THE BANDWIDTH EXTENSION ALGORITHM

In principle, the physical reconstruction of the acoustic bandwidth of (speech) signals can only be feasible if the algorithm has some a priori knowledge about the input signal. For example, if we consider an arbitrary signal that is sampled with a sampling rate of 8 kHz, and if there is no further information available on the kind of the signal, it is impossible due to Nyquist's theorem to tell anything about the signal components beyond the limit frequency of 4 kHz. If, however, a mathematical model of the source of the signal is available, the situation is fundamentally different: both the wideband signal as well as the bandlimited signal are determined by parameters of the common source model. Consequently, exact knowledge of these source parameters would open up the possibility to reconstruct the complete wideband signal as it was originally produced. The parameters of the source, on the other hand, can be estimated from the characteristics of the bandlimited signal.

Because each mathematical model can only be a statistical approximation of the real physical source of a signal, there are several potential drawbacks of such model-based approaches: owing to *simplifications* introduced by the modelling, there will be estimation errors both of the parameters of the source model as well as of the reconstructed wideband signal. In addition, if the characteristics of the actual physical source do not match the characteristics of the source model exactly, that is, if there is a *model mismatch*, the probability of estimation errors and artefacts in the enhanced signal further increases.

Another cause of estimation errors follows from the basic properties of the signal source. In general, it is not possible to estimate the parameters of the source with arbitrary accuracy because of the random attributes of the signal. In this regard, an upper bound on the quality of the estimation of the spectral envelope of the speech signal will be evaluated in Sec. 6.4.3.

The utilization of a particular model for the signal source also imposes fundamental limits on the application areas of the bandwidth extension algorithm. If, for example, the algorithm is based on a model of the process of speech production, the algorithm will naturally not have the capability to extend general audio signals (such as music), or to reconstruct characteristics of the acoustical environment of the speech signal, such as reverberation or background noise.

6.2.1 MODEL OF THE PROCESS OF SPEECH PRODUCTION

For the process of speech production, there exists a well-known source-filter model, which is illustrated in Fig. 6.4. This model has found wide acceptance in many applications of speech signal processing, especially in the areas of speech coding and speech synthesis (see, e.g. Flanagan [71], Rabiner and Schafer [217], Vary *et al.* [286]). According to the physiology of the human vocal tract apparatus, the model can be sub-divided into two parts: first, an *excitation signal* $u(k)$ is produced, which resembles the excitation of the vocal tract as produced by the vocal cords for voiced sounds, or by a constriction of the

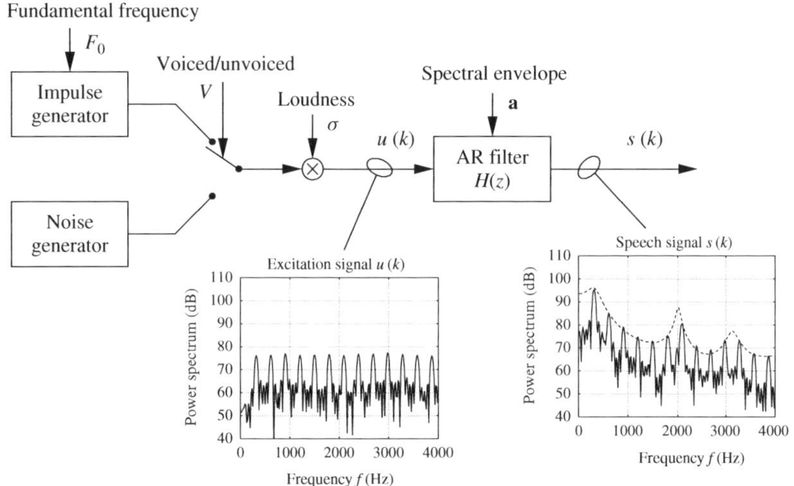

Figure 6.4 Time-discrete, linear source-filter model of the process of speech production. To clarify the principle, exemplarily the power spectra of the excitation and output signal of the model are shown for an idealized voiced speech sound. The spectral envelope of the speech signal is shaped by the auto-regressive (AR) vocal tract filter $H(z)$. The magnitude transfer function of $H(z)$ is illustrated by the dashed line in the right-hand diagram

vocal tract during unvoiced sounds or plosives. The excitation signal $u(k)$ is the input signal to a purely recursive, that is, auto-regressive (AR), digital filter $H(z)$ that models the resonance characteristics of the vocal tract.

All parameters, F_0, V, σ, and $\mathbf{a}$, of the model are basically highly time-variant. However, for speech signals, we can assume that the system is short-term stationary during intervals with a duration of at least 10 to 30 ms. Therefore, in many speech-processing algorithms, the speech signal is processed frame by frame with frame durations between 5 and 30 ms.

Excitation signal In the source-filter model, the excitation signal $u(k)$ is produced by several interacting sub-systems. For voiced sounds (e.g. vowels), a sequence of equidistant impulses with the desired fundamental frequency F_0 of the speech signal is produced by an impulse generator. In the frequency domain, this impulse sequence corresponds to several harmonics, which are positioned at the fundamental frequency F_0 and integer multiples thereof. All harmonics have the same constant amplitude. An example of the short-term power spectrum of the excitation signal of a voiced sound is shown in the left diagram in Fig. 6.4. For unvoiced sounds, the excitation signal is produced by a noise generator that produces white noise with a variance of 1. Note that for both kinds of excitation, the spectrum of the modelled excitation signal is flat. The selection of the particular kind of excitation is performed by a binary switch that is controlled by the voicing parameter V. Finally, the gain of the speech signal is specified with the common scalar gain factor σ.

This simple model of the generation of the excitation signal reflects the real physical process speech production in a very idealized and simplified manner. For example, it is very rare that speech sounds are exclusively of voiced or unvoiced nature. Normally, the excitation signal is a mixture of both kinds of excitation. Further, the excitation of the human vocal tract is not perfectly spectrally flat in reality: the periodic excitation produced by the vocal cords in general has low-pass characteristics; for unvoiced sounds, on the other hand, the spectral characteristics of the excitation signal depend on shape and position of the constriction in the vocal tract that causes the chaotic turbulences of the air. However, such model mismatches of the spectral characteristics of the excitation signal can be taken into account by the subsequent filter $H(z)$ in the source-filter model. Although the model, in a sense, is too simple to describe the complex physical mechanism of speech production, it has proven to be sufficient for most applications of speech processing.

Vocal tract filter In the human vocal tract, the sound-specific spectral envelope of the speech signal is shaped. Its signal-processing model consists of a time-variant auto-regressive (AR) filter

$$H(z) = \frac{1}{A(z)}, \quad \text{with } A(z) = \sum_{i=0}^{N_a} a_i \, z^{-i}. \tag{6.1}$$

The purely recursive structure of the filter $H(z)$ can be motivated by physically modelling the human vocal tract via an idealized, that is, lossless and discretized, acoustic tube with varying diameter (e.g. Flanagan [71], Rabiner and Schafer [217], Vary *et al.* [286]).

Because the human ear is basically insensitive to moderate variations of the signal phase, the vocal tract can be modelled by a minimum-phase AR filter.

According to its role in the decoder of a *linear predictive coding* (LPC) system, the filter modelling the vocal tract is frequently called *LP synthesis filter* in literature. The filter coefficients are combined in the column vector $\mathbf{a} = \left[a_0, a_1, \ldots, a_{N_a}\right]^{\mathrm{T}}$. The first coefficient is normalized to $a_0 = 1$ in general such that

$$\frac{1}{2\pi}\int_{-\pi}^{\pi} \frac{1}{A(\mathrm{e}^{j\Omega})}\,\mathrm{d}\Omega = \frac{1}{2\pi}\int_{-\pi}^{\pi} A(\mathrm{e}^{j\Omega})\,\mathrm{d}\Omega = 1. \tag{6.2}$$

Owing to this normalization, the transfer function of the vocal tract filter is independent of the short-term power (or gain) of the speech signal. Hence, it describes the *shape* of the spectral envelope only. Because of the limited order N_a of the AR filter, it describes a smoothed version of the spectral envelope of the signal. Typical filter orders are $N_a = 8 \ldots 10$ for narrowband speech (sampling frequency $f_\mathrm{s} = 8\,\mathrm{kHz}$), and $N_a = 16 \ldots 18$ for wideband speech ($f_\mathrm{s} = 16$ kHz).

Since the LP synthesis filter $H(z)$ is a minimum-phase filter, there always exists a stable inverse thereof. The inverse $A(z)$ of the LP synthesis filter has a finite impulse response, and it is often called *LP analysis filter*. The analysis filter has an important property for the bandwidth extension application: if the filter coefficients $\mathbf{a}$ are available, that is, if the shape of the spectral envelope of the speech signal is known, applying the LP analysis filter to the speech signal will calculate an estimate of the excitation signal $u(k)$.

The optimal filter coefficients (in the sense of minimizing the power of the estimated excitation signal) for a given segment of speech can be determined by performing a linear prediction analysis of the speech frame (see, e.g. Makhoul [166], Markel and Gray [168], Vary *et al.* [286]). This procedure is commonly performed in two stages: for each speech segment the first $N_a + 1$ short-term auto-correlation coefficients are estimated, which are then transformed into the filter coefficients $\mathbf{a}$, for example, by the recursive Levinson–Durbin algorithm (Roberts and Mullis [228]).

6.2.2 BANDWIDTH EXTENSION ALGORITHM

Most adaptive bandwidth extension algorithms for speech are based on the source-filter model of the speech production process as described in the previous section. The estimation of the missing signal components is performed in a two-stage procedure, indirectly via the model of the source: in the first step the parameters of the wideband source model are estimated from the bandlimited speech signal. These parameters are then used in combination with the model itself to determine an estimate of the wideband speech. This approach is in general well suited for the extension both to high frequencies and to low frequencies.

Below, on the basis of the block diagram in Fig. 6.5, an overview of the principal structure and properties of bandwidth extension algorithms for speech shall be given. According to the structure of the source-filter model from Sec. 6.2.1, the bandwidth extension is performed separately for the excitation signal and for the spectral envelope of the speech signal (Cheng *et al.* [49], Carl [44], Iyengar *et al.*, see Chapter 8). Since these two constituents of the speech signal can be assumed to be mutually independent

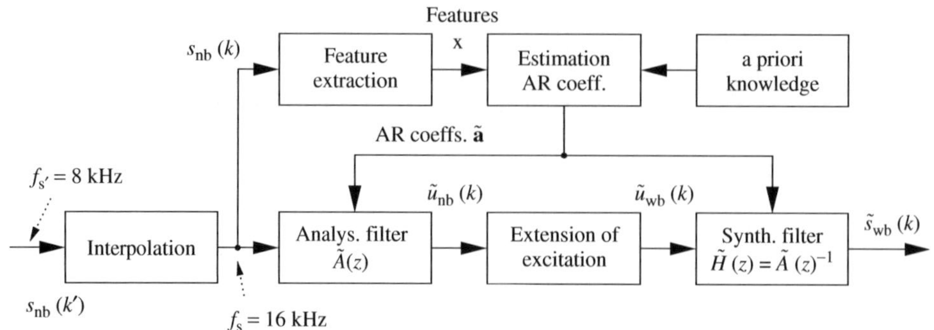

Figure 6.5 Signal flow of an algorithm for the bandwidth extension of speech signals (Jax *et al.* [129]). The final synthesis filter $\tilde{H}(z)$ of the algorithm and certain parts of the sub-system for the extension of the excitation signal reflect the source-filter model from Sec. 6.2.1

to a certain extent, the separate optimization of the two parts of the algorithm leads to an approximation of the global optimum.

The importances of the two sub-tasks are different. For the extension towards high frequencies, the principal problem is posed by the estimation of the wideband spectral envelope. This fact can be verified easily in listening experiments by applying the BWE algorithm utilizing knowledge of the original wideband spectral envelope: by modifying only the excitation signal in the extended frequency bands, the quality of the enhanced speech signal is only slightly inferior to the quality of the original wideband speech (see Sec. 6.3 or Carl [44]). Consequently, the sub-system for the estimation of the spectral envelope has to be designed with special diligence.

For low-frequency BWE, an additional important problem is the correct reconstruction of the pitch information. If the fundamental frequency and/or first overtones thereof are recovered incorrectly, the base-band and extended components will not be grouped to a single auditory stream, see Sec. 1.4.

A detailed description of the two parts of the algorithm concerned with the extension of the excitation signal and of the spectral envelope can be found in Secs. 6.3 and 6.4 ff. respectively.

Interpolation If the sampling rate of the input signal $s_{\text{nb}}(k')$ of the BWE algorithm is not sufficiently high to allow the representation of the extended speech signal, the first step in the BWE system consists of increasing the sampling rate via interpolation (e.g. Oetken and Schüßler [189], Crochiere and Rabiner [53]). In the example that is illustrated in Fig. 6.5, the narrowband input signal is represented with the typical sampling rate (of narrowband speech) of $f_{s'} = 8$ kHz. Hence, to allow the extension of high-frequency components up to a cut-off frequency of 7 kHz, the sampling rate has to be increased to $f_s = 16$ kHz. Note that by the interpolation the signal contents are not modified – the interpolated signal $s_{\text{nb}}(k)$ is still bandlimited in the same manner as the input signal $s_{\text{nb}}(k')$.

All of the subsequent modules are processed with the fixed sampling rate f_s, for example, $f_s = 16\,\text{kHz}$. Furthermore, the processing is mostly performed frame by frame with a frame length of about 20 ms. In the sequel, the frame index is denoted by m. Within each frame, the samples are indexed by the variable κ, with $0 \leq \kappa \leq N_\kappa - 1$, and N_κ being the number of samples per frame (i.e. $N_\kappa = 320$ if $f_s = 16$ kHz).

Estimation of the AR coefficients The actual bandwidth extension starts with the estimation of the coefficient set $\tilde{\mathbf{a}}$, representing the shape of the spectral envelope of the *wideband* speech signal. For this purpose, as much relevant information as possible shall be utilized from the available bandlimited speech. For each signal frame, a vector of features $\mathbf{x}$ of the input speech is calculated, providing the basis for the estimation. A pre-trained statistical model contributes the necessary a priori knowledge on the properties of the process of speech production. A detailed description of the statistical modelling and of different estimation procedures is given in the Secs. 6.4 to 6.9.

Analysis filter The estimated wideband filter coefficient set $\tilde{\mathbf{a}}$ is utilized in an FIR analysis filter $\tilde{A}(z)$, which is applied to the interpolated bandlimited input signal $s_{\text{nb}}(k)$:

$$\tilde{A}(z) = \sum_{i=0}^{N_a} \tilde{a}_i \, z^{-i}, \quad \text{and} \quad \tilde{u}_{\text{nb}}(k) = \sum_{i=0}^{N_a} \tilde{a}_i \, s_{\text{nb}}(k-i). \tag{6.3}$$

Because the analysis filter is the inverse of the corresponding auto-regressive vocal tract filter, the output $\tilde{u}_{\text{nb}}(k)$ of the analysis filter can be interpreted as an approximation of the excitation of the speech. It must be kept in mind, however, that this estimate is bandlimited in the same manner as the input signal of the BWE algorithm.

Extension of the excitation signal The next step in the BWE system consists of substituting the missing frequency components in the excitation signal. Depending on the desired quality of the extended excitation signal, as well as on the admissible complexity of this sub-system, the different parameters, σ, V, or F_0, of the source model can be considered to a greater or lesser extent for this purpose. Owing to the assumed spectral flatness of the excitation signal, and because of the fact that the human ear is quite insensitive to variations of the spectral fine structure at high frequencies, the extension can be realized in a very efficient manner. Different approaches for the extension of the excitation signal are described in Sec. 6.3.

In principle, an extension of low (e.g. below 300 Hz) as well as high components (above 3.4 kHz) of the excitation signal is obtainable. Therefore, the output signal $\tilde{u}_{\text{wb}}(k)$ of this block reflects the desired estimate of the wideband excitation signal.

During the extension of the excitation, it shall be guaranteed that the base-band components of $\tilde{u}_{\text{nb}}(k)$ are not modified – then, the input speech $s_{\text{nb}}(k)$ will be contained transparently in the output signal $\tilde{s}_{\text{wb}}(k)$ of the BWE system.

Synthesis filter So far, both an estimate $\tilde{u}_{\text{wb}}(k)$ of the wideband excitation signal and an approximation $\tilde{\mathbf{a}}$ of the coefficient set of the AR filter representing the spectral envelope of the wideband speech signal have been determined. To finalize the estimate of the wideband speech signal, the two quantities are combined by means of the synthesis filter

$$\tilde{H}(z) = \left(\sum_{i=0}^{N_a} \tilde{a}_i \, z^{-i} \right)^{-1} = \frac{1}{\tilde{A}(z)}. \tag{6.4}$$

Considering the normalization ($\tilde{a}_0 = 1$) of the AR coefficients, the output signal of the bandwidth extension system is computed by

$$\tilde{s}_{\mathrm{wb}}(k) = \tilde{u}_{\mathrm{wb}}(k) - \sum_{i=1}^{N_a} \tilde{a}_i \, \tilde{s}_{\mathrm{wb}}(k-i). \tag{6.5}$$

Note that the transfer function of the synthesis filter is inverse to the transfer function of the employed analysis filter for each signal frame, because the identical coefficient set $\tilde{\mathbf{a}}$ is utilized in both filters.

6.2.3 ALTERNATIVE STRUCTURES

A characteristic property of the algorithm from Sec. 6.2.2 is the fact that both the analysis filter and the synthesis filter are operated at the same sampling rate and, moreover, with the identical coefficient set $\tilde{\mathbf{a}}$. The two filters are exactly mutually inverse. This feature discriminates the algorithm from alternative approaches, which, in a similar manner, perform a separate extension of the excitation signal and the spectral envelope (e.g. Carl [44, 45], Avenando *et al.* [23], Nakatoh *et al.* [183], Enbom and Kleijn [64], Epps and Holmes [66], Miet *et al.* [174], Park and Kim [200], Valin and Lefebvre [279], Fuemmeler *et al.* [78]). In the latter algorithms, narrowband coefficients $\tilde{\mathbf{a}}'_{\mathrm{nb}}$ for the analysis filter are either determined via an LP analysis or taken from a codebook. For the synthesis filter, on the other hand, a different wideband coefficient set $\tilde{\mathbf{a}}_{\mathrm{wb}}$ is utilized, which is estimated or taken from a different codebook (the so-called shadow-codebookcodebook,shadow-codebook). In Fig. 6.6, an example of the structure of such an alternative BWE algorithm is shown. Note that because with the structure from Fig. 6.6 the LP analysis and synthesis filters are not exactly mutually inverse in the base-band, it is important to apply gain correction of the extended speech components.

The two basic approaches from Figs. 6.5 and 6.6 have distinct properties that shall be discussed in the following text.

Transparency in the base-band and mixing An important requirement on algorithms for the bandwidth extension of speech signals is the transparency of the system with respect to the bandlimited input signal. As the input signal provides the best possible speech quality within its limited frequency range, it shall be contained unmodified in the output of the BWE system.

If different coefficient sets (as, e.g., $\tilde{\mathbf{a}}'_{\mathrm{nb}}$ and $\tilde{\mathbf{a}}_{\mathrm{wb}}$ in Fig. 6.6) are used in the analysis and synthesis filter, it is in general necessary, in order to ensure the transparency of the algorithm with respect to the base-band, to mix the original input signal of the BWE algorithm with the band-stop-filtered extended speech to calculate the output signal of

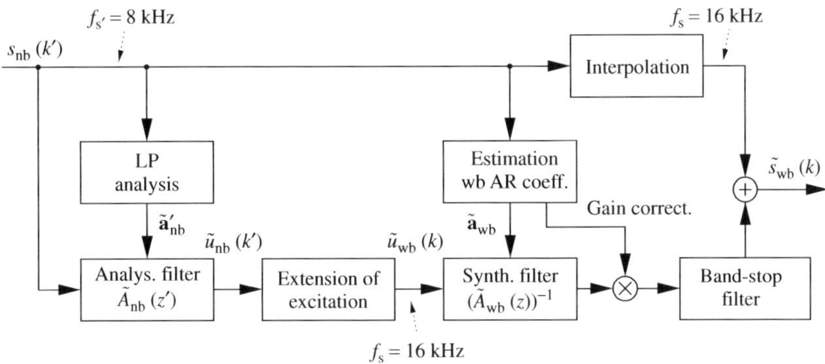

Figure 6.6 Signal flow of an alternative algorithm for bandwidth extension of speech signals (e.g. Park and Kim [200], Fuemmeler *et al.* [78]). The block 'extension of excitation' additionally performs an interpolation of the excitation signal

the BWE system (see Fig. 6.6). Owing to discrepancies in the transfer functions of the analysis and synthesis filter, the base-band speech components are distorted in the lower signal path of Fig. 6.6, and additionally certain artefacts are produced. Further, the relative power of the extended speech is generally altered with this structure. Therefore, prior to mixing the extended signal with the input speech, a correction factor has to be applied to the extended signal (Carl [44], Park and Kim [200], Nilsson and Kleijn [187], Fuemmeler *et al.* [78]). The proper correction factors have to be estimated in addition to the wideband spectral envelope of the speech.

Such measures are not needed in the algorithm from Sec. 6.2.2 because synthesis and analysis filters are mutually inverse: the transparency of the BWE system gets independent of the extension of the spectral envelope. Provided that during the extension of the excitation the narrowband components of the excitation signal are not modified, errors in the estimated spectral envelope of the analysis filter (in the sense of an optimal LPC prediction filtering) are completely compensated by the inverse synthesis filter *within the base-band*. The power of the signal is not modified because of the filtering. However, to achieve transparency of the complete BWE system, the sub-system that is responsible for the extension of the excitation is now required to be transparent with respect to the base-band components (cf. Sec. 6.3).

Impact of estimation errors of the spectral envelope If the estimated coefficient set $\tilde{\mathbf{a}}$, representing the spectral envelope of the wideband speech, is inaccurate, the impact on the quality of the extended frequency bands of the output signal may be two-fold. It is quite obvious that errors of the frequency response of the synthesis filter within the extended frequency bands directly effect the quality of the extended bands: the estimated excitation signal will be shaped by the synthesis filter according to the erroneous spectral envelope.

In the algorithm from Sec. 6.2.2, errors within the base-band of the frequency response of the analysis filter can further impair the quality of the extended bands in an indirect manner. Owing to errors of $\tilde{\mathbf{a}}$, the estimate $\tilde{u}_{nb}(k)$ of the bandlimited excitation signal, that is determined by the erroneous analysis filter, is not spectrally flat as assumed in the source-filter model. During the subsequent extension of the excitation signal, these errors within $\tilde{u}_{\mathrm{nb}}(k)$ will propagate into the extended frequency bands of the estimated wideband excitation $\tilde{u}_{\mathrm{wb}}(k)$. Thus, although base-band transparency is guaranteed by the algorithm, errors in the base-band of the estimated spectral envelope do, nevertheless, effect the extended speech signal $\tilde{s}_{\mathrm{wb}}(k)$. If, for example, the extension of the excitation is performed by spectral translation or folding (see Sec. 6.3.3), the errors of the estimated spectral envelope within the base-band and within the extended band are effectively added up. In Sec. 6.4.1.1, a method that prevents errors in the base-band of the estimated spectral envelope will be described.

Algorithmic delay In real-time speech communication, it is generally desired to keep the signal delay as low as possible. Nevertheless, it is important to apply proper delay compensation in any parallel signal path of a BWE system to ensure that the extended frequency components are psychoacoustically grouped together with the base-band speech (cf. Sec. 1.4).

There are several potential sources of algorithmic delay in the bandwidth extension systems of Figs. 6.5 and 6.6. Firstly, there is always an algorithmic delay due to the frame-based processing of the algorithm: all N_κ samples have to be available before the processing of a frame can start. If the bandwidth extension algorithm is positioned behind a speech decoder, however, in general this source of delay is not relevant since most speech codecs also operate on a per-frame basis. The bandwidth extension system can be merged with the speech decoder.

If the input signal of the BWE algorithm has to be interpolated before applying the analysis filter, an additional delay will be caused by the interpolation low-pass filter. The design criteria of transition bandwidth and stopband attenuation for this filter, however, are not as stringent as for an isolated interpolation system. The high-frequency part of the speech signal will be approximated anyhow by the BWE system. Aliasing errors from non-optimal interpolation may be masked by subsequently added extended frequency components. Consequently, the order and delay of the interpolation filter can be kept rather low.

A further delay of the speech signal might be introduced if any filters are utilized in the sub-system for the extension of the excitation signal to guarantee the base-band transparency of that sub-system (compare Sec. 6.3). Note that in this case also the adjustment of the synthesis filter coefficient set $\tilde{\mathbf{a}}$ has to be delayed accordingly. The processing of the speech signal by the analysis and synthesis filters does not produce any delay of the signal (although both filters are causal) because the two filters are both minimum-phase filters and mutually inverse.

Finally, a delay of the speech signal is necessary if a look-ahead shall be utilized in the estimation of the wideband spectral envelope (compare Sec. 6.9). In this case, the input signal $s_{\mathrm{nb}}(k)$ of the analysis filter has to be delayed in accordance with the implicit delay of the estimated AR coefficients $\tilde{\mathbf{a}}$.

6.3 EXTENSION OF THE EXCITATION SIGNAL

In this section, the sub-system of the BWE algorithm that is responsible for the extension of the excitation signal of the speech (compare Fig. 6.5) is treated. This sub-system gets the bandlimited estimate $\tilde{u}_{\mathrm{nb}}(k)$ of the excitation as its input. The output signal $\tilde{u}_{\mathrm{wb}}(k)$ on the other hand serves as the input to the final synthesis filter of the BWE system and reflects an estimate of the wideband excitation signal. The task of the extension of the excitation signal is the recovery of the *spectral fine structure* of the speech signal.

Potential algorithms that can be employed for the extension of the excitation signal benefit both from the quite simple structure of the excitation signal according to the source-filter model of speech production (compare Sec. 6.2.1) as well as from insensitivities of the human auditory system with regard to certain distortions of the spectral fine structure at high respectively low frequencies. In this chapter, several algorithms from literature are described and evaluated. The different methods for the extension of the excitation either reuse the signal components of the estimated bandlimited excitation signal $\tilde{u}_{\mathrm{nb}}(k)$, for example, by spectral translation (Sec. 6.3.3) or pitch scaling (Sec. 6.3.4), or they generate new components via explicit signal generation (Sec. 6.3.1) or by non-linear distortion (Sec. 6.3.2).

An important requirement that has to be demanded for the estimated wideband excitation signal $\tilde{u}_{\mathrm{wb}}(k)$ is that it transparently contains the estimated bandlimited excitation signal $\tilde{u}_{\mathrm{nb}}(k)$ – in this case, the complete BWE system becomes transparent with respect to the narrowband input speech (see Sec. 6.2.3). To guarantee this transparency, it is necessary for some of the following methods to mix the original bandlimited excitation $\tilde{u}_{\mathrm{nb}}(k)$ with an appropriately high-pass respectively low-pass-filtered version of the extended excitation.

Because the vast majority of publications on the topic of bandwidth extension of speech signals to date is concerned primarily with the extension of the spectral envelope of the speech, most of the known methods for the extension of the excitation signal have been adopted from the field of speech coding. Especially, techniques from so-called *base-band codecs* are used. In these speech codecs, only a part of the frequency components of the LPC residual signal is coded and transmitted while the remaining components are recovered at the receiving site via *high-frequency regeneration* (HFR, e.g. Makhoul and Berouti [167], Kroon *et al.* [150], Taori *et al.* [265], McCree *et al.* [171]). A prominent representative of this category of speech codecs is the GSM full-rate codec [69], Vary *et al.* [285].

6.3.1 EXPLICIT SIGNAL GENERATION

The most straightforward solution to extend the excitation signal consists of the explicit generation of the missing signal components. Basically, by this approach the excitation part of the source model from Sec. 6.2.1 is implemented directly. The method therefore strongly depends on estimates of the source parameters, that is, on estimates of the voiced/unvoiced state V, the gain factor σ, and the fundamental frequency F_0 of the speech (compare Fig. 6.7).

According to the admissible computational complexity and to the desired accuracy of the simulated source model, there are several prevalent approaches:

- *Noise only*: The missing components of the excitation signal are produced by a noise generator and a subsequent band-stop filter. Further, the gain $\tilde{\sigma}$ of the noise signal $\tilde{u}_{\mathrm{mb}}(k)$ has to be adapted to match the gain of the base-band excitation signal $\tilde{u}_{\mathrm{nb}}(k)$. The procedure is reflected by the block diagram of Fig. 6.7 if the voiced/unvoiced

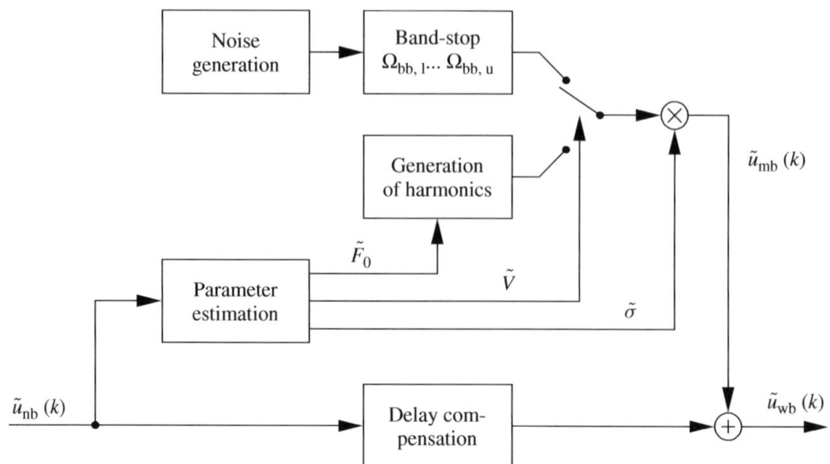

Figure 6.7 Extension of the excitation signal via explicit signal generation

switch (parameter $\tilde{V}$) is invariably set to the upper (unvoiced) position. This approach can be motivated by the fact that the main contribution of high-frequency components of the speech is during unvoiced sounds.

In fact, the addition of noise components yields very good results if the extended frequency band is rather narrow. For example, this approach has been used successfully in several wideband speech codecs for the coding of the frequency components above 6 or 6.4 kHz and up to 7 kHz, for example, Paulus [206], Schnitzler [238], Erdmann *et al.* [67], 3GPP TS 26.190 [3]. Regarding the quite wide missing frequency band above 3.4 kHz in the artificial bandwidth extension of narrowband telephone speech, however, both the extended excitation signal as well as the extended speech signal (after the AR synthesis filter) sound quite noisy. Especially during voiced speech segments, the noisy signal components added at high frequencies are then well audible and annoying.

- *Noise and/or sinusoids*: The algorithm can be refined by distinguishing between voiced and unvoiced segments of the speech. During unvoiced phases a noise generator is utilized, and during voiced sounds a tonal excitation is produced in the missing frequency band. The techniques for sine-wave generation resemble those from *sinusoidal* or *harmonic coding* (Griffin and Lim [100], Carl [44], McAulay and Quatieri [170]). The voiced/unvoiced switching can either be 'hard', allowing either a noisy or a tonal extended excitation $\tilde{u}_{mb}(k)$ at a time, or the kind of excitation is specified individually for different frequency bands. In the latter case, the approach corresponds to the *harmonic plus noise model* (HNM) from Griffin and Lim [100], Abrantes *et al.* [13], Stylianou [258]. In the bandwidth extension system, the newly generated signal components $\tilde{u}_{mb}(k)$ are mixed with the original bandlimited excitation $\tilde{u}_{nb}(k)$.

  In informal listening experiments, it can be found that a very good estimation of the fundamental speech frequency F_0 is crucial for the generation of tonal speech components: if the estimate $\tilde{F}_0$ is inaccurate, the objectionable impression is produced that an interfering simultaneous speaker with a slightly different pitch frequency is added to the speech signal. This problem can be circumvented if the excitation signal is substituted completely, that is, if also the base-band components are regenerated (Chan and Hui [47, 48]). This, however, may also introduce artefacts in the base-band frequency range of $\tilde{s}_{\mathrm{wb}}(k)$.

6.3.2 NON-LINEAR PROCESSING

The first approach to the artificial bandwidth extension of speech signals to our knowledge was the application of non-linear distortions to the narrowband speech $s_{\mathrm{nb}}(k)$ as proposed by Schmidt [236]. The same basic method can also be used to extend the excitation signal of the speech: an estimate of the wideband excitation signal is determined by applying a non-linear function $g(\cdot)$ to the bandlimited excitation $\tilde{u}_{\mathrm{nb}}(k)$

$$\tilde{u}_{\mathrm{nl}}(k) = g\left(\tilde{u}_{\mathrm{nb}}(k)\right). \tag{6.6}$$

Owing to the non-linear processing, (harmonic) distortions that reflect the desired new signal components in the missing frequency bands are created. The signal $\tilde{u}_{\mathrm{nl}}(k)$ denotes a generalized extended excitation signal here, that is, it can correspond to the high- or low-frequency band of the speech signal, respectively.

There is an unlimited number of possible non-linear functions $g(\cdot)$, and it is quite difficult to find that particular function that yields the best results in the bandwidth extension application. Non-linear functions have been used in bandwidth extension algorithms mainly for the generation of low-frequency speech components to date (Schmidt [236], Croll [54], Patrick *et al.* [201], Valin and Lefebvre [279], Kornagel [147]). The utilized non-linearities $g(\cdot)$ have been, for example, quadratic, cubic, or saturation functions, and half-wave respectively full-wave rectification. Note that here in contrast to Chapters 2 and 3 the non-linearities are applied to a signal containing more than only one harmonic. Unfortunately, the effects of the non-linear function $g(\cdot)$ are very difficult to predict, as any modification of the input signal $\tilde{u}_{\mathrm{nb}}(k)$ (e.g. scaling, the addition of signal components, application of phase distortions, or a simple addition of a constant value) can significantly effect the properties of the distorted signal $\tilde{u}_{\mathrm{nl}}(k)$. In general, either the narrowband signal $\tilde{u}_{\mathrm{nb}}(k)$ has to be pre-processed (normalized) prior to applying the non-linearity, or the distorted signal $\tilde{u}_{\mathrm{nl}}(k)$ has to be post-processed.

The possibility of a post-processing of the distorted signal is illustrated in Fig. 6.8. The shape of the envelope as well as the gain of the distorted signal $\tilde{u}_{\mathrm{nl}}(k)$ are corrected (and adapted to the base-band excitation) to match the assumption of a spectrally flat wideband excitation signal from the source-filter model of Sec. 6.2.1: first, an LP analysis of the signal is performed. The corresponding adaptive LP analysis filter is applied, yielding a whitening of the signal. Afterwards, the gain of the signal is adjusted to match the gain of the base-band excitation $\tilde{u}_{\mathrm{nb}}(k)$. As the non-linear distortion effects the whole frequency range of the distorted signal, measures have to be taken to guarantee base-band transparency. The base-band components have to be removed from $\tilde{u}_{\mathrm{eb}}(k)$ by band-stop filtering, or by high-pass respectively low-pass filtering if only an extension towards high respectively low frequencies is desired. The filtered signal $\tilde{u}_{\mathrm{mb}}(k)$ is mixed with the narrowband excitation signal to determine the estimated wideband excitation signal $\tilde{u}_{\mathrm{wb}}(k)$.

In experiments, very good results were achieved with a simple quadratic non-linear distortion, that is, with the function $\tilde{u}_{\mathrm{nl}}(k) = (\tilde{u}_{\mathrm{nb}}(k))^2$. By this non-linearity, both

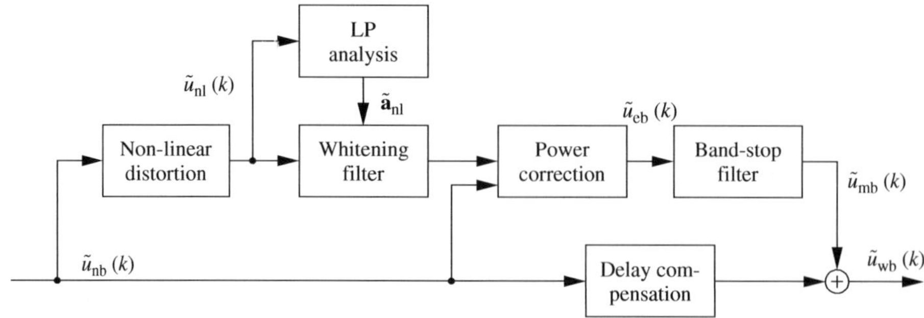

Figure 6.8 Extension of the excitation signal by non-linear distortion. To compensate for hardly controllable adverse effects of the non-linearity, sophisticated post-processing is required. The algorithmic delay of the band-stop filter has to be compensated for in the path of the base-band signal

low-frequency and high-frequency components can be generated. The upper cut-off frequency of the distorted signal $\tilde{u}_{\mathrm{nl}}(k)$ is twice the upper band limit of the narrowband excitation $\tilde{u}_{\mathrm{nb}}(k)$. Since only *harmonic* distortions are produced by the squaring operation, the tonal components in the enhanced excitation signal $\tilde{u}_{\mathrm{wb}}(k)$ always match the harmonic structure of the bandlimited excitation $\tilde{u}_{\mathrm{nb}}(k)$ during voiced sounds.

For low-frequency bandwidth extension, simple full-wave rectification has proven successful. It has the advantage that the signal level is not altered (a full-wave rectifier is a homogeneous system, see Sec. 1.1.1) such that it can easily be implemented.

6.3.3 MODULATION IN THE TIME DOMAIN

In this section, we consider algorithms that are based on a modulation of the bandlimited excitation signal $\tilde{u}_{\mathrm{nb}}(k)$ (Carl [44], Fuemmeler *et al.* [77], Kornagel [147], Jax and Vary [130]). Because a modulation in the time domain corresponds to a translation in the frequency domain, the input signal is virtually reused by 'shifting' it into the missing frequency band(s). Several well-known methods for the extension of the excitation signal, such as spectral folding or spectral translation, are special cases of the more general modulation concept (Carl [44]).

The straightforward implementation of a spectral translation would be based on the analytical signal of the bandlimited excitation. The product of the analytic signal with a complex-valued modulation function directly yields the desired extended signal. However, the determination of the analytic signal by Hilbert transformation either in the time- or frequency domain is quite complex (e.g. Schuessler [242], Marple [169]). In general, equivalent results as with the analytic signal can also be achieved by modulation of the input signal with a *real-valued* modulation function. However, in this case the shifted spectra cause mutual overlappings that have to be removed by subsequent frequency-selective filtering as illustrated in Fig. 6.9.

In the following, the modulation shall be performed using a real-valued cosine function:

$$\tilde{u}_{\mathrm{M}}(k) = \tilde{u}_{\mathrm{nb}}(k) \cdot \zeta \, \cos(\Omega_{\mathrm{M}}\, k). \tag{6.7}$$

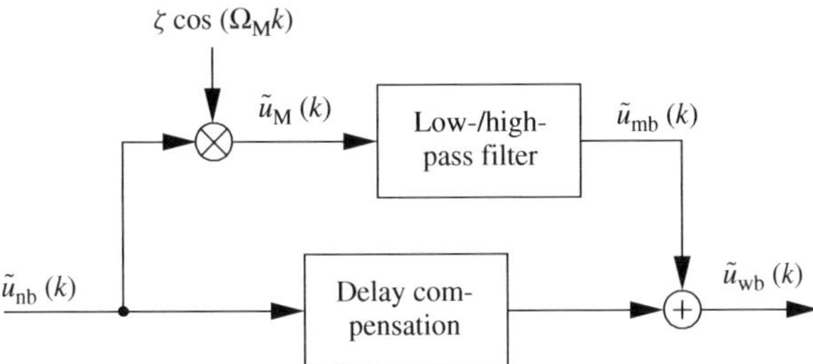

Figure 6.9 Extension of the excitation signal by modulation. The algorithmic delay of the high-pass filter has to be compensated for in the path of the base-band signal $\tilde{u}_{\mathrm{nb}}(k)$

Depending on the particular modulation frequency Ω_{M}, the fixed scalar factor ζ in Eqn. 6.7 has to be chosen from $\zeta \in \{1, 2\}$ to obtain the correct power of the extended excitation signal. The argument of the cosine function consists of the phase function $\Omega_{\mathrm{M}}\, k$, which is linear in time if the modulation frequency Ω_{M} is fixed. By the multiplication of the input signal with the cosine signal in time domain, two shifted copies of the original spectrum $\tilde{U}_{\mathrm{nb}}(e^{j\Omega})$ are generated

$$\tilde{U}_{\mathrm{M}}(e^{j\Omega}) = \frac{\zeta}{2}(\tilde{U}_{\mathrm{nb}}(\mathrm{e}^{j(\Omega-\Omega_{\mathrm{M}})}) + \tilde{U}_{\mathrm{nb}}(\mathrm{e}^{j(\Omega+\Omega_{\mathrm{M}})})). \tag{6.8}$$

The two shifted spectra may overlap in different frequency ranges. Whether such overlappings occur, and to which extent, depends on the lower- and upper-frequency limits $\Omega_{\mathrm{bb,l}}$ and $\Omega_{\mathrm{bb,u}}$ of the bandlimited speech signal as well as on the modulation frequency Ω_{M}.

The modulation approach to the extension of the excitation signal is especially suited for the extension of high frequencies, because the frequency range to be extended above the upper band limit $\Omega_{\mathrm{bb,u}}$ of the input speech is – in contrast to the lower extended frequency range – in general larger than the bandwidth of the base-band. This property has consequences for the design criteria of the high-pass respectively low-pass filter from Fig. 6.9: if the bandwidth of the shifted spectrum is greater than the width of the extended frequency range, the implemented filter shall have very steep slopes.

Owing to the importance of the fundamental frequency of the speech in the low-frequency excitation, only a pitch-adaptive approach is applicable to the extension of the missing low-frequency band.

6.3.3.1 Spectral Folding

The method of spectral folding reflects a special case of the modulation method that is exclusively suitable for the extension of high-frequency components. The modulation frequency is specified to be equal to the Nyquist frequency $\Omega_{\mathrm{M}} = \pi$ (corresponding to $f_{\mathrm{M}} = 8\,\mathrm{kHz}$ for the (interpolated) excitation signal $\tilde{u}_{\mathrm{nb}}(k)$). Thereby, the modulation function is simplified to a sequence of alternating signs $\cos(\Omega_{\mathrm{M}}\, k) = (-1)^k$. The two shifted spectra are superimposed constructively such that $\zeta = 1$. By the modulation, a 'folded' version of the input signal is generated, the spectrum of which is mirrored at $\Omega = \pi/2$, that is, the half of the Nyquist frequency.

Because the folded spectrum is bandlimited in the same way as the spectrum of the input signal, the high-pass filter from Fig. 6.9 can be omitted. The combination of the folded signal with the input signal yields the most efficient method for an extension of the excitation signal towards high frequencies:

$$\begin{aligned}\tilde{u}_{\mathrm{wb}}(k) &= \tilde{u}_{\mathrm{nb}}(k) + \tilde{u}_{\mathrm{nb}}(k)\ \cos(\Omega_{\mathrm{M}}\, k) \\ &= \tilde{u}_{\mathrm{nb}}(k)\,(1 + (-1)^k). \end{aligned} \tag{6.9}$$

The base-band of the input signal is preserved transparently if the upper band limit of the base-band speech is $\Omega_{\mathrm{bb,u}} < \pi/2$. Because of its efficiency, the method of spectral folding is used very frequently in bandwidth extension algorithms. Similar approaches can also be found in base-band speech codecs such as the GSM full-rate speech codec [69].

By the use of the spectral folding method, some systematic errors are produced, as can be observed in Fig. 6.10 (a). Since the fundamental frequency of the speech is not considered, the reproduced discrete structure in the extended frequency band is inconsistent during voiced sounds – the discrete frequency components are not correctly placed at integer multiples of the fundamental frequency, resulting in a metallic sound or 'ringing' of the enhanced speech $\tilde{s}_{\mathrm{wb}}(k)$. Further, the position of the extended frequency band is invariably determined by the sampling rate and the band limits of the input signal. In general, a spectral gap is created in the frequency range $\Omega_{\mathrm{bb,u}} < \Omega < \pi - \Omega_{\mathrm{bb,u}}$. For typical telephone speech (300 Hz to 3.4 kHz), for example, there will be a large gap between 3.4 and 4.6 kHz. In addition, the upper band limit of the folded signal is determined by the lower band limit of the input speech. For telephone speech, the upper limit of the extended speech is at 7.7 kHz. Serious artefacts are produced if there is a DC component in the input signal: the folded DC component yields a strong stationary sinusoid at the Nyquist frequency.

6.3.3.2 Spectral Translation

In this section, the modulation shall be performed with a *fixed* modulation frequency as well. Now, the modulation frequency Ω_{M} is specified by the bandwidth of the bandlimited input speech (lower band limit $\Omega_{\mathrm{bb,l}}$ and upper band limit $\Omega_{\mathrm{bb,u}}$)

$$\Omega_{\mathrm{M}} = \Omega_{\mathrm{bb,u}} - \Omega_{\mathrm{bb,l}}. \tag{6.10}$$

Owing to this setting of Ω_{M} the spectrum $\tilde{U}_{\mathrm{nb}}(\mathrm{e}^{j(\Omega-\Omega_{\mathrm{M}})})$, shifted towards high frequencies (see Eqn. 6.8), starts in continuation of the base-band spectrum $\tilde{U}_{\mathrm{nb}}(e^{j\Omega})$, that is, there is no gap in the spectrum of the extended speech. The upper band limit of the extended speech depends on the band limits of the base-band signal. It is defined by the frequency $\Omega_{\mathrm{bb,u}} + \Omega_{\mathrm{M}} = 2\,\Omega_{\mathrm{bb,u}} - \Omega_{\mathrm{bb,l}}$ (corresponding to 6.5 kHz for telephone speech).

Prior to the mixing of the extended excitation with the base-band excitation signal, the frequency range in which the downwardly shifted spectrum is situated shall be removed by high-pass filtering (see Fig. 6.9). The cut-off frequency of the employed high-pass filter shall be equal to the upper band limit $\Omega_{\mathrm{bb,u}}$ of the base-band which is, due to the

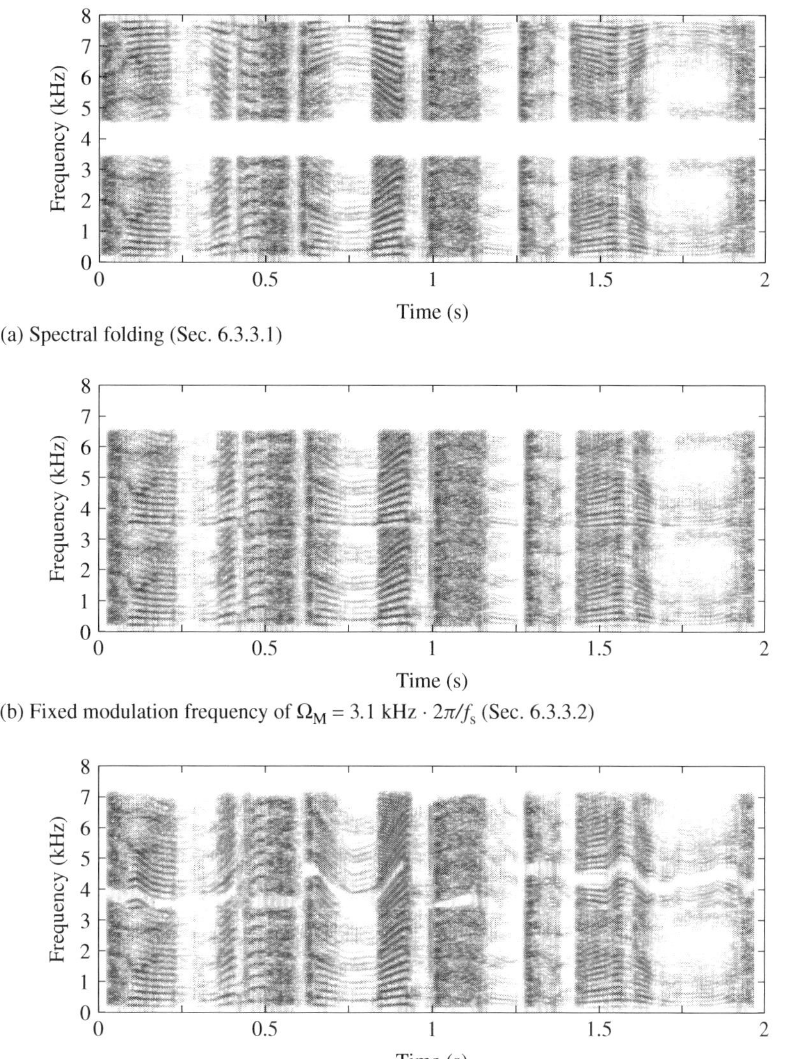

(a) Spectral folding (Sec. 6.3.3.1)

(b) Fixed modulation frequency of $\Omega_M = 3.1$ kHz $\cdot\, 2\pi/f_s$ (Sec. 6.3.3.2)

(c) Pitch-adaptive modulation and low-pass filtering at 7 kHz (Sec. 6.3.3.3)

Figure 6.10 Spectrograms of the excitation signal $\tilde{u}_{\text{wb}}(k)$, extended via modulation-based techniques. Black regions reflect a large short-term power spectrum. The sentence 'to administer medicine to animals' is spoken by a female voice. Note that there is a pitch estimation error in the pitch-adaptive approach (c) after about 1.5 s

particular choice of the modulation frequency, also the lower limit of the spectrum shifted towards high frequencies. The desired stopband attenuation of the filter shall be obtained for frequencies below $\max\left(\Omega_{\text{bb,l}}, \Omega_{\text{bb,u}} - 2\,\Omega_{\text{bb,l}}\right)$, for example, below 2.8 kHz if typical telephone speech is the input of the BWE system.

6.3.3.3 Pitch-adaptive Modulation

The modulation schemes with fixed modulation frequencies that have been described so far share the disadvantage that the discrete spectral structure of the extended excitation signal $\tilde{u}_{\mathrm{wb}}(k)$ during voiced sounds is inconsistent. To achieve a better performance, a further possibility to control the modulation frequency Ω_{M} that takes the pitch frequency Ω_{p} of the current speech frame into account shall be studied. The method has been developed independently by Fuemmeler and Hardie [77], Kornagel [147], and Jax and Vary [130], all in 2001. The idea to utilize information on the fundamental frequency of the speech for the extension of an excitation signal was first proposed by Makhoul and Berouti [167].

The basis of the *pitch-adaptive modulation* (PAM) in the time domain is an estimate $\tilde{\Omega}_{\mathrm{p}} = 2\pi\,\tilde{F}_0/f_{\mathrm{s}}$ of the fundamental frequency in the currently processed frame of the speech signal (e.g. Hess [109]). The modulation frequency Ω_{M} is then adjusted in dependence of the estimate $\tilde{\Omega}_{\mathrm{p}}$, such that the shifted tonal components of the base-band excitation correspond to proper harmonics of the fundamental frequency within the extended frequency band

$$\Omega_{\mathrm{M}} = n_{\mathrm{M}}\,\tilde{\Omega}_{\mathrm{p}} \qquad \text{with } n_{\mathrm{M}} \in \mathbb{N}^{+} \text{ and } \Omega_{\mathrm{M,l}} \leq \Omega_{\mathrm{M}} \leq \Omega_{\mathrm{M,u}}. \tag{6.11}$$

In this way, for example, the qth harmonic of the fundamental frequency of the speech is shifted to the position of the $(q + n_{\mathrm{M}})$th harmonic. The integer-valued factor n_{M} is an adjustable parameter that has to be specified for each signal frame depending on the estimated fundamental frequency such that the resulting modulation frequency is between $\Omega_{\mathrm{M,l}}$ and $\Omega_{\mathrm{M,u}}$. By the limitation of the range of values of Ω_{M}, it shall be prevented that the bandwidth of the extended speech signal fluctuates strongly because of the variations of the fundamental frequency of the speech. The adaptive calculation of $n_{\mathrm{M}}(m)$ can, for example, be performed by the following method

$$n_{\mathrm{M}}(m) = \begin{cases} \left\lceil \dfrac{\Omega_{\mathrm{M,l}}}{\tilde{\Omega}_{\mathrm{p}}(m)} \right\rceil, & \text{if } n_{\mathrm{M}}(m-1)\,\tilde{\Omega}_{\mathrm{p}}(m) < \Omega_{\mathrm{M,l}} \\ n_{\mathrm{M}}(m-1), & \text{if } \Omega_{\mathrm{M,l}} \leq n_{\mathrm{M}}(m-1)\,\tilde{\Omega}_{\mathrm{p}}(m) \leq \Omega_{\mathrm{M,u}} \\ \left\lfloor \dfrac{\Omega_{\mathrm{M,u}}}{\tilde{\Omega}_{\mathrm{p}}(m)} \right\rfloor, & \text{if } n_{\mathrm{M}}(m-1)\,\tilde{\Omega}_{\mathrm{p}}(m) > \Omega_{\mathrm{M,u}}. \end{cases} \tag{6.12}$$

The basic principle of Eqn. 6.12 is to keep the factor $n_{\mathrm{M}}(m-1)$ that has been used in the preceding frame, if possible, thereby minimizing the number of switchings. If the reuse of the factor $n_{\mathrm{M}}(m)$ would lead to an under- or overshooting of the minimum or maximum modulation frequencies $\Omega_{\mathrm{M,l}}$ and $\Omega_{\mathrm{M,u}}$, respectively, the value of $n_{\mathrm{M}}(m)$ is corrected such that the new modulation frequency is just within the admissible range. The described procedure to control $n_{\mathrm{M}}(m)$ implies that the difference $\Omega_{\mathrm{M,u}} - \Omega_{\mathrm{M,l}}$ is greater than the maximum possible fundamental speech frequency such that there exists a valid factor n_{M}, fulfilling the requirements from Eqn. 6.11, for each potential estimate $\tilde{\Omega}_{\mathrm{p}}$.

If the input signal of the bandwidth extension system has the typical telephone bandwidth (300 Hz to 3.4 kHz), the minimum modulation frequency should be specified by the bandwidth of the input speech $\Omega_{\mathrm{M,l}} = 3.1\,\mathrm{kHz} \cdot 2\pi/f_{\mathrm{s}}$. The upper limit $\Omega_{\mathrm{M,u}}$ of the modulation frequency can, for example, be adjusted to $\Omega_{\mathrm{M,u}} = 4.6\,\mathrm{kHz} \cdot 2\pi/f_\mathrm{s}$ such that the maximum upper band limit of the extended speech is about 7 kHz. In the example of Fig. 6.10 (c), the maximum modulation frequency is set to $\Omega_{\mathrm{M,u}} = 3.6\,\mathrm{kHz} \cdot 2\pi/f_\mathrm{s}$, and the variations of the upper band limit of the extended excitation signal $\tilde{u}_{\mathrm{wb}}(k)$ are suppressed by low-pass filtering the modulated signal $\tilde{u}_{\mathrm{M}}(k)$ with a cut-off frequency of 7 kHz.

Since an absolutely accurate estimate of the fundamental frequency of the speech cannot by expected from any realizable pitch estimation algorithm, we have evaluated the impacts of typical estimation errors on the performance of the pitch-adaptive modulation method. A quite typical error of pitch estimation algorithms is *pitch doubling*, that is, $\tilde{\Omega}_\mathrm{p} = 2\,\Omega_\mathrm{p}$. Pitch-doubling errors, however, do not have any effect on the results of the pitch-adaptive modulation approach because the resulting modulation frequency Ω_M is again an integer multiple of the true fundamental frequency Ω_p. In the case of relatively small deviations of the estimated pitch frequency $\tilde{\Omega}_\mathrm{p}$, on the other hand, the PAM algorithm exhibits a quite strong susceptibility because any such error is significantly increased by the multiplication with the large factor n_M. Consequently, a very accurate pitch estimation algorithm is needed for the estimation of $\tilde{\Omega}_\mathrm{p}$. Otherwise, the positions of discrete tonal components in the extended frequency bands will be inconsistent, and the performances of the PAM algorithm and the fixed spectral translation (Sec. 6.3.3.2) will be to a certain extent alike. If, on the other hand, a sufficiently accurate pitch estimation algorithm is used with the pitch-adaptive modulation approach, the resulting speech signal $\tilde{s}_{\mathrm{wb}}(k)$ will sound more natural in comparison. The artefacts (metallic sound, 'ringing') that are produced by mis-aligned harmonics are reduced noticeably.

6.3.4 PITCH SCALING

Finally, a new method for the extension of the excitation signal towards high frequencies that is based on frequency scaling of the original bandlimited speech signal shall be described. Figure 6.11 illustrates the concept of the approach by a block diagram. The application of the method for an exemplary speech signal is shown in the spectrograms in Fig. 6.12: first, by doubling the pitch frequency (pitch doubling), a version $\tilde{u}_{\mathrm{eb}}(k)$ of the excitation signal is produced, which has a doubled upper band limit in comparison to the bandlimited excitation signal $\tilde{u}_{\mathrm{nb}}(k)$. Comparing the spectrograms of the narrowband excitation $\tilde{u}_{\mathrm{nb}}(k)$ from Fig. 6.12 (a) and of the signal $\tilde{u}_{\mathrm{eb}}(k)$ after pitch

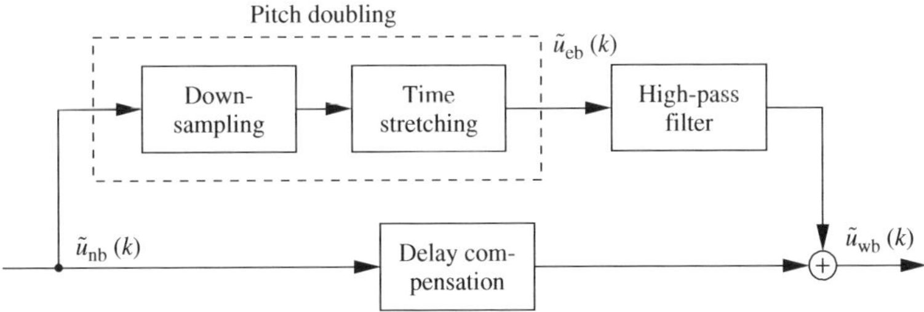

Figure 6.11 Extension of the excitation signal via pitch scaling. The pitch doubling is realized by a downsampling by a factor of 2 and subsequent time stretching

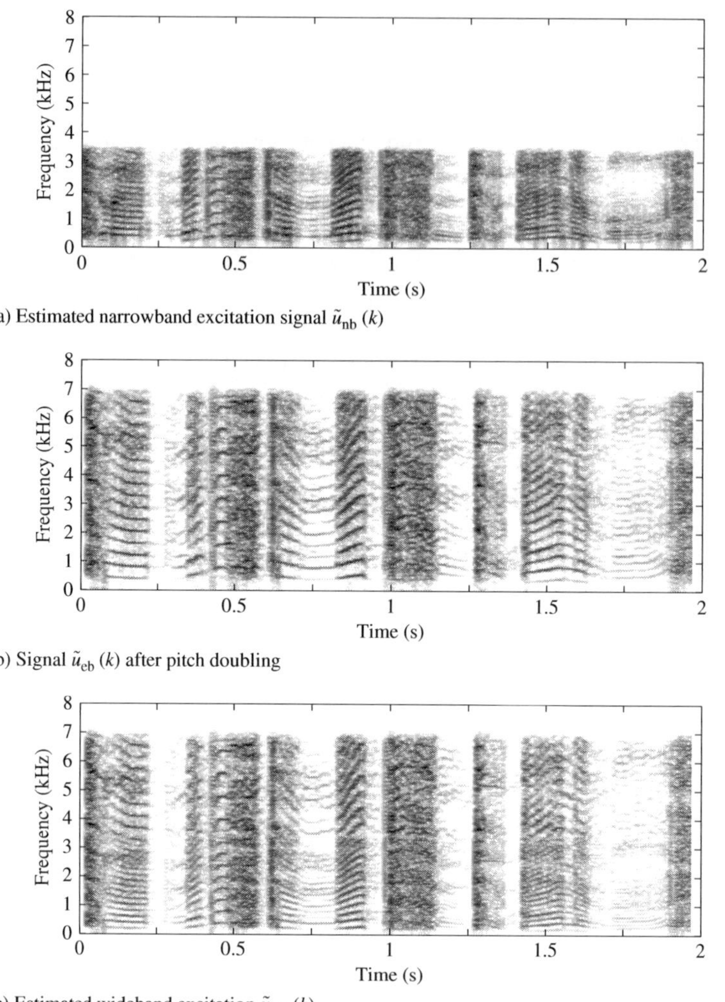

(a) Estimated narrowband excitation signal $\tilde{u}_{\mathrm{nb}}(k)$

(b) Signal $\tilde{u}_{\mathrm{eb}}(k)$ after pitch doubling

(c) Estimated wideband excitation $\tilde{u}_{\mathrm{wb}}(k)$

Figure 6.12 Spectrograms of intermediate signals for the pitch-scaling approach to the extension of the excitation signal. Black regions reflect a large short-term power spectrum. The sentence 'to administer medicine to animals' is spoken by a female voice

doubling in Fig. 6.12 (b), it can be observed that the latter signal has frequency components up to a cut-off frequency of about 6.8 kHz. During voiced sounds, the signal $\tilde{u}_{\mathrm{eb}}(k)$ contains tonal components only at *even* integer multiples of the fundamental frequency of the bandlimited excitation $\tilde{u}_{\mathrm{nb}}(k)$ from Fig. 6.12 (a).

The pitch doubling can, for example, be performed as depicted in Fig. 6.11: first, the sampling rate of the narrowband excitation $\tilde{u}_{\mathrm{nb}}(k)$ is reduced by a factor of 2[1],

[1] In the design of the downsampling method, an advantage can be taken here from the fact that the bandlimited excitation signal $\tilde{u}_{\mathrm{nb}}(k)$ already has an upper band limit that is lower than half of the Nyquist frequency. Therefore, a downsampling by a factor of 2 can simply be performed by omitting every other sample of the signal $\tilde{u}_{\mathrm{nb}}(k)$.

thereby effectively halving the length of signal segments (as expressed in a number of samples) In the second step, the downsampled signal is elongated conversely by employing time-scaling techniques (e.g. Verhelst and Roelands [289], Moulines and Verhelst [181], Verhelst [288]), that is, retaining the pitch frequency of the *downsampled* excitation signal. Since the impacts of the downsampling and time-scaling operations on the length of the speech segments as measured in samples per frame compensate each other, the serial concatenation of the two operations yields a doubling of the pitch of the signal components in $\tilde{u}_{\text{nb}}(k)$.

After pitch doubling, the signal $\tilde{u}_{\text{eb}}(k)$ is high-pass filtered and added to the properly delayed bandlimited excitation signal $\tilde{u}_{\text{nb}}(k)$, thus yielding the estimated wideband excitation signal $\tilde{u}_{\text{wb}}(k)$. It can be seen in Fig. 6.12 (c) that in those speech phases in which the bandlimited speech represents a harmonic complex tone, the pitch-scaling approach has the advantage that tonal signal components in the extended frequency band are at integer multiples of the fundamental frequency of the speech. Accordingly, there are no metallic or ringing artefacts in the enhanced speech. Note, however, that only harmonics with an even order are present in the extended band (above 3.4 kHz) due to the pitch doubling.

The method in general succeeds in regenerating the proper harmonic structure in the estimated wideband excitation signal. In some cases, particularly for speakers with a very low pitch frequency, the impression is produced by the algorithm that a second simultaneous speaker is present in the background of the enhanced speech who speaks with a doubled pitch in comparison to the original speaker. A particular advantage of this algorithm is the fact that it does not need any explicit estimate of any parameter, F_0, V, or σ, of the excitation part of the source model from Sec. 6.2.1. Accordingly, the method has a very high robustness.

6.3.5 DISCUSSION

To evaluate the performances of the different methods for the extension of the excitation signal, extensive informal listening tests have been performed. To produce the speech samples for these tests, the set-up from Fig. 6.13 was used. The block diagram resembles our bandwidth extension system from Sec. 6.2.2, except that the estimation of the wideband spectral envelope from the upper signal path of Fig. 6.5 is replaced by an LP analysis of the *original* wideband speech. Thus, the AR coefficients $\mathbf{a}_{\text{wb}}$ can be assumed

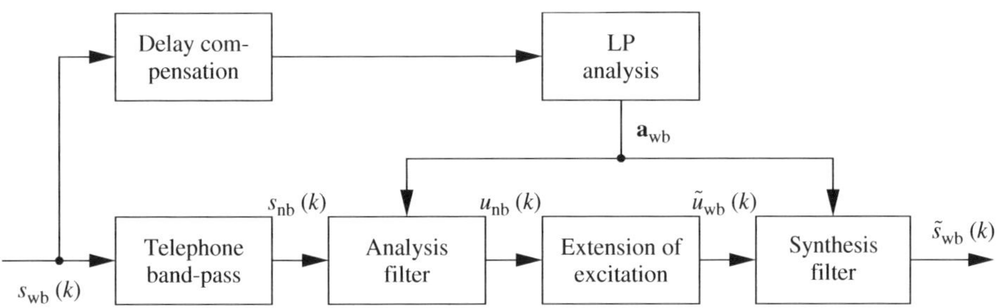

Figure 6.13 System for the generation of speech samples for evaluating the quality of different approaches for the extension of the excitation signal via informal listening tests

to be optimal (in the sense of estimating the narrowband excitation signal $u_{\mathrm{nb}}(k)$), and potential artefacts in $\tilde{s}_{\mathrm{wb}}(k)$ are solely due to the extension of the excitation signal.

We have performed many informal listening tests that have shown that – on the precondition that the bandwidth extension of the spectral envelope works well – the human ear is amazingly insensitive to distortions of the excitation signal at high frequencies above 3.4 kHz. For example, spectral gaps of moderate width as produced by band-stop filters are almost inaudible. Further, inconsistencies of the harmonic structure of speech at high frequencies do not significantly degrade the subjective quality of the enhanced speech signal. The above comments particularly apply if the extended speech signal is played back via the physically constrained acoustical front-end of, for example, a mobile handset. As any such front-end in general has low-pass characteristics, the audibility of artefacts in the spectral fine structure of the enhanced speech $\tilde{s}_{\mathrm{wb}}(k)$ is reduced even further.

Owing to the beneficial properties of the human auditory system at high frequencies, all of the described methods for the extension of the excitation signal towards high frequencies perform well or very well if a good estimate of the wideband spectral envelope is available. A reasonable compromise between the maximization of the subjective quality of the output signal and the computational complexity is given by the modulation with the fixed modulation frequency of $\Omega_{\mathrm{M}} = \Omega_{\mathrm{bb,u}} - \Omega_{\mathrm{bb,l}}$.

The extension of the excitation signal towards low frequencies, on the other hand, is more difficult. The low-frequency components (e.g. below 300 Hz) are especially dominant during voiced sounds, and the human ear is rather sensitive to variations of the harmonic structure in this frequency range. Because most of the methods that are capable of extending the excitation towards low frequencies are based on a pitch estimation algorithm (with limited accuracy), the regenerated low-frequency harmonics often do not fit the harmonics within the base-band of the speech. This produces the distracting impression that a second simultaneous speaker is contained in the enhanced speech signal.

Nevertheless, the subjective quality of the speech signals generated by the system from Fig. 6.13, that is, with knowledge of the true spectral envelope of the wideband speech, is reasonably well, particularly for the extension towards high frequencies. This observation conforms to the results of previous investigations, where it was found that the quality of the estimated wideband spectral envelope is far more important for the subjective quality of the bandwidth-extended speech signal than the extension of the excitation signal (Carl [44]). This has also been recognized in high-frequency BWE for audio applications, see Sec. 5.5.

6.4 ESTIMATION OF THE WIDEBAND SPECTRAL ENVELOPE

The essential step in bandwidth extension algorithms is the estimation of the spectral envelope of the wideband speech signal. This task corresponds to the upper signal path in the block-diagram Fig. 6.5 of the bandwidth extension algorithm.

In most adaptive bandwidth extension algorithms, statistical estimation methods are used, which are to a certain extent similar to approaches from pattern recognition or speech recognition (see e.g. Fukunaga [79], Rabiner and Juang [216]). The estimation of the spectral envelope is in general performed in several consecutive steps as illustrated in Fig. 6.14. The three steps are executed for each frame of the speech signal and for each

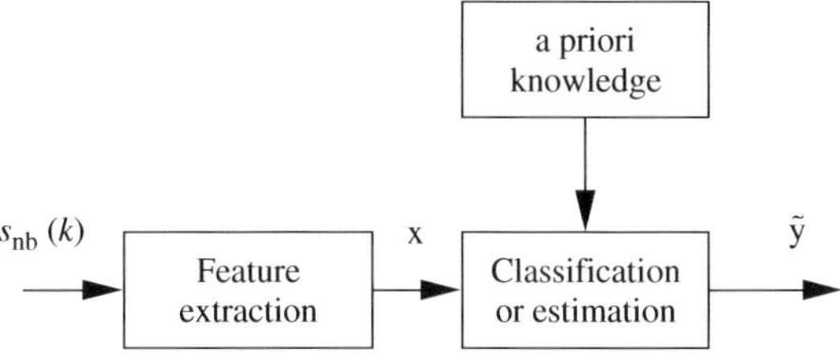

Figure 6.14 Intermediate steps in the estimation of the representation $\tilde{\mathbf{y}}$ of the wideband spectral envelope

missing frequency band. Note that the frame index m will be omitted in the following sections if it is not essential for the understanding of the particular topic.

Feature extraction From each signal frame of the narrowband input speech $s_{nb}(k)$, several features $\mathbf{x}$ are extracted, which carry information on the state of the source model, that is, indirectly on the estimated spectral envelope of the missing frequency band. By the feature extraction algorithm, the dimension and complexity of the estimation problem are reduced significantly. The art is to find a *compact* set of features, that is, the dimension $\dim \mathbf{x}$ of the feature vectors $\mathbf{x}$ shall be low although the features shall carry much information to allow a proper estimation of $\mathbf{y}$. A more detailed description of the feature extraction step will be given in Sec. 6.5.

A priori knowledge The extracted features $\mathbf{x}$ are compared with a priori knowledge, comprising information on the joint behaviour of the features $\mathbf{x}$ and the unknown quantity $\mathbf{y}$. Several representations are possible such as linked codebooks (tables), transformation matrices, reflecting linear correlation, or statistical models, for example, of the joint PDF $p(\mathbf{x}, \mathbf{y})$. The utilized representation of the a priori knowledge is strongly linked with the employed estimation method.

In general, the a priori knowledge has to be acquired during an off-line training phase before applying bandwidth extension. For this, a larger amount of wideband speech data is utilized. The model parameters will be stored for later use in the application phase of the BWE system.

Classification or estimation The final step is the estimation of the representation $\mathbf{y}$ of the spectral envelope. This step can be based on different classification or estimation concepts, the most prominent of which are codebook mapping, linear or piecewise-linear mapping, and Bayesian estimation, according to the possible representations of the a priori knowledge as listed above. These approaches will be described together with the employed a priori knowledge in Secs. 6.6 to 6.9. In literature, sporadically also other approaches may be found such as, for example, neural networks (Tanaka and Hatazoe [262], Uncini *et al.* [277], Iser and Schmidt [119]). In some papers, fixed spectral envelopes are used (Larsen *et al.* [157]).

Note that commonly the required AR coefficients $\tilde{\mathbf{a}}$ of the wideband linear prediction filters are not estimated directly but some other mathematical representation of the spectral envelope is determined first. To keep our presentation abstract, we will denote the estimator output by the quantity $\tilde{\mathbf{y}}$ in the sequel, assuming that $\tilde{\mathbf{y}}$ can directly be converted into the coefficient vector $\tilde{\mathbf{a}}$ to be applied in the analysis and synthesis filters from Fig. 6.5. Common representations will be described in Sec. 6.4.1.

To make things even more complicated, $\tilde{\mathbf{y}}$ may also stand for the shape and relative gain of the spectral envelope within a particular missing sub-band only. In this case, the wideband spectral envelope coefficients $\tilde{\mathbf{a}}$ have to be computed by evaluating $\tilde{\mathbf{y}}$ in addition to the available narrowband speech signal $s_{\mathrm{nb}}(k)$. This approach will be discussed in Sec. 6.4.1.1.

6.4.1 REPRESENTATIONS OF THE ESTIMATED SPECTRAL ENVELOPE

It is an open question, what is the best representation $\mathbf{y}$ of the wideband spectral envelope to be estimated by the framework from Fig. 6.14. The spectral envelope can be represented in many different forms. From an algorithmic view, the most natural representation would be to directly use the AR coefficients $\mathbf{a}$ (i.e. $\tilde{\mathbf{y}} = \tilde{\mathbf{a}}$). In fact, this representation can be used if the estimation algorithm performs a hard classification, for example, by the codebook mapping approach (Carl [44], Carl and Heute [45], Yoshida and Abe [301]). Using just slightly more sophisticated estimation methods, for example, by averaging over the most likely codebook entries, there will be the problem that the stability of the LP synthesis filter $1/A(z)$ cannot be guaranteed. Therefore, direct estimation of AR coefficients is not often used for bandwidth extension.

The most frequently used representations of the spectral envelope in bandwidth extension of speech are line spectral frequencies (LSF), for example, in Enbom and Kleijn [64], Miet *et al.* [174], Chennoukh *et al.* [50]. They are defined as the roots of two symmetric and anti-symmetric polynomials reflecting the transfer function of the linear prediction filter as given in Eqn. 6.4 (Itakura [120]). The conversion from AR coefficients to LSF vectors and vice versa is unique. The outstanding advantage of the LSF representation is that there is a very simple rule to guarantee stable LP synthesis filters: the elements of the LSF vector have to be sorted in ascending order, and their values must be between zero and π. For detailed information on properties of LSF vectors, the reader is referred to the speech-coding literature, for example, Markel and Gray [168], Paliwal and Kleijn [197].

Other interesting representations of the wideband spectral envelope are cepstral coefficients. The real cepstrum of a signal frame is computed by an inverse discrete Fourier transform (DFT) of its logarithmized amplitude spectrum. Analogously, the log amplitude frequency characteristics of an AR filter can be approximated by a series of cepstral coefficients

$$\ln \frac{\sigma^2}{|A(\mathrm{e}^{j\Omega})|^2} = \sum_{i=-\infty}^{\infty} c_i \,\mathrm{e}^{-ji\Omega}, \tag{6.13}$$

where ln denotes the natural logarithm to the base of e, and σ is a scalar gain factor. The cepstral coefficients c_i are real valued and even ($c_{-i} = c_i$) owing to the minimum-phase frequency response of the all-pole filter $1/A(\mathrm{e}^{j\Omega})$ (Hagen [103]). The cepstral coefficients $c_0, c_1, \ldots$ can be calculated directly from the AR coefficients $\mathbf{a}$ and the gain factor σ via a simple recursive formula given, for example, in Markel and Gray [168]

$$c_0 = \ln \sigma^2$$
$$c_i = -a_i - \sum_{n=1}^{i-1} \frac{n}{i} c_n a_{i-n} \qquad \text{for } i > 0, \tag{6.14}$$

with $a_i = 0$ for $i > N_a$. Note that only the first $N_a + 1$ cepstral coefficients derived in this way are non-redundant, while the remaining ones can be determined from these foremost coefficients.

A particular advantage of the cepstral representation of AR coefficients is that a minimum mean square error (MMSE) solution for the estimated coefficients $\tilde{\mathbf{y}} = \tilde{\mathbf{c}}$ corresponds to a minimization of the log spectral distortion (LSD) measure

$$d_{\text{LSD}}^2 = \left(\frac{10}{\ln 10}\right)^2 \sum_{i=-\infty}^{\infty} (c_i - \tilde{c}_i)^2$$
$$= \frac{1}{2\pi} \int_{-\pi}^{\pi} \left(20 \log_{10} \frac{\sigma}{|A(\mathrm{e}^{j\Omega})|} - 20 \log_{10} \frac{\tilde{\sigma}}{|\tilde{A}(\mathrm{e}^{j\Omega})|}\right)^2 \mathrm{d}\Omega. \tag{6.15}$$

The interest in this distortion measure is motivated by the fact that in speech coding the log spectral distortion correlates reasonably well with the subjective speech quality. Therefore, it has found wide acceptance in speech coding to assess the quality of quantizers of representations (i.e. parameters/coefficients) of the spectral envelope. Cepstral coefficients have been used for bandwidth extension of speech, for example, in Avendano *et al.* [23], Park and Kim [200], and Nilsson and Kleijn [187].

6.4.1.1 Sub-band-based Assembly of the Wideband Spectral Envelope

If we consider the estimation of the wideband spectral envelope, we must distinguish between missing and available sub-bands: information on the spectral envelope within the frequency range of the narrowband input speech $s_{\text{nb}}(k)$ of the BWE algorithm can be determined by conventional linear prediction techniques. For the missing frequency band(s), on the other hand, more or less sophisticated estimation methods as illustrated in Fig. 6.14 have to be used.

This distinction leads to the concept of subdividing the estimation of the (single) wideband spectral envelope according to the different frequency bands in the wideband speech (Jax [128]). The approach is illustrated in Fig. 6.15. First, the estimation of the shape and (relative) gain of the spectral envelope of the missing frequency band(s) is performed individually, according to Fig. 6.14. Each estimator of an individual sub-band spectral envelope can be tuned optimally to the specific properties of the particular frequency band. The separate estimators output their results in a short-term power spectrum domain. Then, the assembly of a joint description for the full frequency range of the wideband speech signal can be performed by simple concatenation of the estimated power spectra from the

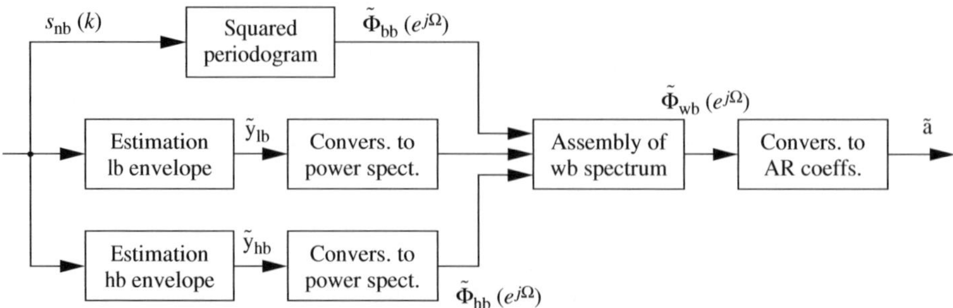

Figure 6.15 Block diagram of the sub-band approach to estimate the AR coefficient set $\tilde{\mathbf{a}}$ of the wideband speech signal. The sub-band estimators for each missing frequency band consist of individual statistical estimators as shown in Fig. 6.14

sub-bands. In a final step, the assembled smoothed power spectrum of the wideband signal is converted into the corresponding AR coefficients $\tilde{\mathbf{a}}$ (e.g. applying an inverse DFT and the Levinson–Durbin algorithm), which can then be used in the wideband analysis and synthesis filters of the BWE algorithm.

In general, the aforementioned procedure can be performed with an arbitrary number of sub-bands. In Fig. 6.15, the algorithm is depicted for three sub-bands, that is, in addition to the base-band there are two missing frequency bands at low and high frequencies. Further details on the sub-band-based assembly of the wideband spectral envelope can be found in Jax [128], Jax and Vary [133].

As illustrated in the lower diagram of Fig. 6.16 (b), the resulting AR coefficients constitute a good estimate of the wideband spectral envelope. In the base-band, the frequency response of the estimated AR filter (solid line) strongly matches the frequency response of the optimal AR filter as derived from the original wideband speech (dashed line). There are only slight deviations visible at the band edges due to errors in the estimated sub-band power spectra within the neighbouring missing frequency bands. In the extended frequency bands, there are some errors in the formant structure. Nevertheless, the ample run of the frequency response is estimated correctly. The common modelling of the spectral envelope of the base-band and extended bands has the advantage that the spectral envelope of the enhanced speech $\tilde{s}_{wb}(k)$ is smoothed in the transition regions between the bands.

6.4.2 INSTRUMENTAL PERFORMANCE MEASURE

Before discussing various solutions for the sub-blocks of Fig. 6.14 in the remainder of this chapter, we will now define an instrumental performance measure needed to evaluate the different alternatives. Later, we are interested in what we can expect from bandwidth extension algorithms: will it be possible to perform as good as true wideband speech codecs by extending telephone speech or are there fundamental limits? To answer this question, BWE of speech shall be examined from an information theoretic perspective in Sec. 6.4.3.

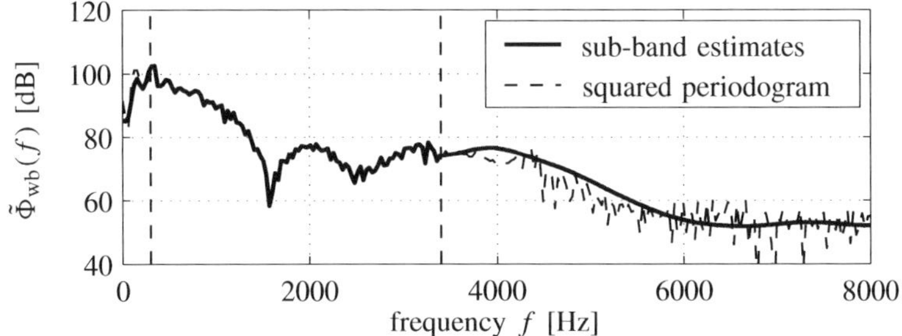

(a) Concatenated short-term power spectrum

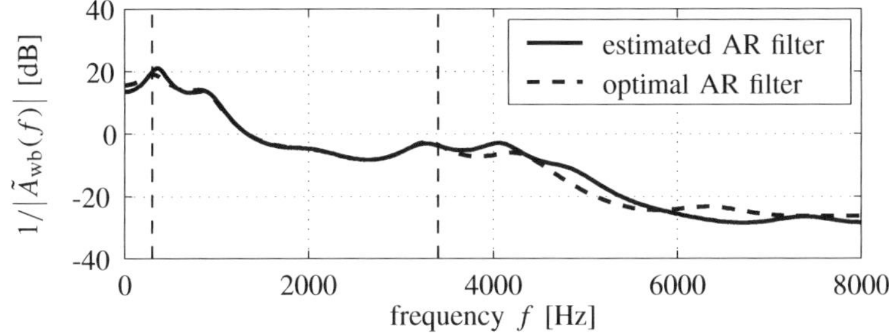

(b) Estimated wideband spectral envelope

Figure 6.16 Example for the procedure of the sub-band-based estimation of the spectral envelope. In this example, there are two missing frequency bands, below 300 Hz and beyond 3.4 kHz. For comparison, the respective quantities as derived from the corresponding signal frame of the original wideband speech signal are shown by the dashed curves

6.4.2.1 Auto-regressive Modelling of the Missing Sub-band Spectral Envelope

As described earlier, the spectral envelope of the enhanced speech signal is the principal key for a high subjective quality of the output signal of a bandwidth extension system. In Sec. 6.2.3, it was further pointed out that the bandlimited input signal $s_{\mathrm{nb}}(k)$ shall be contained transparently in the extended output signal $\tilde{s}_{\mathrm{wb}}(k)$. This can either be guaranteed implicitly by the structure of the algorithm as in our approach from Sec. 6.2.3, or the bandlimited input speech has to be considered explicitly as shown in Fig. 6.6. With both structures, there will only be errors within the spectral envelope of the extended frequency band of the output speech $\tilde{s}_{\mathrm{wb}}(k)$, and a performance measure should accordingly consider only distortions in this extended frequency band. Since a significant source of distortion is the attenuation or amplification of the spectral envelope in the extended frequency band with respect to the base-band spectral envelope, the *shape* as well as the *gain* of the spectral envelope of the extended band shall be investigated. The gain shall be expressed with respect to the base-band signal components, that is, it shall be a *relative* gain as specified later (compare Nilsson *et al.* [185], Park and Kim [200], Nilsson and Kleijn [187]).

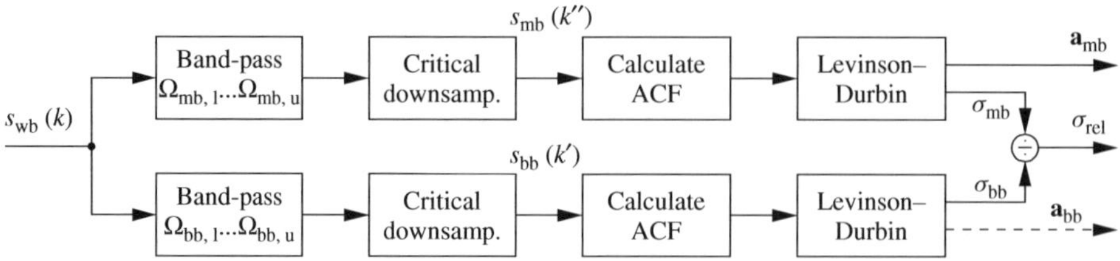

Figure 6.17 Modelling of the spectral envelope of the missing (respectively extended) frequency band with respect to the base-band signal via two sub-band signals $s_{\text{mb}}(k'')$ and $s_{\text{bb}}(k')$ (non-general scheme: see footnote 3). For each sub-band signal, the parameters of an auto-regressive model are obtained from the estimated auto-correlation function (ACF) via a Levinson–Durbin recursion here. Note that the AR coefficient set $\mathbf{a}_{\text{bb}}$ of the base-band signal is not utilized

Without loss of generality, it shall be assumed in the following that the wideband speech signal is constituted from two sub-band signals: the base-band signal that corresponds to the input signal $s_{\text{nb}}(k)$ of the bandwidth extension system, and the sub-band signal containing the missing respectively extended frequency components[2]. The missing frequency band will be denoted by the subscript mb in the following text. It covers the frequencies between the lower band edge $\Omega_{\text{mb,l}}$ and the upper band edge $\Omega_{\text{mb,u}}$. The base-band of the speech signal starts at the lower cut-off frequency $\Omega_{\text{bb,l}}$ and ends at the upper cut-off frequency $\Omega_{\text{bb,u}}$ of the bandlimited input signal.

For the evaluation of an instrumental performance measure, it is presumed that the wideband speech signal $s_{\text{wb}}(k)$ is available. To define the spectral distortion measure, first the wideband speech signal is split into the two aforementioned sub-band signals (see Fig. 6.17). This is accomplished by band-pass filtering the wideband signal $s_{\text{wb}}(k)$, using two filters with lower and upper cut-off frequencies of $\Omega_{\text{bb,l}}$ and $\Omega_{\text{bb,u}}$, and $\Omega_{\text{mb,l}}$ and $\Omega_{\text{mb,u}}$, respectively. The band-pass-filtered signals are further critically downsampled such that the respective frequency components cover the whole frequency range of $\Omega = 0 \ldots \pi$ of the downsampled signals[3]. The two resulting sub-band signals are denoted by $s_{\text{bb}}(k')$ for the base-band signal, and by $s_{\text{mb}}(k'')$ for the signal containing the missing frequency band of the wideband speech. The sampling rates of the downsampled sub-band signals $s_{\text{bb}}(k')$ and $s_{\text{mb}}(k'')$ are

$$f_{s'} = f_s \frac{1}{\pi}(\Omega_{\text{bb,u}} - \Omega_{\text{bb,l}}) \quad \text{and} \quad f_{s''} = f_s \frac{1}{\pi}(\Omega_{\text{mb,u}} - \Omega_{\text{mb,l}}), \tag{6.16}$$

[2] Note that there are applications in which there are *two* (or even more) missing frequency bands, for example, at low (<300 Hz) and high (>3.4 kHz) frequencies. The performances of the two sub-band estimators can then be described individually using the approach of this chapter.

[3] Employing this procedure, it is inherently assumed that the ratios between the sampling rates of the wideband speech and of the downsampled sub-band signals are integer valued and that the band limits of the sub-bands are at suitable frequencies. In a more general scenario, an additional modulation of the sub-band signals is necessary to ensure that the band limits (e.g. $\Omega_{\text{mb,l}}$ and $\Omega_{\text{mb,u}}$) of the sub-bands in the wideband speech are mapped to the band limits 0 and π of the critically downsampled signals.

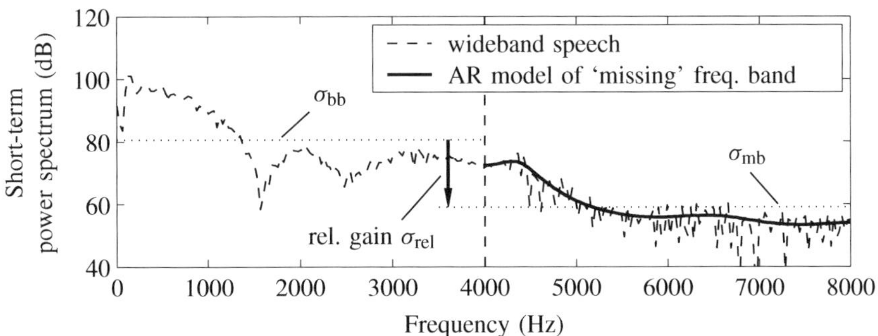

Figure 6.18 Auto-regressive modelling of the spectral envelope of the sub-band signal containing the missing frequency band. In this example, the missing frequency band ranges from 4 to 8 kHz

respectively. The corresponding time indices or angular frequencies will be marked with one (for the base-band) or two (for the missing frequency band) apostrophes in the following.

In the next step, two individual auto-regressive models are fitted to frames of the two sub-band signals. For example, the model used for representing the missing frequency band spectrum $|S_{\text{mb}}(e^{j\Omega''})|^2$ is defined by (compare Fig. 6.18)

$$|S_{\text{mb}}(e^{j\Omega''})|^2 \approx \left|\frac{\sigma_{\text{mb}}}{A_{\text{mb}}(e^{j\Omega''})}\right|^2 = \left|\frac{\sigma_{\text{mb}}}{A_{\text{mb}}(z'')}\right|^2_{z''=e^{j\Omega''}} \tag{6.17}$$

with

$$A_{\text{mb}}(z'') = 1 + \sum_{i=1}^{N_{a,\text{mb}}} a_{\text{mb},i}\,(z'')^{-i}. \tag{6.18}$$

The parameters of the models are estimated by conventional LP analysis, for example, by the Levinson–Durbin algorithm (Markel and Gray [168]). This is done individually for the two sub-band signals $s_{\text{bb}}(k')$ and $s_{\text{mb}}(k'')$. The results of the LP analysis are the coefficient set $\mathbf{a}_{\text{mb}}$, representing the spectral envelope of the missing frequency band, as well as two gain factors σ_{bb} and σ_{mb} of the base-band and the missing frequency band, respectively. Since the *relative* gain of the extended frequency band shall be measured, we define $\sigma_{\text{rel}} = \sigma_{\text{mb}}/\sigma_{\text{bb}}$. The order of $\mathbf{a}_{\text{mb}} = \left[a_{\text{mb},1}, a_{\text{mb},2}, \ldots a_{\text{mb},N_{a,\text{mb}}}\right]^{\text{T}}$ is $N_{a,\text{mb}}$.

The parameters of the auto-regressive model of the missing frequency band of wideband speech that was defined in the previous paragraphs can alternatively be determined using *selective linear prediction* (SLP) techniques as described in Markel and Gray [168, Sec. 6.4]. Similar to the procedure described above (Fig. 6.17), the SLP approach allows to fit an auto-regressive model to a sub-band of the short-term spectrum of a signal. The resulting model corresponds to a critically sampled sub-band signal. Since the splitting of the sub-bands is performed in the frequency domain, the SLP method can flexibly be adapted to any bandwidth extension scenario (Jax [128]).

6.4.2.2 Sub-band Log Spectral Distortion Measure

The performance of the estimation of the wideband spectral envelope shall be defined in terms of the *log spectral distortion* (LSD) of the missing frequency band. The squared LSD measure is specified in the frequency domain by (e.g. Markel and Gray [95], Gray, Buzo, Gray, and Matsuyama [96], Vary *et al.* [286])

$$d_{\text{LSD}}^2 = \frac{1}{2\pi} \int_{-\pi}^{\pi} \left(20 \log_{10} \frac{\sigma_{\text{rel}}}{|A_{\text{mb}}(\mathrm{e}^{j\Omega''})|} - 20 \log_{10} \frac{\tilde{\sigma}_{\text{rel}}}{|\tilde{A}_{\text{mb}}(\mathrm{e}^{j\Omega''})|} \right)^2 \mathrm{d}\Omega''. \tag{6.19}$$

Here, the quantities $A_{\text{mb}}(\mathrm{e}^{j\Omega''})$ and σ_{rel} refer to the modelled frequency spectrum and relative gain of the missing frequency band of original wideband speech, and $\tilde{A}_{\text{mb}}(\mathrm{e}^{j\Omega''})$ and $\tilde{\sigma}_{\text{rel}}$ denote the corresponding estimated parameters as determined by a bandwidth extension system. Note that, because the LSD measure is evaluated for the critically downsampled sub-band signal $s_{\text{mb}}(k'')$ containing only the missing frequency band, the integration range of $-\pi$ to π in Eqn. 6.19 covers the missing frequency range in the original wideband speech signal. The unit of d_{LSD} is dB.

Unfortunately, the evaluation of the LSD measure in the frequency domain in general is quite complicated. Therefore, an alternative representation by a mean-square error criterion in the cepstral domain, following the definition from Eqn. 6.13, will be used in the following

$$\ln \frac{\sigma_{\text{rel}}^2}{|A_{\text{mb}}(\mathrm{e}^{j\Omega''})|^2} = \sum_{i=-\infty}^{\infty} c_i \, \mathrm{e}^{-ji\Omega''}. \tag{6.20}$$

With this definition, for a sequence of speech frames the *root mean square* (RMS) average of the LSD is given by

$$\overline{d}_{\text{LSD}} = \frac{\sqrt{2}\,10}{\ln 10} \sqrt{E\left\{ \frac{1}{2}(c_0 - \tilde{c}_0)^2 + \sum_{i=1}^{\infty}(c_i - \tilde{c}_i)^2 \right\}}. \tag{6.21}$$

Here, the function $E\{\cdot\}$ denotes the expectation operation.

Now, the output representation $\tilde{\mathbf{y}}$ of the estimation shall be defined in such a manner that the estimation performance can be determined by a mean-square error criterion. For this, the quantity $\mathbf{y}$ is defined as a *weighted* cepstral representation of the missing frequency band. It can be determined from the cepstral coefficients $c_0, c_1, \ldots$ that represent the AR model of the spectral envelope of the missing frequency band

$$y_i = \begin{cases} \frac{1}{\sqrt{2}} c_i, & \text{if } i = 0 \\ c_i, & \text{if } 1 \le i < \text{ d}. \end{cases} \tag{6.22}$$

The scalar values y_i constitute the d-dimensional vector $\mathbf{y} = [y_0, y_1, \ldots y_{d-1}]^{\mathrm{T}}$. The dimension d of $\mathbf{y}$ should be at least equal to $N_{a,\text{mb}}$ such that all non-redundant cepstral coefficients are considered. Inserting the definition of Eqn. 6.22 into Eqn. 6.21 yields the relationship

$$d_{\mathrm{LSD}}^2 \approx \left(\frac{\sqrt{2}\,10}{\ln 10}\right)^2 \sum_{i=0}^{d-1} (y_i - \tilde{y}_i)^2. \tag{6.23}$$

Note that the term on the right-hand side of Eqn. 6.23 is only an approximation of the log spectral distortion (Eqn. 6.19) because only the first d summands of the sum in Eqn. 6.21 are considered.

To get a notion of the admissible sub-band log spectral distortion in bandwidth extension, the LSD performances of several wideband speech codecs (G.722 and AMR-WB at several bit rates) were investigated in Jax [128]. It was found that the wideband codecs achieve a near-transparent subjective speech quality even with an RMS LSD of more than 2 dB for the low-frequency band from 50 to 300 Hz, and with an RMS LSD of more than 3 dB for the high-frequency band from 3.4 to 7 kHz. Thus, it is conjectured that it is also possible in the BWE application to obtain a 'near-transparent' speech quality for RMS LSD values about 2 to 3 dB.

6.4.3 THEORETICAL PERFORMANCE BOUND

It is plausible that an extension of the bandwidth of speech signals is only possible, if there are sufficient dependencies between the available bandlimited speech signal and the missing frequency components. The fact that the narrowband speech and the missing signal components are results of the same physical speech production process gives rise to the assumption that there are such dependencies in speech signals. This assumption is supported by the success of many BWE methods published throughout the last two decades. There are only few publications, however, that shade some light onto the information theoretic background of artificial bandwidth extension (Nilsson *et al.* [185, 186], Nordén *et al.* [188], Jax and Vary [131], Yang *et al.* [297], Epps [65]).

In digital signal processing, *linear* dependencies between signals are commonly described in terms of *correlation* factors. In an information theoretic perspective, the dependencies between different signals are described by their mutual information (MI), for example, Cover and Thomas [52]. In contrast to the correlation measure, mutual information covers all kinds of linear and non-linear dependencies. The aim of this section is to investigate the relationship between an upper bound on the achievable quality of a BWE algorithm (measured in terms of the instrumental performance measure from the previous section) on one hand, and the mutual information between representations of the bandlimited speech (feature vector $\mathbf{x}$) and of the missing frequency components (sub-band spectral envelope $\mathbf{y}$) on the other hand.

The quantity $\tilde{\mathbf{y}}$ as defined in Eqn. 6.22 can be calculated for *any* BWE algorithm, either from the extended speech signal or directly from the estimated representation of the wideband spectral envelope (e.g. as $\tilde{\mathbf{a}}$ in Fig. 6.5) if applicable. Therefore, an upper bound on the performance of a generalized BWE algorithm – measured in terms of the RMS LSD or the mean-square error of $\tilde{\mathbf{y}}$, respectively – also constitutes a bound on the performance of any other BWE algorithm with a differing representation of the wideband spectral envelope.

If we assume a memoryless, deterministic estimation[4] of $\tilde{\mathbf{y}} = f(\mathbf{x})$ of the missing spectral envelope from the feature vector $\mathbf{x}$, a relation between the mutual information $I(\mathbf{x};\mathbf{y})$, expressed in *nats* (Cover and Thomas [52]), and the minimum possible mean square estimation error of $\tilde{\mathbf{y}}$ can be formulated (Jax and Vary [131], Jax [128])

$$E\{\|\mathbf{y}-\tilde{\mathbf{y}}\|^2\} \geq \frac{\mathrm{d}}{2\pi \mathrm{e}} \exp\left(\frac{2}{\mathrm{d}}\big(h(\mathbf{y}) - I(\mathbf{x};\mathbf{y})\big)\right). \tag{6.24}$$

The relation depends on the differential entropy $h(\mathbf{y}) = -E\{\ln p(\mathbf{y})\}$, which comprises statistical properties of the estimated quantity $\mathbf{y}$. Further, $d = \dim \mathbf{y}$.

Owing to the weighting of the representation $\mathbf{y}$ of the missing frequency band, the mean-square error $E\{\|\mathbf{y}-\tilde{\mathbf{y}}\|^2\}$ resembles a truncated version of the cepstral distance within the square root of Eqn. 6.21. Because the truncated elements are non-negative, we find the inequality

$$\begin{aligned}\bar{d}_{\mathrm{LSD}} &\geq \frac{\sqrt{2}\,10}{\log 10}\sqrt{E\{\|\mathbf{y}-\tilde{\mathbf{y}}\|^2\}} \\ &\geq \frac{\sqrt{2}\,10}{\log 10}\sqrt{\frac{\mathrm{d}}{2\pi \mathrm{e}}} \exp\left(\frac{1}{\mathrm{d}}\big(h(\mathbf{y}) - I(\mathbf{x};\mathbf{y})\big)\right).\end{aligned} \tag{6.25}$$

which gives a lower bound on the achievable RMS log spectral distortion in dependence of the mutual information $I(\mathbf{x};\mathbf{y})$ and differential entropy $h(\mathbf{y})$. While the differential entropy $h(\mathbf{y})$ only depends on statistical properties of the cepstral representation $\mathbf{y}$, the mutual information $I(\mathbf{x};\mathbf{y})$ additionally depends on the chosen feature set $\mathbf{x}$.

In the following, we will give examples of bounds defined by Eqn. 6.25 for the bandwidth extension of telephone speech. Two applications will be considered: extension towards low frequencies below 300 Hz and extension to high frequencies above 3.4 up to 7 kHz. In Jax [128], differential entropies $h(\mathbf{y})$ have been approximated for these applications by numerical simulations using the large SI100 speech corpus from the *Bayerisches Archiv für Sprachsignale* (BAS) (Schiel [234]). The SI100 corpus contains more than 35 hours of continuous wideband German speech spoken by 101 male and female speakers. The resulting bounds for the two aforementioned applications are illustrated in Fig. 6.19 in dependence of the mutual information. Similar investigations have been reported for the TIMIT database in Jax and Vary [131]. Details on the mutual information obtained for different feature sets will be presented in Sec. 6.5.

Unfortunately, it is in the nature of the information theoretic bound that it does not point out a particular strategy to design estimators that are optimal in the sense of minimizing the log spectral distortion measure. However, from the dependency of the performance bounds from the mutual information $I(\mathbf{x};\mathbf{y})$, it can be concluded that it is advantageous to select the elements of the utilized feature vector $\mathbf{x}$ such as to maximize the mutual information $I(\mathbf{x};\mathbf{y})$ (compare Sec. 6.5). To obtain the best possible quality with the BWE

[4] Note that we do not assume anything about the particular realization of the estimator: the function $f(\mathbf{x})$ can be interpreted as a generalized description of any kind of linear or non-linear classification or estimation with arbitrary complexity–including the approaches commonly used for BWE, for example, codebook mapping, linear mapping, statistical estimation, and neural networks.

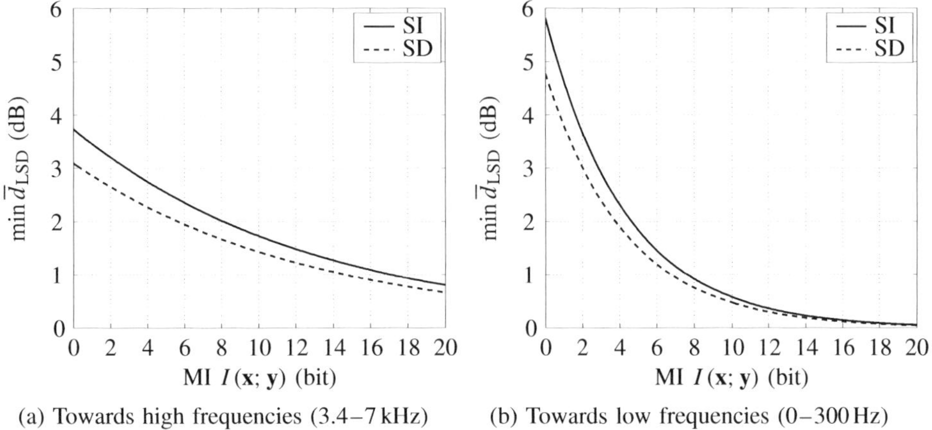

(a) Towards high frequencies (3.4–7 kHz) (b) Towards low frequencies (0–300 Hz)

Figure 6.19 Lower bounds on the RMS log spectral distortion $\overline{d}_{\rm LSD}$ for memoryless bandwidth extension of telephone speech (300 Hz–3.4 kHz). The bounds are given for speaker-independent (SI) and speaker-dependent (SD) solutions

system, it appears favourable – at least if the available mutual information $I(\mathbf{x};\mathbf{y})$ is low – to aim at speaker-dependent approaches.

Since the theoretical bound of Eqn. 6.25 is not tight, however, high mutual information $I(\mathbf{x};\mathbf{y})$ only is a necessary but not a sufficient condition for achieving a high performance with a specific feature vector $\mathbf{x}$. Thus, by observing the bound in Fig. 6.19, we cannot acquire knowledge on what we can expect from BWE algorithms but only on what we can *not* expect. We will see at the end of Sec. 6.5 that by selection of proper elements to include in the feature vector $\mathbf{x}$ it was to date only possible to achieve a mutual information of about 3 bit at best. Inserting this value into Eqn. 6.25 results in lower bounds of about 3 dB for speaker-independent solutions in both considered applications. Consequently, we can safely conclude that it is not possible, at least with the investigated features, to outperform the quality of wideband speech codecs by bandwidth extension.

6.5 FEATURE SELECTION

In this section, the focus shall be on the feature extraction block that is preceding the step of estimating the wideband spectral envelope (see Fig. 6.14). In general, feature extraction and estimation of the wideband spectral envelope are performed on a frame-by-frame basis with frame lengths of about 10 to 30 ms. The feature extraction reduces the dimensionality of each frame of the narrowband signal $s_{\rm nb}(k)$ such that the subsequent estimation of the spectral envelope representation is feasible and computationally efficient. The result is the feature vector $\mathbf{x} = [x_1, \ldots x_b]^{\rm T}$ with the dimension $b = \dim \mathbf{x}$. Usually, representations of the spectral envelope of the narrowband signal $s_{\rm nb}(k)$ are used as features, for example, LPC or LSF vectors or cepstral coefficients. In some contributions, additional features such as voicing criteria are taken into account. Here, we want to find some measures that help in finding the best composition of the feature vector $\mathbf{x}$.

The optimal feature extraction method (in the sense of high-quality bandwidth extension) for a fixed dimension of the feature vector allows the BWE algorithm to achieve the best subjective performance as compared to all other possible mappings with the same dimension. Unfortunately, evaluation and comparison of the subjective performances for a large number of alternative algorithms and/or feature sets $\mathbf{x}$ is very time consuming. Therefore, other means have to be used to assess the 'quality' of single features or feature vectors – instrumental measures are needed that provide suggestive hints for the selection of the best feature set.

In the next two sub-sections, two instrumental measures from information theory and statistics will be reviewed. In Sec. 6.5.3, the *linear discriminant analysis* (LDA) that results from an optimization of the separability measure will be introduced. With an LDA, the dimension of a feature vector can be reduced while the maximum discriminating power of the features is retained. In the last sub-sections, the usability of different features, well-tried and new ones, for the BWE problem will be evaluated using the introduced measures and procedures. The insights and results of this section are mostly independent from the particular approach used for estimating the wideband spectral envelope.

6.5.1 MUTUAL INFORMATION

Shannon's *mutual information* (MI) $I(\mathbf{x};\mathbf{y})$ gives the mean information we gain on the estimated wideband spectral envelope representation $\mathbf{y}$ by knowledge of the feature vector $\mathbf{x}$. Mutual information can be regarded as an indication of the feasibility of the estimation task (Nilsson *et al.* [186], Jax and Vary [131]): in Sec. 6.4.3, it has been shown that for a specific mutual information $I(\mathbf{x};\mathbf{y})$ the minimum achievable mean-square estimation error $E\{\|\mathbf{y}-\tilde{\mathbf{y}}\|^2\}$ is lower bounded. The larger the MI, the lower is the bound. Hence, a large mutual information $I(\mathbf{x};\mathbf{y})$ is a *necessary* condition for high-quality estimation of $\mathbf{y}$ from the observations $\mathbf{x}$. Now, we want to investigate the mutual information for different features $\mathbf{x}$ of the narrowband speech signal.

For estimating the mutual information $I(\mathbf{x};\mathbf{y})$, we have to use a parametric approach because of the high dimension of the continuous vectors $\mathbf{x}$ and $\mathbf{y}$. The joint probability density function (PDF) $p(\mathbf{x},\mathbf{y})$ is approximated by a Gaussian mixture model (GMM) $\tilde{p}(\mathbf{x},\mathbf{y})$, that is, a sum of L weighted multivariate Gaussian densities $\mathcal{N}(\cdot)$ with mean vectors μ_l and covariance matrices $\mathbf{V}_l$

$$\tilde{p}(\mathbf{x},\mathbf{y}) = \sum_{l=1}^{L} \rho_l \, \mathcal{N}(\mathbf{x},\mathbf{y};\mu_l,\mathbf{V}_l) \approx p(\mathbf{x},\mathbf{y}). \tag{6.26}$$

The scalar weights ρ_l and the parameters μ_l and $\mathbf{V}_l$ of the individual Gaussians are trained by the expectation-maximization (EM) algorithm[5]. Then, the mutual information is estimated numerically from the parameters of the GMM (Hedelin and Skoglund [106])

$$I(\mathbf{x};\mathbf{y}) \approx E_{\tilde{p}(\mathbf{x},\mathbf{y})}\left\{\log \frac{\tilde{p}(\dot{\mathbf{x}},\dot{\mathbf{y}})}{\tilde{p}(\dot{\mathbf{x}})\,\tilde{p}(\dot{\mathbf{y}})}\right\}$$

[5] Further details on Gaussian mixture models can be found in Sec. 6.8.

$$\approx \frac{1}{M} \sum_{\nu=1}^{M} \log \frac{\tilde{p}(\dot{\mathbf{x}}(\nu), \dot{\mathbf{y}}(\nu))}{\tilde{p}(\dot{\mathbf{x}}(\nu))\, \tilde{p}(\dot{\mathbf{y}}(\nu))}. \tag{6.27}$$

In the computation of Eqn. 6.27, the vector pairs $\dot{\mathbf{x}}(\nu), \dot{\mathbf{y}}(\nu)$ are generated synthetically according to the model PDF $\tilde{p}(\mathbf{x}, \mathbf{y})$. In our investigations presented in Sec. 6.5.5, we have used $L = 256$ Gaussians with full covariance matrices. The numerical evaluation of Eqn. 6.27 was performed with $M = 10^6$ synthetic vector pairs (Jax [128]).

From the definition of mutual information, for example, Cover and Thomas [52], the following important properties of this measure for feature selection can be found:

- If the relation between two different feature vectors is defined by a bijective mapping, the MI is identical for both feature vectors. In this case, the MI measure does not provide any hint on which feature set shall be preferred.
- If several parameters of the narrowband speech (say x_A, x_B and x_C) form a Markov chain $x_A \rightarrow x_B \rightarrow x_C$, that is, if x_C is calculated from x_B, and x_B is calculated from x_A, it appears favourable to select the very first element x_A of the chain as a feature. Owing to the data-processing inequality (Cover and Thomas [52]), MI is maximized by this choice.
- For combined feature vectors, the MI cannot be simply added. In general, the MI has to be estimated again for the new vector.

6.5.2 SEPARABILITY

From the field of pattern recognition, the *separability* is known as a measure of the quality of a particular feature set for a classification problem (Fukunaga [79]). In the BWE application, the class definitions should best be adopted to the method used to estimate the wideband spectral envelope: for example, if codebook mapping is used (Carl [44]), the classes should correspond to the correct codebook indices as computed from true wideband speech. For an HMM-based approach (Jax and Vary [133]), the classes should be the true HMM state information.

The separability measure can be calculated from a labelled set of training data, that is, for each feature vector in the set the corresponding class must be known. Let Ξ_i denote the set of feature vectors $\mathbf{x}$ assigned to the ith class. The number of feature vectors in the ith set is $N_{\Xi_i} = |\Xi_i|$. The constant N_S denotes the number of classes. Then, the total number of frames in the training data is given by $N_m = \sum_{i=1}^{N_S} N_{\Xi_i}$. From the labelled training data, the *within-class* covariance matrix

$$\mathbf{V_x} = \frac{1}{N_m} \sum_{i=1}^{N_S} \sum_{\mathbf{x} \in \Xi_i} (\mathbf{x} - \mu_i)(\mathbf{x} - \mu_i)^T \tag{6.28}$$

and the *between-class* covariance matrix

$$\mathbf{B_x} = \sum_{i=1}^{N_S} \frac{N_{\Xi_i}}{N_m} (\mu_i - \mu)(\mu_i - \mu)^T \tag{6.29}$$

are calculated, where

$$\mu_i = \frac{1}{N_{\Xi_i}} \sum_{\mathbf{x}\in\Xi_i} \mathbf{x} \quad \text{and} \quad \mu = \sum_{i=1}^{N_S} \frac{N_{\Xi_i}}{N_m}\, \mu_i. \tag{6.30}$$

The separability measure shall be larger if the between-class covariance gets larger and/or if the within-class covariance gets smaller. Accordingly, the separability measure is empirically defined by the term $\mathbf{J}_\mathbf{x} = \mathbf{V}_\mathbf{x}^{-1}\,\mathbf{B}_\mathbf{x}$. To obtain a scalar measure for the separability of the classes, a trace criterion is used (Fukunaga [79])

$$\zeta(\mathbf{x}) = \text{tr}\ \mathbf{J}_\mathbf{x} = \text{tr}\left(\mathbf{V}_\mathbf{x}^{-1}\,\mathbf{B}_\mathbf{x}\right). \tag{6.31}$$

The separability depends on the definition of the classes. Comparing $\zeta(\mathbf{x})$ for different feature vectors $\mathbf{x}$ with the same class definitions, a larger value indicates a better suitability of the corresponding feature vector for classification and estimation.

The separability measure has the following properties:

- The definition of the separability measure is based on the implicit assumption of a normal distribution of the feature vectors that are assigned to each class. If this assumption is not valid, the significance of the separability measure is reduced.
- By the separability measure, all classes are treated alike. Therefore, the separability of two very similar classes (w.r.t. the represented speech sound) is rated like the separability of two very different classes. Hence, maximizing the separability does not necessarily lead to the optimum achievable estimation performance (e.g. in the MMSE sense) of the subsequent estimation rule.
- In general, the values of the separabilities cannot be added up if several features are assembled to a composite feature vector. In this case, the separability of the composite feature vector must be measured anew.

6.5.3 LINEAR DISCRIMINANT ANALYSIS

The purpose of the *linear discriminant analysis* (LDA) is to obtain a feature vector with maximal compactness (Fukunaga [79]): starting from a high-dimensional 'super-vector' $\mathbf{x}_0$, the dimension of the feature vector $\mathbf{x}$ shall be reduced, while the discriminating power shall be retained or decreased as little as possible. The reduction of dimension is performed (in the BWE application phase) by means of a linear transformation

$$\mathbf{x} = \mathbf{H}^\text{T}\,\mathbf{x}_0, \tag{6.32}$$

where the matrix $\mathbf{H}$ is a $\beta \times b$ matrix with $b = \dim \mathbf{x} < \beta = \dim \mathbf{x}_0$. The column vectors in $\mathbf{H}$ shall be linearly independent.

The matrix $\mathbf{H}$ is optimized such that the separability of $\mathbf{x}$ is maximized (Fukunaga [79])

$$\mathbf{H} = \arg\max_{\mathbf{H}} \zeta(\mathbf{x}), \quad \text{where} \tag{6.33}$$

$$\zeta(\mathbf{x}) = \text{tr}\left(\mathbf{V}_\mathbf{x}^{-1}\mathbf{B}_\mathbf{x}\right) = \text{tr}\left(\mathbf{H}^\text{T}\mathbf{V}_{\mathbf{x}_0}^{-1}\mathbf{B}_{\mathbf{x}_0}\mathbf{H}\right).$$

The solution to Eqn. 6.33 is achieved by composing the matrix $\mathbf{H}$ from the eigenvectors $\Phi_1, \Phi_2 \ldots \Phi_b$ that are assigned to the b largest eigenvalues $\lambda_1 \geq \lambda_2 \geq \ldots \geq \lambda_b$ of $\mathbf{V}_{\mathbf{x}_0}^{-1}\mathbf{B}_{\mathbf{x}_0}$. The computationally complex preparation of the transformation matrix $\mathbf{H}$ is performed only once, off-line during the training phase of the BWE algorithm.

The LDA makes it possible to take many primary features of the bandlimited speech signal into account, using a high-dimensional super-vector $\mathbf{x}_0$. Nevertheless, the dimension of $\mathbf{x}$ can be small – without loosing too much discriminating power – such that the computational complexity and memory consumption of the subsequent estimation algorithm are low.

6.5.4 PRIMARY FEATURES

In the following text, brief definitions and descriptions of features of the narrowband speech $s_{\text{nb}}(k)$ typically used in bandwidth extension algorithms for speech are given. For further particulars on specific features, the reader is referred to Jax [128] and the cited literature.

Coefficients of the auto-correlation function (ACF) are often used for the voiced/unvoiced classification of speech segments (Campbell and Tremain [43], Wang [294]). Especially, the normalized first coefficient of the ACF is used because it reflects the spectral tilt of the signal spectrum. The normalized coefficients of the ACF can be estimated as follows

$$x_{\text{acf}}(\lambda) = \frac{\sum\limits_{\kappa=\lambda}^{N_\kappa-1} s_{\text{nb}}(\kappa - \lambda)\, s_{\text{nb}}(\kappa)}{\sum\limits_{\kappa=0}^{N_\kappa-1} (s_{\text{nb}}(\kappa))^2}, \tag{6.34}$$

where λ denotes the index of the ACF coefficient, and N_κ is the number of samples per frame. Later, we will investigate the first ten coefficients ($\lambda = 1 \ldots 10$) and the ACF at the pitch lag (see below).

Coefficients of a linear prediction filter (LPC) In *linear predictive coding* (LPC) of speech signals, FIR prediction filters with the coefficients $a_i, i = 1 \ldots N_a$ are used (compare Sec. 6.2.1). The prediction filter is described by the difference equation

$$\tilde{s}_{\text{nb}}(k) = -\sum_{i=1}^{N_a} a_i\, s_{\text{nb}}(k - i). \tag{6.35}$$

The optimal (in the sense of minimizing the power of the error signal $s_{\text{nb}}(k) - \tilde{s}_{\text{nb}}(k)$) filter coefficients a_i are derived from the first coefficients of the auto-correlation function of the speech signal $s_{\text{nb}}(k)$ utilizing, for example, the Levinson–Durbin algorithm (Makhoul [166], Markel and Gray [168]).

The line spectral frequencies (LSFs) are an alternative representation of the LPC coefficients from the previous paragraph (Itakura [120]). The LSF coefficients have several advantageous properties with regard to coding and interpolation (Paliwal [196], Vary *et al.* [286]), compare Sec. 6.4.1. Therefore, this representation of the LPC coefficients is often used in speech codecs. It has been used for bandwidth extension in Miet *et al.* [174], Chennoukh *et al.* [50].

LPC-derived cepstral coefficients (LPC-cepstrum) The transfer function of an LP synthesis filter can be represented by an infinite sequence of cepstral coefficients (e.g. Sec. 6.4.1, Markel and Gray [168], Hagen [103]). The according cepstral coefficients c_i are calculated from the prediction coefficients a_i by the simple recursive formula from Eqn. 6.14. The cepstral representation has the advantage of a good decorrelation of the coefficients. This is advantageous for modelling the PDF $p(\mathbf{x})$. LPC-derived cepstral coefficients have been used for BWE first in Abe and Yoshida [11] and Avendano *et al.* [23].

Linear cepstral coefficients A cepstral representation of the spectral envelope of the speech signal can alternatively be calculated from the magnitude spectrum of the signal frame. For this purpose, the speech frame is transformed into the frequency domain via a *discrete Fourier transform* (DFT). Then, the logarithm is applied to the magnitude spectrum, and the result is transformed to the cepstral domain with an inverse DFT (Oppenheim and Schafer [194]). If the cepstrum is truncated, the coefficients represent a cepstrally smoothed version of the magnitude spectrum of the input speech signal.

The mel-frequency cepstral coefficients (MFCC) are based on the perceptually motivated mel-scale filter-bank. This representation is frequently used in speech recognition (Davis and Mermelstein [57], Rabiner and Juang [216]). The MFCC have been utilized for BWE in Enbom and Kleijn [64] and Nilsson and Kleijn [187].

The calculation of the MFCC vector for a signal frame is performed in several steps: first, a pre-emphasis filter is applied to the input speech. Then, the Fourier coefficients are calculated via windowing (Hamming window), zero-padding and DFT. The Fourier coefficients are combined into 31 filter-bank outputs according to the mel-scale filter-bank as, for example, defined by Davis and Mermelstein [57]. The inverse discrete cosine transform is applied to the logarithms of the 31 filter-bank outputs, which yields 31 cepstral coefficients. The first cepstral coefficients are finally combined in the MFCC feature vector. Note that the MFCC representation includes information on the gain or power of the input speech in the feature vector because the signal is not normalized during the calculation of the MFCC.

The normed frame energy Energy-based criteria provide a quite robust indication for voice activity, for example, Rabiner and Schafer [217], at least if the signal-to-noise ratio is sufficiently high. The signal power further differs for distinct speech sounds: in general the short-term power of the signal is greater for voiced sounds, while it is lower for unvoiced sounds. To become independent of long-term variations of the signal power, the frame energy has to be normalized adaptively.

For the mth frame, the normed logarithmic frame energy may, for example, be calculated by

$$x_{\mathrm{nrp}}(m) = \frac{\log E(m) - \log E_{\min}(m)}{\log \overline{E}(m) - \log E_{\min}(m)}, \tag{6.36}$$

with

$$E(m) = \sum_{\kappa=0}^{N_\kappa - 1} s_{\mathrm{nb}}^2(\kappa)$$

$$E_{\min}(m) = \min_{\mu=0}^{N_{\min}} E(m-\mu)$$

$$\overline{E}(m) = \alpha\, \overline{E}(m-1) + (1-\alpha)\, E(m).$$

A reasonable setting for telephone speech sampled with $f_s = 8\,\mathrm{kHz}$ and with a frame rate of 50 frames/sec is to use a forgetting factor of $\alpha = 0.96$ and a size of $N_{\min} = 200$ for the minimum search window.

The gradient index has been proposed for the voiced/unvoiced classification of speech segments (Paulus [207, 206]). The measure is based on the sum of magnitudes of the gradient of the speech signal at each change of direction

$$x_{\mathrm{gi}} = \sum_{\kappa=2}^{N_\kappa} \frac{\Psi(\kappa)\,|s_{\mathrm{nb}}(\kappa) - s_{\mathrm{nb}}(\kappa-1)|}{\sqrt{\frac{1}{N_\kappa} E(m)}}. \tag{6.37}$$

$\Psi(\kappa)$ is an indicator function for the 'change of direction' of the signal. That is, $\Psi(\kappa) = 1/2\,|\psi(\kappa) - \psi(\kappa-1)|$, where the variable $\psi(\kappa)$ denotes the sign of the gradient $s_{\mathrm{nb}}(\kappa) - s_{\mathrm{nb}}(\kappa-1)$, that is, $\psi(\kappa) \in \{-1, 1\}$.

The zero-crossing rate counts the number of times the signal crosses the zero level within each frame (Sec. I.3.2.4 and Eqn. I.1). The zero-crossing rate has been used extensively in speech recognition (Rabiner and Schafer [217]) and as a voicing criterion (Atal and Rabiner [22], Campbell and Tremain [43], and Wang [294]).

The pitch period in the current speech frame depends on the instantaneous fundamental frequency F_0 of the speech signal. Its calculation is based on the auto-correlation function of the signal (see above): the position of the local maximum of the ACF in a limited range of reasonable time lags, for example, corresponding to pitch frequencies between 60 and 400 Hz, gives the estimated pitch period (e.g. Hess [109], Vary *et al.* [286]).

The kurtosis is a measure from higher-order statistics that is based on the fourth and second-order moments of the signal. Here, we use an estimate of the *local* kurtosis

(Krishnamachari [149])

$$x_{\mathrm{k}} = \frac{\frac{1}{N_\kappa}\sum_{\kappa=0}^{N_\kappa-1}(s_{\mathrm{nb}}(\kappa))^4}{\left(\frac{1}{N_\kappa}\sum_{\kappa=0}^{N_\kappa-1}(s_{\mathrm{nb}}(\kappa))^2\right)^2}. \tag{6.38}$$

The kurtosis is a measure of the 'Gaussianity' of a random signal, and it is a dimensionless parameter. A Gaussian random variable has a kurtosis of 3. In the short-term, it can be observed that the local kurtosis is less than 3 for most voiced speech sounds. There are substantial peaks in the local kurtosis measure at the onset of plosives and of strong vowels.

The spectral centroid is defined as the 'centre of gravity' of the magnitude spectrum of the bandlimited speech, reflecting its 'brightness' (see Sec. 1.4.6 on timbre)

$$x_{\mathrm{sc}} = \frac{\sum_{i=0}^{N_i/2} i \cdot \left|S_{\mathrm{nb}}(\mathrm{e}^{j\Omega_i})\right|}{\left(\frac{N_i}{2}+1\right)\sum_{i=0}^{N_i/2}\left|S_{\mathrm{nb}}(\mathrm{e}^{j\Omega_i})\right|}. \tag{6.39}$$

The quantity $S_{\mathrm{nb}}(\mathrm{e}^{j\Omega_i})$ labels the ith coefficient of a *discrete Fourier transformation* (DFT) of the length N_i of the input signal frame. The spectral centroid is mainly around 1500 Hz (corresponding to $x_{\mathrm{sc}} \approx 0.35$ for a sampling rate of 8 kHz) for voiced speech sounds and increases significantly for unvoiced speech segments (Heide and Kang [107], Abdelatty Ali *et al.* [17], Abdelatty Ali and Van der Spiegel [16]). Note that this definition of spectral centroid x_{sc} differs from the spectral centroid C_S (Eqn. 1.95).

The spectral flatness is defined as the ratio between the geometric and arithmetic mean of the estimated power spectrum

$$x_{\mathrm{sfm}} = \frac{\sqrt[N_i]{\prod_{i=0}^{N_i-1}\left|S_{\mathrm{nb}}(\mathrm{e}^{j\Omega_i})\right|^2}}{\frac{1}{N_i}\sum_{i=0}^{N_i-1}\left|S_{\mathrm{nb}}(\mathrm{e}^{j\Omega_i})\right|^2}. \tag{6.40}$$

Because the arithmetic mean of a sequence of non-negative values is always greater than (or equal to) its geometric mean, the spectral flatness is between zero and one. The spectral flatness has, for example, been used to measure the tonality of signal segments (Johnston [136]).

6.5.5 EVALUATION

In this section, feature extraction for the typical application of bandwidth extension of telephone speech will be considered. That is, the narrowband speech signal $s_{\mathrm{nb}}(k)$ has frequency components in the range of 0.3–3.4 kHz. By the BWE algorithm, a wideband signal $s_{\mathrm{wb}}(k)$ with frequency components from 50 Hz up to 7 kHz shall be produced. We distinguish between the applications of low-frequency extension (below 300 Hz) and high-frequency extension (3.4–7 kHz).

For measuring mutual information and separability, speech signals are sub-divided into frames with a length of 20 ms. For each signal frame, all of the primary features defined in Sec. 6.5.4 and (from corresponding wideband speech) the vector $\mathbf{y}$ is determined. The vector $\mathbf{y}$ consists of weighted cepstral coefficients representing the gain and shape of the spectral envelope within the missing frequency band (50–300 Hz respectively 3.4–7 kHz) according to Eqn. 6.22. All of the measurements were performed using the BAS SI100 speech corpus (Schiel [234]).

Mutual information and separability The estimated mutual information and separabilities between $\mathbf{y}$ and the investigated primary features are listed in Table 6.1. It can be observed that the features describing the spectral envelope of the bandlimited speech in fact play a major role for the bandwidth extension. Both the mutual information and the separability measures are maximal for these features. It must be taken into account, however, that the dimension of the primary features from this group is ten times higher than those of the scalar features. The mutual information and separabilities for the MFCC

Table 6.1 Estimates of mutual information $I(\mathbf{x};\mathbf{y})$ and separability $\zeta(\mathbf{x})$ for bandwidth extension of telephone speech (0.3–3.4 kHz). For calculating the separability, the 16 classes were defined by vector quantizing the true wideband spectral envelope representation $\mathbf{y}$ (Jax and Vary [133])

Feature vector $\mathbf{x}$	dim $\mathbf{x}$	Towards high frequencies		Towards low frequencies	
		$I(\mathbf{x};\mathbf{y})$ [bit/frame]	$\zeta(\mathbf{x})$ (16 classes)	$I(\mathbf{x};\mathbf{y})$ [bit/frame]	$\zeta(\mathbf{x})$ (16 classes)
ACF	10	2.6089	1.6349	2.7530	2.3977
LPC	10	2.3054	1.5295	2.1100	1.7901
LSF	10	2.3597	1.5596	2.2125	2.5817
LPC-cepstrum	10	2.2401	1.4282	2.1778	2.3879
Cepstrum	10	2.3075	1.5483	1.9398	2.5473
MFCC	10	2.3325	2.2659	3.0771	6.6142
ACF (1)	1	0.7514	1.1237	0.7324	1.1065
ACF (pitch period)	1	0.4450	0.4058	0.5441	0.6745
Frame energy	1	0.9285	1.0756	1.3968	4.2328
Gradient index	1	0.8011	1.2520	0.5403	0.6983
Zero-crossing rate	1	0.7453	1.0795	0.7456	1.1685
Pitch period	1	0.2451	0.0530	0.4823	0.1122
Local kurtosis	1	0.2037	0.0225	0.2979	0.0809
Spectral centroid	1	0.7913	1.0179	0.6630	0.9276
Spectral flatness	1	0.4387	0.3538	0.4201	0.4648

feature vector are highest because this vector already incorporates some information on the power level of the narrowband speech.

To achieve the best results with the BWE algorithm, it can further be motivated from Table 6.1 to additionally include certain scalar features in the feature vector. Particularly, the consideration of the frame energy as well as the gradient index, zero-crossing rate, and/or spectral centroid is very promising.

Linear discriminant analysis To evaluate the impact of a linear discriminant analysis, the estimation quality obtained with the transformed feature vectors was determined. The HMM-based MMSE estimation rule from Sec. 6.9 (Jax and Vary [133]) was used with $N_S = 64$ HMM states and $L = 16$ mixture components in the state-specific GMMs. Both speaker-dependent and speaker-independent models were investigated. The results are expressed in terms of the *root mean square log spectral distortion* (RMS LSD) of the estimated spectral envelope within the missing frequency band, according to the definition of the performance measure from Sec. 6.4.2. The 15-dimensional feature super-vector $\mathbf{x}_0$ consisted of the first 10 normalized auto-correlation coefficients, the zero-crossing rate, the normed frame energy, the gradient index, the local kurtosis, and the spectral centroid.

In Fig. 6.20, the mean performances that were obtained both without LDA and with the application of LDAs for the dimensions $b = 1 \ldots 5$ are depicted. As expected, the distortions of the estimates are decreased by increasing the dimension of the LDA transform. Remarkably, the achieved performances with a dimension of the LDA transform of $b = 5$ are even superior to those of the estimator that uses the original non-transformed feature vectors with a dimension of $\beta = 15$. This effect is the result of the improved compactness of the feature vectors: if the dimension of the feature vectors $\mathbf{x}$ is reduced significantly, the quality of the statistical modelling is enhanced. Thus, by utilizing a linear discriminant analysis, the performance and robustness of the bandwidth extension system can be improved, yet simultaneously reducing the computational complexity of the estimation algorithm substantially.

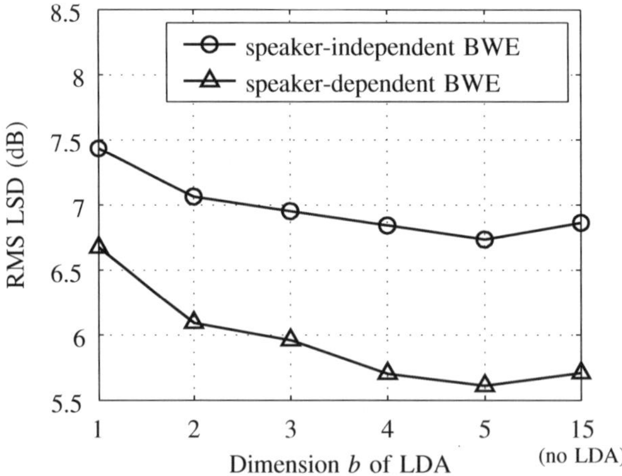

Figure 6.20 Impact of a linear discriminant analysis on the estimation performance

6.6 CODEBOOK MAPPING

The first and most commonly used method for estimating the wideband spectral envelope is the codebook mapping approach (Carl and Heute [45], Yoshida and Abe [301], Carl [44], Yasukawa [299]). The principle of this class of algorithms is based on the observation that there occur only a limited number of typical sounds (i.e. typical shapes of the spectral envelope) in speech signals. Accordingly, the codebook mapping approach is based on a pair of coupled codebooks that contain representations of the spectral envelopes of the narrowband and wideband speech, respectively.

The basic algorithm is depicted in Fig. 6.21. For each signal frame, the spectral envelope of the narrowband speech signal, represented by the feature vector $\mathbf{x}$, is compared to a list of typical narrowband spectral envelopes that are stored in a pre-trained codebook. The most similar codebook entry is selected. In parallel to the searched primary narrowband codebook, there exists a second codebook, the so-called shadow codebook, which contains corresponding wideband spectral envelope representatives. Hence, the estimate $\tilde{\mathbf{y}}$ of the wideband spectral envelope is simply the entry of the shadow codebook that is assigned to the previously selected codebook entry of the narrowband codebook.

The estimates $\tilde{\mathbf{y}}$ are confined to the discrete entries $\hat{\mathbf{y}}$ of the shadow codebook. This is beneficial, on one hand, because it is guaranteed that the estimate yields stable LP synthesis filters in the BWE algorithm framework. On the other hand, the performance of the codebook mapping method is restricted by the number and quality of the entries in the corresponding codebooks.

To improve the performance of the codebook mapping approach, some authors have proposed to use interpolation methods. Then, instead of a simple table lookup, the estimate $\tilde{\mathbf{y}}$ is determined by a weighted sum of all or the most probable codebook entries

$$\tilde{\mathbf{y}} = \sum_{i=1}^{N_S} w_i \, \hat{\mathbf{y}}_i, \tag{6.41}$$

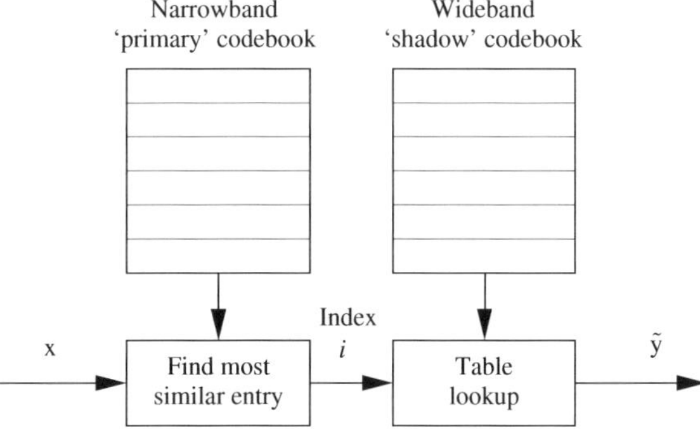

Figure 6.21 Estimation of the spectral envelope representation $\mathbf{y}$ via codebook mapping. Corresponding entries with the same index in both codebooks reflect properties of the same typical speech sound

where $\hat{\mathbf{y}}_i$ denotes the ith entry of the shadow codebook. The weights $0 \leq w_i \leq 1$ have to be normed such that $\sum_{i=1}^{N_S} w_i = 1$. The individual weights w_i are, for example, inverse proportional to the distance of the feature vector $\mathbf{x}$ to the respective codebook entry $\hat{\mathbf{x}}_i$ (Epps [65]).

6.6.1 VECTOR QUANTIZATION AND TRAINING OF THE PRIMARY CODEBOOK

In the following paragraph, the entries of the primary narrowband codebook shall be defined by vector quantization (VQ) of the feature vector $\mathbf{x}$. Each code vector of the VQ represents the properties of a typical speech sound. For a comprehensive introduction into vector quantization see, for example, Gersho and Gray [87] and Gray and Neuhoff [97].

Vector quantization of the b-dimensional feature vectors $\mathbf{x} = [x_0, \ldots x_{b-1}]^{\mathrm{T}}$ is described by the mapping $Q : \mathbb{R}^b \rightarrow \mathcal{C}_{\mathbf{x}}$ from the b-dimensional Euclidian space into the finite sub-space $\mathcal{C}_{\mathbf{x}}$. This sub-space is defined by a codebook $\mathcal{C}_{\mathbf{x}}$ in which all possible representatives $\hat{\mathbf{x}}_i$ are combined, that is, $\mathcal{C}_{\mathbf{x}} = \{\hat{\mathbf{x}}_1, \ldots \hat{\mathbf{x}}_{N_S}\}$. The number of representatives is denoted by N_S.

The quantization mapping Q is defined such as to minimize some error criterion $d(\mathbf{x}, \hat{\mathbf{x}}_i)$ between the input vectors $\mathbf{x}$ and the entries $\hat{\mathbf{x}}_i$, $i = 1 \ldots N_S$ of the vector codebook

$$Q(\mathbf{x}) = \arg \min_{\hat{\mathbf{x}}_i \in \mathcal{C}_{\mathbf{x}}} \mathrm{d}(\mathbf{x}, \hat{\mathbf{x}}_i). \tag{6.42}$$

Note that quantization with this mapping in general requires an extensive codebook search. There exist computationally more efficient, albeit sub-optimal, schemes that are commonly used for large codebooks, for example, *multi-stage vector quantization* (MSVQ) (Gersho [87]).

By the mapping from Eqn. 6.42, a region Υ_i of the b-dimensional Euclidian space, the quantizer cell, is assigned to each code vector $\hat{\mathbf{x}}_i$

$$\Upsilon_i = \left\{\mathbf{x} \in \mathbb{R}^b : \ \mathrm{d}(\mathbf{x}, \hat{\mathbf{x}}_i) < \ \mathrm{d}(\mathbf{x}, \hat{\mathbf{x}}_j), \forall \hat{\mathbf{x}}_j \in \mathcal{C}_{\mathbf{x}} \setminus \{\hat{\mathbf{x}}_i\}\right\}. \tag{6.43}$$

The set union of all quantizer regions Υ_i fills the entire Euclidian space $\mathbb{R}^b$ without overlappings, that is, $\bigcup_{i=1}^{N_S} \Upsilon_i = \mathbb{R}^b$, and $\Upsilon_i \cap \Upsilon_j = \emptyset$ for any $j \neq i$.

During training of the vector quantizer, the objective is to minimize the mean quantization distortion. For a fixed number N_S of code vectors, this is accomplished by modifying the code vectors according to

$$\begin{aligned} \mathcal{C}_{\mathbf{x}} &= \arg \min_{\mathcal{C}} E\{\ \mathrm{d}(\mathbf{x}, Q(\mathbf{x}))\} \\ &\approx \arg \min_{\mathcal{C}} \frac{1}{N_m} \sum_{m=0}^{N_m - 1} \min_{\hat{\mathbf{x}} \in \mathcal{C}} \ \mathrm{d}(\mathbf{x}(m), \hat{\mathbf{x}}). \end{aligned} \tag{6.44}$$

This task is usually performed by an iterative refinement procedure based on a large set of training vectors $\mathbf{x}(m), m = 0 \ldots N_m - 1$. Commonly, the well-known LBG algorithm (Linde *et al.* [162]) is used, which is a variant of the generalized Lloyd algorithm (Lloyd [163]). By the training, a clustering of the training data is obtained.

6.6.2 TRAINING OF THE SHADOW CODEBOOK

Because there is a fixed relationship between the entries of the primary codebook and of the shadow codebook, the shadow codebook cannot be trained until having obtained the primary codebook. The entries $\hat{\mathbf{y}}_i$ of the shadow codebook $\mathcal{C}_{\mathbf{y}}$ are then defined by clustering of the training data w.r.t. the primary codebook $\mathcal{C}_{\mathbf{x}}$.

Since the mapping between $\mathbf{x}$ and $\mathbf{y}$ is not unique, the 'quantizer cells', defined by VQ of $\mathbf{x}$ but situated in the Euclidian space of $\mathbf{y}$, will in general be overlapping. Individual cells may even be discontiguous. Accordingly, to describe the regions assigned to the ith code vector $\hat{\mathbf{y}}_i$, it is not sufficient to use a minimum distortion criterion like in Eqn. 6.43. Instead, the d-dimensional conditional probability density function $p(\mathbf{y}|\mathbf{x} \in \Upsilon_i)$ with $\mathbf{y} = [y_0 \dots y_{d-1}]^{\mathrm{T}}$ has to be employed. With this, we can define the optimal code vectors as

$$\begin{aligned}\hat{\mathbf{y}}_i &= \arg\min_{\hat{\mathbf{y}} \in \mathbb{R}^d} E\left\{ \mathrm{d}(\mathbf{y}, \hat{\mathbf{y}}) \middle| \mathbf{x} \in \Upsilon_i \right\} \\ &= \arg\min_{\hat{\mathbf{y}} \in \mathbb{R}^d} \int_{\mathbb{R}^d} p(\mathbf{y}|\mathbf{x} \in \Upsilon_i)\ \mathrm{d}(\mathbf{y}, \hat{\mathbf{y}})\, \mathrm{d}\mathbf{y}. \end{aligned} \tag{6.45}$$

If, for example, the mean-square error criterion $d(\mathbf{y}, \hat{\mathbf{y}}) = \|\mathbf{y} - \hat{\mathbf{y}}\|^2$ is used, then Eqn. 6.45 is solved by the conditional expectation $\hat{\mathbf{y}}_i = E\{\mathbf{y}|\mathbf{x} \in \Upsilon_i\}$. The ith expectation can be determined using a large number of pairs of training vectors $\{\mathbf{x}(m), \mathbf{y}(m)\}$, $m = 1 \dots N_m$ by averaging the vectors $\mathbf{y}$ extracted from those signal frames for which $\mathbf{x}(m) \in \Upsilon_i$.

The performance of the vector quantization approach strongly depends on the choice of representations of $\mathbf{x}$ and $\mathbf{y}$, and on the chosen distortion measures $d(\mathbf{x}, \hat{\mathbf{x}})$ respectively $d(\mathbf{y}, \hat{\mathbf{y}})$. Since in most implementations of the codebook mapping method for BWE the feature vector $\mathbf{x}$ reflects some representation of the spectral envelope of the narrowband speech, usually distortion measures used in speech coding for optimizing the quantization of LPC coefficients are employed, for example, Gray and Markel [95] and Gray *et al.* [96]. In principle, it is not necessary that the distortion measures for the primary and shadow codebook are identical.

Besides the distortion measure(s), the performance of codebook mapping depends on the sizes of the primary and/or shadow codebook. The estimation distortion is lower, the higher the number of codebook entries. Several authors have found that the codebook mapping performance in bandwidth extension for telephone speech saturates for codebook sizes greater than about 256 (Carl [44], Epps [65]). In Epps [65], algorithms have been developed that allow to decrease the size of the shadow codebook (with a fixed primary codebook) without sacrificing lots of performance.

6.7 LINEAR MAPPING

Another approach that has been used successfully to estimate the representation $\mathbf{y}$ of the wideband spectral envelope from a feature vector $\mathbf{x}$ is by linear mapping or piecewise-linear mapping, for example, Nakatoh *et al.* [183], Epps and Holmes [66], and Miet *et al.* [174]. With linear mapping, an estimate of the unknown quantity $\mathbf{y}$ is derived from the observed feature vector $\mathbf{x}$ by the transformation

$$\tilde{\mathbf{y}} = \mathbf{A}^{\mathrm{T}} \cdot \mathbf{x}. \tag{6.46}$$

The dimension of the transformation matrix $\mathbf{A}$ is $b \times d$ with $b = \dim \mathbf{x}$ and $d = \dim \mathbf{y}$. The whole a priori knowledge on dependencies between $\mathbf{x}$ and $\mathbf{y}$ is contained in the matrix $\mathbf{A}$, which is derived and stored during the off-line training phase of the BWE system (see below). Therefore, there are no large memory requirements with this approach. Further, the estimation rule in Eqn. 6.46 is very simple to implement and computationally efficient.

A problem of the linear mapping algorithm is that it is in general not possible to strictly confine the estimates $\tilde{\mathbf{y}}$ to a reasonable and admissible range of values. Accordingly, mainly depending on the chosen representations of $\mathbf{x}$ and $\mathbf{y}$, applying the linear mapping rule sometimes results in an instability of the LP synthesis filter in the BWE system.

To prevent strong artefacts, such severe estimation errors are commonly concealed by mostly heuristically derived countermeasures. For example, in Chennoukh *et al.* [50], wideband LSF vectors ($\mathbf{y}$) are estimated from the narrowband LSF vectors ($\mathbf{x}$). This sometimes results in LSF elements larger than π, which makes it necessary to scale down the estimated vector so far that all elements are well below π (cf. Sec. 6.4.1). Although preventing the worst, such measures impair the mean estimation performance.

Another flavour of linear mapping was used in Avendano *et al.* [23]. This contribution has the distinctive feature that the estimate $\tilde{\mathbf{y}}(m)$ depends not only on the features $\mathbf{x}(m)$ extracted from the current frame m but also on the features from a number of preceding and following signal frames. In this case, the concept of linear mapping corresponds to multi-dimensional filtering.

6.7.1 TRAINING PROCEDURE

During the training phase of the bandwidth extension system, the transformation matrix $\mathbf{A}$ has to be found. For this purpose, a database of true wideband speech is needed. By band-pass filtering, the corresponding narrowband speech is produced, and both signals are cut into time-aligned signal frames. From each pair of wideband and narrowband signal frames, the vectors $\mathbf{y}$ and $\mathbf{x}$ are extracted and collected in two large matrices $\mathbf{F}_{\mathbf{y}}$ and $\mathbf{F}_{\mathbf{x}}$. The rows of the matrix $\mathbf{F}_{\mathbf{y}}$ consist of the d-dimensional training vectors $\mathbf{y}(m)$, $m = 0 \ldots N_m$ computed from the wideband speech, and the rows of $\mathbf{F}_{\mathbf{x}}$ consist of b-dimensional feature vectors $\mathbf{x}(m)$ extracted from the corresponding narrowband speech frames

$$\mathbf{F}_{\mathbf{y}} = \begin{pmatrix} \mathbf{y}^{\mathrm{T}}(0) \\ \mathbf{y}^{\mathrm{T}}(1) \\ \vdots \\ \mathbf{y}^{\mathrm{T}}(N_m - 1) \end{pmatrix} \quad \text{and} \quad \mathbf{F}_{\mathbf{x}} = \begin{pmatrix} \mathbf{x}^{\mathrm{T}}(0) \\ \mathbf{x}^{\mathrm{T}}(1) \\ \vdots \\ \mathbf{x}^{\mathrm{T}}(N_m - 1) \end{pmatrix}. \tag{6.47}$$

The number N_m denotes the number of signal frames in the training data set.

Now, the transformation matrix $\mathbf{A}$ shall be optimized such as to minimize the model error $\mathbf{y} - \mathbf{A}^{\mathrm{T}}\mathbf{x}$ for the complete training data set. This training procedure results from a least squares approach, leading to minimization of the trace criterion

$$e^2 = \mathrm{tr}\left[(\mathbf{F_y} - \mathbf{F_x A})^{\mathrm{T}}(\mathbf{F_y} - \mathbf{F_x A})\right]. \tag{6.48}$$

This is the Frobenius norm of the error $\mathbf{F_y} - \mathbf{F_x A}$, namely the sum of squares of all differences (estimation errors) $y_i(m) - \tilde{y}_i(m)$. Derivation with respect to one element a_{ij} of the transformation matrix $\mathbf{A}$ delivers (Scharf [233])

$$\frac{\partial e^2}{\partial a_{ij}} = -2\,\mathrm{tr}\left[(\mathbf{F_y} - \mathbf{F_x}\,\mathbf{A})^{\mathrm{T}}\mathbf{F_x}\frac{\partial \mathbf{A}}{\partial a_{ij}}\right]. \tag{6.49}$$

For the minimization of e^2, the zero point of the derivative (Eqn. 6.49) has to be found, that is, Eqn. 6.49 has to be solved for the unknown matrix $\mathbf{A}$. This leads to the condition

$$(\mathbf{F_y} - \mathbf{F_x A})^{\mathrm{T}}\mathbf{F_x} \equiv 0. \tag{6.50}$$

Because this condition is independent of the position of the element a_{ij}, solving it is sufficient for finding all elements of the transformation matrix $\mathbf{A}$.

A unique solution for the least squares problem can be found if and only if the inverse of the Gram matrix $\mathbf{F_x^T}\,\mathbf{F_x}$ exists[6]. Then, we arrive at the training algorithm

$$\mathbf{A} = \left(\mathbf{F_x^T}\,\mathbf{F_x}\right)^{-1}\mathbf{F_x^T}\,\mathbf{F_y}. \tag{6.51}$$

Note that the calculation of Eqn. 6.51 is not trivial. It requires computation and inversion of the Gram matrix $\mathbf{F_x^T}\,\mathbf{F_x}$, which is a huge $N_m \times N_m$ square matrix.

To verify the solution of Eqn. 6.51, we have to investigate the second derivative of the trace criterion (Eqn. 6.48). With $\frac{\partial^2 \mathbf{A}}{\partial a_{ij}^2} = 0$, we get (Scharf [233])

$$\frac{\partial}{\partial a_{ij}}\left(\frac{\partial e^2}{\partial a_{ij}}\right) = \mathrm{tr}\left[\left(\mathbf{F_x}\frac{\partial \mathbf{A}}{\partial a_{ij}}\right)^{\mathrm{T}}\mathbf{F_x}\frac{\partial \mathbf{A}}{\partial a_{ij}}\right] \geq 0 \tag{6.52}$$

which is the Frobenius norm of the matrix $\mathbf{F_x}\frac{\partial \mathbf{A}}{\partial a_{ij}}$ and therefore always non-negative. Accordingly, we can be sure that the solution of Eqn. 6.51 indeed minimizes Eqn. 6.48.

6.7.2 *PIECEWISE-LINEAR MAPPING*

The basic problem of the linear mapping approach described above is that the statistical model of a linear dependency between feature vector $\mathbf{x}$ and spectral envelope representation $\mathbf{y}$ in general is too simple to describe the true relationship. Consequently, the linear mapping approach has been extended by a preceding classification stage (Nakatoh *et al.* [183], Chennoukh *et al.* [50]), to better reflect the possibly non-linear relationship between $\mathbf{x}$ and $\mathbf{y}$: first, a classification of the feature vector $\mathbf{x}(m)$ is performed. Then, if the ith

[6] The inverse of the Gram matrix $\mathbf{F_x^T}\,\mathbf{F_x}$ exists iff the columns of $\mathbf{F_x}$ are mutually independent.

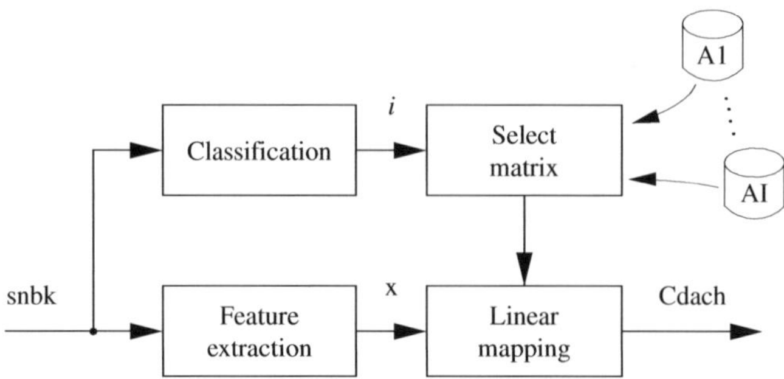

Figure 6.22 Estimation of the wideband spectral envelope representation $\mathbf{y}$ via piecewise-linear mapping

class has been detected, a specific matrix $\mathbf{A}_i$ is used to determine the estimated spectral envelope representation $\tilde{\mathbf{y}}$ by linear mapping as described in Eqn. 6.46. The method is illustrated in Fig. 6.22.

For the first stage in Fig. 6.22, any classification method can be used. In literature, for example, vector quantization of the feature vector $\mathbf{x}$ has been utilized in Nakatoh *et al.* [183], or it was aimed at detecting certain phonemes via thresholding the reflection factors of the narrowband speech in Chennouk *et al.* [50]. Instead of a hard classification, the first stage in Fig. 6.22 can be enhanced to a soft decision scheme. Then, the final estimate $\tilde{\mathbf{y}}$ is determined by a weighted sum of individual mappings obtained for all classes (Nakatoh *et al.* [183]).

For the training of a piecewise-linear mapping method, the mapping matrix $\mathbf{A}_i$ for the ith class is computed by Eqn. 6.51 using only those signal frames from the training data set for which the ith class has been detected by the classification rule. It appears advantageous that the same classification rule is used during the training and application phase of the BWE system. Otherwise, there might be a model mismatch in the class-specific linear mapping matrices $\mathbf{A}_i$.

6.8 GAUSSIAN MIXTURE MODEL

The linear mapping approach described above has the disadvantage that the statistical model is principally limited to multivariate normal distributions. To employ more sophisticated statistically optimized estimation schemes, a more exact model of the joint PDF $p(\mathbf{x}, \mathbf{y})$, describing the joint behaviour of the two multi-dimensional random variables $\mathbf{x}$ and $\mathbf{y}$, is necessary (Park and Kim [200], Raza and Chan [220]). Then, even non-linear dependencies of $\mathbf{x}$ and $\mathbf{y}$ may be exploited.

Because the dimension $\dim \mathbf{x} + \dim \mathbf{y}$ of the joint PDF $p(\mathbf{x}, \mathbf{y})$ may be fairly large, prohibiting the use of histograms owing to memory constraints, it is necessary in practice to approximate the PDF by some parametric model $\tilde{p}(\mathbf{x}, \mathbf{y})$. Then, only the model parameters have to be obtained and stored in the off-line training phase.

For the definition of the joint PDF in the following, the two vectors $\mathbf{x} = [x_0, \ldots x_{b-1}]^{\mathrm{T}}$ and $\mathbf{y} = [y_0, \ldots y_{d-1}]^{\mathrm{T}}$ shall be combined in the column vector $\mathbf{z} = [\mathbf{x}^{\mathrm{T}}\, \mathbf{y}^{\mathrm{T}}]^{\mathrm{T}}$. A common way to model unknown high-dimensional real-world probability density functions is the approximation with Gaussian mixture models (GMM, see e.g. Reynolds and Rose [222], Vaseghi [287]). In these parametric models, the PDF is approximated by the sum of weighted multivariate Gaussian distributions

$$p(\mathbf{x},\mathbf{y}) \approx \tilde{p}(\mathbf{x},\mathbf{y}) = \tilde{p}(\mathbf{z}) = \sum_{l=1}^{L} \rho_l \,\mathcal{N}(\mathbf{z};\mu_{\mathbf{z},l},\mathbf{V}_{\mathbf{z},l}) \tag{6.53}$$

with mean vectors $\mu_{\mathbf{z},l}$ and covariance matrices $\mathbf{V}_{\mathbf{z},l}$. The individual $(b+d)$-dimensional joint Gaussian densities (with $\dim \mathbf{z} = b+d$, $b = \dim \mathbf{x}$, $d = \dim \mathbf{y}$) are given by

$$\mathcal{N}(\mathbf{z};\mu_{\mathbf{z},l},\mathbf{V}_{\mathbf{z},l}) = \frac{\sqrt{\det \mathbf{A}_{\mathbf{z},l}}}{(2\pi)^{(b+d)/2}} \exp\left(-\frac{1}{2}(\mathbf{z}-\mu_{\mathbf{z},l})^{\mathrm{T}} \mathbf{A}_{\mathbf{z},l}\,(\mathbf{z}-\mu_{\mathbf{z},l})\right), \tag{6.54}$$

where $\mu_{\mathbf{z},l} = [\mu_{\mathbf{x},l}^{\mathrm{T}} \mu_{\mathbf{y},l}^{\mathrm{T}}]^{\mathrm{T}}$ and $\mathbf{A}_{\mathbf{z},l}$ is the inverse of the covariance matrix $\mathbf{V}_{\mathbf{z},l}$

$$\mathbf{A}_{\mathbf{z},l} = \begin{pmatrix} \mathbf{A}_{\mathbf{xx},l} & \mathbf{A}_{\mathbf{xy},l} \\ \mathbf{A}_{\mathbf{yx},l} & \mathbf{A}_{\mathbf{yy},l} \end{pmatrix} = \mathbf{V}_{\mathbf{z},l}^{-1} = \begin{pmatrix} \mathbf{V}_{\mathbf{xx},l} & \mathbf{V}_{\mathbf{xy},l} \\ \mathbf{V}_{\mathbf{yx},l} & \mathbf{V}_{\mathbf{yy},l} \end{pmatrix}^{-1}. \tag{6.55}$$

The scalar weighting factors ρ_l in Eqn. 6.53 define the relative contribution of the lth Gaussian distribution to the modelled PDF. The model represents a true PDF if the weighting factors meet the constraints

$$0 \le \rho_l \le 1 \qquad \text{and} \qquad \sum_{l=1}^{L} \rho_l = 1. \tag{6.56}$$

The parameters of the GMM are combined in the set $\Theta = \{\Theta_l;\ l = 1, 2, \ldots L\}$ with the subsets $\Theta_l = \{\rho_l, \mu_{\mathbf{z},l}, \mathbf{V}_{\mathbf{z},l}\}$ of the respective parameters of the individual Gaussian mixture components.

It has been shown that any smooth continuous probability density function can be approximated arbitrarily closely by increasing the model order L (Sorenson [254]). To show the qualitative behaviour, Gaussian mixture models with different orders L for an exemplary two-dimensional random process are illustrated in Fig. 6.23.

6.8.1 MINIMUM MEAN SQUARE ERROR ESTIMATION

The *minimum mean square error* (MMSE) estimation rule shall take the joint PDF $p(\mathbf{x},\mathbf{y})$ into consideration. The aim of the MMSE criterion is the minimization of

$$\begin{aligned} \mathcal{D}_{\text{MSE}}(\mathbf{y},\tilde{\mathbf{y}}|\mathbf{x}) &= E\left\{\|\mathbf{y}-\tilde{\mathbf{y}}\|^2 \middle| \mathbf{x}\right\} \\ &= \int_{\mathbb{R}^d} p(\mathbf{y}|\mathbf{x})\, \|\mathbf{y}-\tilde{\mathbf{y}}\|^2 \, d\mathbf{y} \end{aligned} \tag{6.57}$$

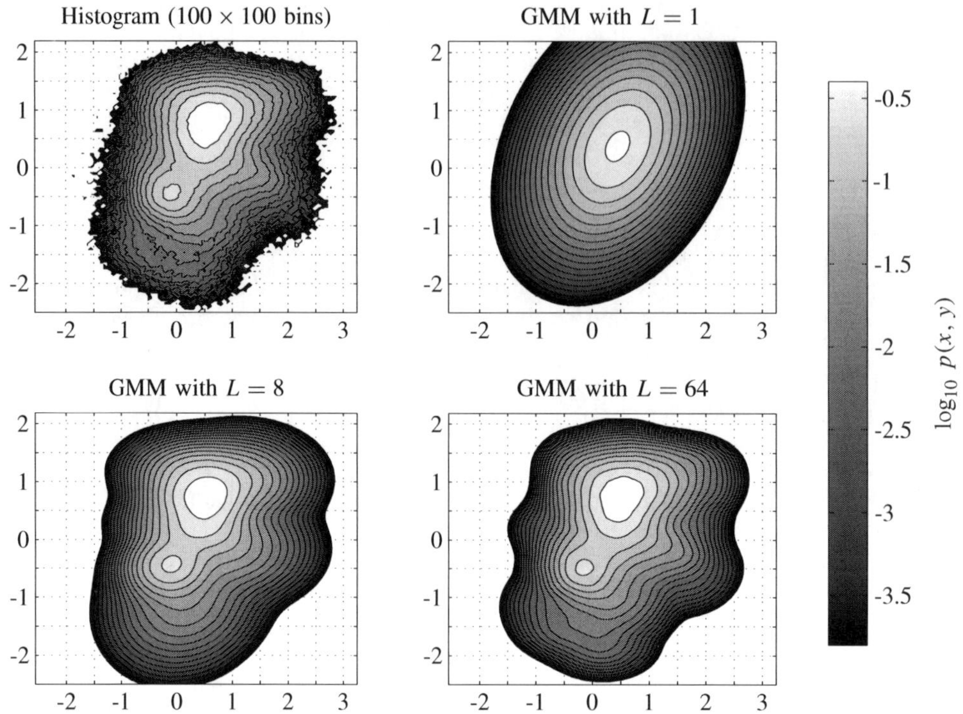

Figure 6.23 Contour diagrams of Gaussian mixture models for two-dimensional exemplary data. In the upper left diagram, a measured histogram of the training data is shown. The other diagrams illustrate GMMs of this data with 1, 8, and 64 multivariate normal distributions, respectively

The solution is the conditional expectation

$$\tilde{\mathbf{y}} = E\{\mathbf{y}|\mathbf{x}\} = \int_{\mathbb{R}^d} \mathbf{y}\, p(\mathbf{y}|\mathbf{x})\, \mathrm{d}\mathbf{y}, \tag{6.58}$$

which can be calculated in closed form from the parameters of a GMM of the joint PDF $p(\mathbf{x}, \mathbf{y})$ (Park and Kim [200], Jax [128])

$$\tilde{\mathbf{y}} = \sum_{l=1}^{L} \rho_{\mathbf{y}|\mathbf{x},l} \left(\mu_{\mathbf{y},l} - \left((\mathbf{x} - \mu_{\mathbf{x},l})^{\mathrm{T}} \mathbf{A}_{\mathbf{xy},l} \mathbf{A}_{\mathbf{yy},l}^{-1} \right)^{\mathrm{T}} \right). \tag{6.59}$$

The 'new' weighting factors $\rho_{\mathbf{y}|\mathbf{x},l}$ are defined by

$$\rho_{\mathbf{y}|\mathbf{x},l} = \frac{\rho_l\, \mathcal{N}(\mathbf{x}; \mu_{\mathbf{x},l}, \mathbf{V}_{\mathbf{xx},l})}{\sum_{l=0}^{L} \rho_l\, \mathcal{N}(\mathbf{x}; \mu_{\mathbf{x},l}, \mathbf{V}_{\mathbf{xx},l})}. \tag{6.60}$$

The parameters $\mu_{\mathbf{x},l}$ and $\mathbf{V}_{\mathbf{xx},l}$ of the marginal Gaussian densities $\mathcal{N}(\mathbf{x};\, \mu_{\mathbf{x},l}, \mathbf{V}_{\mathbf{xx},l})$ can be determined from the parameters of the GMM using the definitions from Eqn. 6.55.

Bandwidth extension using raw Gaussian mixture models has been introduced in Park and Kim [200]. GMMs are also often used as parts of *hidden Markov models* (HMMs), which will be treated in the next section.

6.8.2 TRAINING BY THE EXPECTATION-MAXIMIZATION ALGORITHM

The parameters of the Gaussian mixture model are determined and stored during an off-line training phase. For the training of a GMM, a variety of algorithms have been proposed, which are based on different optimization criteria, for example, Bahl *et al.* [24], Gopalakrishnan *et al.* [94], Valtchev *et al.* [280], Schlüter and Macherey [235], Hedelin and Skoglund [106], Povey and Woodland [212], and Yang and Zwolinksi [298]. Most of these training algorithms have been applied to the training of statistical models for speech recognition.

Here, the *expectation-maximization* (EM) algorithm shall be outlined, which is prevalent in the GMM literature (e.g. Dempster *et al.* [59], Reynolds and Rose [222], Moon [175], Vaseghi [287]). The optimization criterion in the EM algorithm is the maximization of the log-likelihood function

$$\mathcal{L}(\Theta) = \log\left(\prod_{\mathbf{z}\in\Xi} \tilde{p}(\mathbf{z};\,\Theta)\right) = \sum_{\mathbf{z}\in\Xi} \log\left(\sum_{l=1}^{L} \rho_l\, \mathcal{N}(\mathbf{z};\,\Theta_l)\right). \qquad (6.61)$$

Consequently, the method realizes a maximum likelihood (ML) optimization of the parameters Θ of the model, corresponding to a minimization of the Kullback-Leibler distance between the PDF $p(\mathbf{x}, \mathbf{y})$ and its model $\tilde{p}(\mathbf{x}, \mathbf{y})$ (Cover and Thomas [52]). The training of the GMM is based on a set Ξ of training vectors that are taken from the original random process $\{\mathbf{x}(m), \mathbf{y}(m);\ m = 1 \ldots N_m\}$. The number of data vectors in the training set is denoted by $N_\Xi = |\Xi|$.

Unfortunately, the log-likelihood term (Eqn. 6.61) contains the logarithm of a sum such that a closed-form analytical solution for the maximization of $\mathcal{L}(\Theta)$ cannot be formulated. Instead, the EM approach leads to an iterative numerical training algorithm. The parameters ρ_l, $\mu_{\mathbf{z},l}$, and $\mathbf{V}_{\mathbf{z},l}$ of the GMM are refined in each iteration step (with the iteration index ν) by the following update equations

$$\rho_l^{(\nu+1)} = \frac{1}{N_\Xi} \sum_{\mathbf{z}\in\Xi} \psi_1^{(\nu)}(\mathbf{z}),$$

$$\mu_{\mathbf{z},l}^{(\nu+1)} = \frac{\sum\limits_{\mathbf{z}\in\Xi} \psi_1^{(\nu)}(\mathbf{z}) \cdot \mathbf{z}}{\sum\limits_{\mathbf{z}\in\Xi} \psi_1^{(\nu)}(\mathbf{z})},$$

$$\mathbf{V}_{\mathbf{z},l}^{(\nu+1)} = \frac{\sum\limits_{\mathbf{z}\in\Xi} \psi_1^{(\nu)}(\mathbf{z}) \cdot \left(\mathbf{z} - \mu_{\mathbf{z},l}^{(\nu+1)}\right)\left(\mathbf{z} - \mu_{\mathbf{z},l}^{(\nu+1)}\right)^{\mathrm{T}}}{\sum\limits_{\mathbf{z}\in\Xi} \psi_1^{(\nu)}(\mathbf{z})}. \qquad (6.62)$$

The variable $\psi_1(\mathbf{z})$ is defined by the a posteriori probability

$$\psi_1^{(\nu)}(\mathbf{z}) = \frac{\rho_l^{(\nu)}\,\mathcal{N}(\mathbf{z};\,\mu_{\mathbf{z},l}^{(\nu)},\,\mathbf{V}_{\mathbf{z},l}^{(\nu)})}{\sum_{l=1}^{L}\rho_l^{(\nu)}\,\mathcal{N}(\mathbf{z};\,\mu_{\mathbf{z},l}^{(\nu)},\,\mathbf{V}_{\mathbf{z},l}^{(\nu)})}. \tag{6.63}$$

For a detailed derivation of these terms, refer to the literature, for example, Dempster *et al.* [59], Moon [175], and Vaseghi [287]. Note that by simple modification of the above update rules, certain structures of the model parameters can be enforced, for example, diagonal covariance matrices.

For the initialization of the model prior to applying the EM algorithm, the training data set is sub-divided into clusters, for example, using the well-known binary-split LBG algorithm (Linde *et al.* [162], cf. Sec. 6.6.1). The centroids and covariances of the feature vectors that are assigned to the individual clusters are then used as the initial parameters $\mu_{\mathbf{z},l}^{(0)}$ and $\mathbf{V}_{\mathbf{z},l}^{(0)}$ of the model. The weighting factors $\rho_l^{(0)}$ shall be proportional to the number of feature vectors in the lth cluster.

It is a property of the EM algorithm that, provided the same large training data set is used for each iteration, the log-likelihood function increases strictly monotonically with every iteration step of the EM algorithm (Dempster *et al.* [59], Vaseghi [287]), that is, $\mathcal{L}(\Theta^{(\nu+1)}) \geq \mathcal{L}(\Theta^{(\nu)})$. The training is continued until the relative increase of the log-likelihood between two iterations falls below a predefined value ε, that is, the stop condition is

$$\frac{\mathcal{L}(\Theta^{(\nu+1)}) - \mathcal{L}(\Theta^{(\nu)})}{\left|\mathcal{L}(\Theta^{(\nu)})\right|} \leq \varepsilon. \tag{6.64}$$

Owing to the monotonical increase of the log-likelihood function during the training, it is guaranteed that the EM algorithm approaches a local maximum of $\mathcal{L}(\Theta)$. However, it can, in general, not be guaranteed that the *global* maximum is found.

6.9 HIDDEN MARKOV MODEL

In this section, the source-filter model from Sec. 6.2 shall be extended by a *hidden Markov model* (HMM). A HMM is a discrete-time *composite source model* (CSM), consisting of a finite number of independent sub-sources that are controlled by a switch, compare Fig. 6.24 (a). Each setting of the switch defines a state of the model, and for each state of the model the statistical properties of the output signal of the CSM correspond to the statistical properties of the selected sub-source. It is further assumed that the position of the switch is governed by a Markov chain. In this case, the model is referred to as a hidden Markov model in literature. The statistical characteristics of the state sequence can be described by a matrix of transition probabilities. Further details on hidden Markov models can be found in the literature, for example, Rabiner [215], Papoulis [199], Rabiner and Juang [216], and Vaseghi [287].

The application of a hidden Markov model to the process of speech production is illustrated in Fig. 6.24 (a). The state $\mathcal{S}_i$ of the source is represented by a switch that redirects

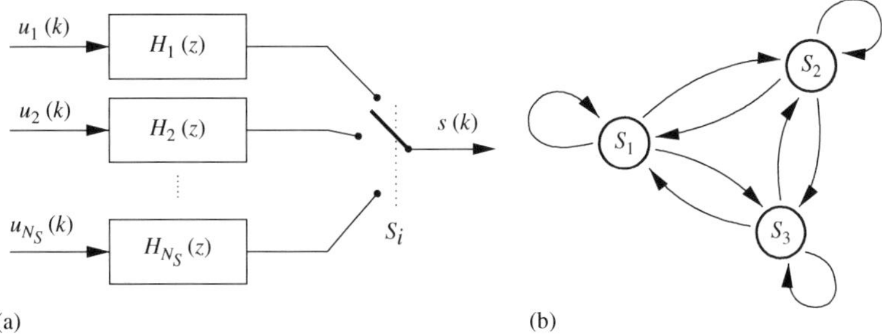

Figure 6.24 (a) Hidden Markov model of the process of speech generation. The AR filters $H_i(z)$ and excitation signals $u_i(k)$ represent typical (wideband) speech sounds for each state. (b) State transition diagram of an ergodic first-order Markov chain with $N_\mathcal{S} = 3$ states

the output of one of the sub-sources to the output $s(k)$ of the HMM. The sub-sources of the HMM correspond to individual source-filter models as introduced in Sec. 6.2.1. Each of the sub-sources is assumed to be *stationary* and represents the characteristics of one particular *wideband* speech sound. To simplify matters, state transitions are only allowed at the boundaries of signal frames in the following. Further, the transition from any state to any other state shall be possible, in which case the HMM is called ergodic[7].

For the bandwidth extension application, the states of the HMM are defined by vector quantization of the spectral envelope representation $\mathbf{y}$ (Jax *et al.* [129]). Every state $\mathcal{S}_i$ of the HMM corresponds to one entry $\hat{\mathbf{y}}_i$ of the VQ codebook $\mathcal{C}_\mathbf{y} = \{\hat{\mathbf{y}}_1 \ldots \hat{\mathbf{y}}_{N_\mathcal{S}}\}$ such that the number of states $N_\mathcal{S}$ in the HMM is the same as the number of entries in the codebook. The training of the VQ codebook is performed off-line with real wideband speech, using the LBG algorithm (Linde [162]).

Per definition, if wideband speech is available, the *true* state of the source in the mth frame can be determined by vector quantization of the spectral envelope representation $\mathbf{y}(m)$

$$\mathcal{S}_{\text{true}}(m) = \mathcal{S}_{i_{\text{opt}}(m)} \quad \text{where} \quad i_{\text{opt}}(m) = \arg \min_{i=1}^{N_\mathcal{S}} \|\mathbf{y}(m) - \hat{\mathbf{y}}_i\|^2, \tag{6.65}$$

The true state sequence is needed during the off-line training phase to obtain the parameters of the statistical model describing the Markov states.

Note that, with the above definition, the Markov state is *not hidden* during training. Nevertheless, wideband speech is only available in the training phase whereas in the application phase of the BWE system the states $\mathcal{S}_i$ have to be identified from the features $\mathbf{x}$ of the narrowband speech. Corresponding estimators will be described in Sec. 6.9.2.

6.9.1 STATISTICAL MODEL OF THE MARKOV STATES

For each possible state $\mathcal{S}_i$ of the hidden Markov model, the features $\mathbf{x}$ as well as the wideband spectral envelope representation $\mathbf{y}$ exhibit characteristic statistical properties.

[7] In certain applications, for example, for speech recognition, it is useful to restrict the possible state transitions to get directed state diagrams (e.g. a left–right model).

To describe these properties, a statistical model that consists of three parts is employed: the state and transition probabilities of the Markov chain, the probability density function (PDF) of the feature vectors $\mathbf{x}$ of the bandlimited speech (observation probability), and the PDF or expectation of the estimated quantity $\mathbf{y}$ (emission probability). We will elaborate on these three parts of the statistical model below. The information that is contained in the statistical model is needed as a priori knowledge for the subsequent estimation methods.

The training of the statistical model is performed off-line using a training database of corresponding wideband and bandlimited speech signals. The wideband speech is needed to calculate the true state sequence of the HMM. According to the true state $\mathcal{S}_{i_{\text{opt}}}$ of the HMM, as defined by Eqn. 6.65, the vectors $\mathbf{x}$ and $\mathbf{y}$ are assigned to the corresponding sets Ξ_i, $i = 1 \dots N_{\mathcal{S}}$).

Observation Probability

As the observation probability, we define the conditional probability density function $p(\mathbf{x}|\mathcal{S}_i)$ of the feature vectors $\mathbf{x}$. The conditioning is with respect to the state $\mathcal{S}$ such that there exists a separate PDF $p(\mathbf{x}|\mathcal{S}_i)$ for each HMM state $\mathcal{S}_i$. In accordance with the definition of the HMM, it is assumed that the observation $\mathbf{x}(m)$ for each frame only depends on the state $\mathcal{S}_{\text{true}}(m)$ of the Markov chain during that particular frame.

The modelling of the observation PDFs is complicated by the fact that the features $\mathbf{x}$ are continuous variables. Further, they constitute multi-dimensional vectors with the dimension b for each signal frame. Therefore, we use parametric *Gaussian mixture models* (GMMs). The observation probability density is expressed by

$$p(\mathbf{x}|\mathcal{S}_i) \approx \tilde{p}(\mathbf{x}|\mathcal{S}_i) = \sum_{l=1}^{L} \rho_{il}\,\mathcal{N}(\mathbf{x};\, \mu_{il}, \mathbf{V}_{il}). \tag{6.66}$$

For each state $\mathcal{S}_i$ of the hidden Markov model, there exists an individual GMM with the parameter set $\Theta_i = \{\rho_{il}, \mu_{il}, \mathbf{V}_{il}; l = 1, 2, \dots L\}$. The EM training procedure to determine the parameters Θ_i as well as some details on the structure and parameterization of GMMs has already been described in Sec. 6.8.

The observation probability constitutes the connection between the state of the HMM and the observed characteristics (features $\mathbf{x}$) of the bandlimited speech signal $s_{\text{nb}}(k)$. Consequently, the observation probability is the decisive element of the statistical model of the speech production process for detecting the momentary HMM state, and the modelling of $p(\mathbf{x}|\mathcal{S}_i)$ has to be implemented with special care.

Emission Probability

The emission probability describes the statistical characteristics of the variable $\mathbf{y}$ representing the estimated spectral envelope of the missing frequency band. The modelling of the emission probability depends on the type of estimation rule that will be described in Sec. 6.9.2.

If the estimation of $\mathbf{y}$ is based solely on the detection of the actual state of the HMM, for example, by a ML (*maximum likelihood*), MAP (*maximum a posteriori*) or MMSE *soft* classification (MMSE variant I), the only information on $\mathbf{y}$ that can be extracted from the conditional PDF $p(\mathbf{y}|\mathcal{S}_i)$ is the conditional expectation $E\{\mathbf{y}|\mathcal{S}_i\}$. Thus, it is sufficient to store the vectors $\hat{\mathbf{y}}_i = E\{\mathbf{y}|\mathcal{S}_i\}$ in a codebook. Note that this codebook is identical to the codebook $\mathcal{C}_\mathbf{y}$ of the constituting vector quantizer of the HMM.

In more sophisticated estimation rules, state-specific mutual dependencies between the variables $\mathbf{x}$ and $\mathbf{y}$ are additionally taken into account. For such estimators the emission probability is described by models of the conditional joint PDFs $p(\mathbf{x}, \mathbf{y}|\mathcal{S}_i)$. Since both $\mathbf{x}$ and $\mathbf{y}$ are multi-dimensional, continuous variables, state-specific Gaussian mixture models can be utilized to model these joint PDFs (compare Sec. 6.8).

Parameters of the Markov Chain

The dependencies between the states of consecutive frames shall be considered in the statistical model. These dependencies are reproduced by the parameters of the Markov chain in the HMM. In the sequel, depending on the modelling and exploitation of the Markov chain parameters, the utilized a priori knowledge will be labelled as follows:

AK0: Only the probability of occurrence of the states is considered, that is, it is assumed that the probability of a state is independent from the state of the source at preceding or following frame instants.

AK1: A first-order ergodic Markov chain is assumed, that is, transition probabilities between consecutive states of the source are taken into account.

State probability (AK0) The scalar value $P(\mathcal{S}_i)$ describes the (non-conditional) state probability, that is, the a priori probability that the HMM is in state $\mathcal{S}_i$ without incorporating any further observation or a priori knowledge, for example, of the feature vector $\mathbf{x}$, or of the state in preceding or following frames.

The state probabilities can easily be estimated by computing the true state sequence for the wideband training material and counting the number of occurrences $N_{\Xi_i} = |\Xi_i|$ of each state $\mathcal{S}_i$. The ratio between the number of occurrences of state $\mathcal{S}_i$ and the total number N_m of speech frames in the training set gives the estimated state probability $\tilde{P}(\mathcal{S}_i) = N_{\Xi_i}/N_m$. The resulting probability values are stored in a table such that the actual bandwidth extension algorithm can later on access the a priori state probabilities by simple table lookups.

Transition probabilities (AK1) The *transition probability* $P(\mathcal{S}_i(m+1)|\mathcal{S}_j(m))$ describes the conditional probability that the state of the source changes from state $\mathcal{S}_j$ in one signal frame to state $\mathcal{S}_i$ in the next frame.

Because the true state sequence is known during the training phase of the BWE algorithm, the transition probabilities can be estimated as the ratio between the counted number of occurrences of a particular transition from $\mathcal{S}_j$ to $\mathcal{S}_i$ and the total number of occurrences of state $\mathcal{S}_j$. Because, in general, transitions from any state to any other state are possible owing to the ergodicity of the Markov chain, a $N_\mathcal{S} \times N_\mathcal{S}$ matrix (i.e. a two-dimensional table) is necessary to store the transition probabilities for the later bandwidth extension.

Higher-order Markov modelling is possible, but the straightforward implementation yields huge lookup tables to store the transition probabilities, and the computational complexity of estimation rules increases exponentially with the model order.

6.9.2 ESTIMATION RULES

The actual classification or estimation constitutes the final step towards the determination of the coefficients $\tilde{\mathbf{y}}$ representing the spectral envelope of the missing frequency band. In this section, three well-known estimation methods are described, and their application in the context of the estimation of the spectral envelope is depicted. Two of the three methods are based on a posteriori probabilities that will first be defined in the next section. The most important advantage of the HMM-based estimation rules using a posteriori probabilities is that they explicitly use memory, that is they take observations from adjacent signal frames into account.

6.9.2.1 Calculation of A Posteriori Probabilities

To be able to utilize a priori knowledge on the temporal dependencies of the states of the hidden Markov model, we define the *observation sequence* $\mathbf{X}(m_k) = \{\mathbf{x}(1)\ldots\mathbf{x}(m_k)\}$ containing all feature vectors that have been observed up to the m_kth frame. Note that the index m_k of the most recently observed feature vector is allowed to be greater than the frame index m of the currently processed signal frame, corresponding to an interpolation or look-ahead. The a posteriori probabilities shall be expressed with respect to all observed signal frames

$$P(\mathcal{S}_i(m)|\mathbf{X}(m_k)) = P(\mathcal{S}_i(m)|\mathbf{x}(1), \mathbf{x}(2), \ldots \mathbf{x}(m), \ldots \mathbf{x}(m_k)). \tag{6.67}$$

The definition and calculation of the a posteriori probabilities $P(\mathcal{S}_i(m)|\mathbf{X}(m_k))$ depends on the kind of a priori knowledge that shall be utilized. According to the definitions from Sec. 6.9.1, the two cases AK0 and AK1 will be distinguished in the following paragraphs.

No consideration of transition probabilities (AK0) If only the state probabilities of the Markov chain shall be considered (AK0), it is assumed that the state of the source for the mth frame of the signal only depends on the features $\mathbf{x}(m)$ observed for that frame. Then, the a posteriori PMFs from Eqn. 6.67 can be simplified

$$P(\mathcal{S}_i(m)|\mathbf{x}(1), \mathbf{x}(2), \ldots \mathbf{x}(m), \ldots \mathbf{x}(m_k)) = P(\mathcal{S}_i(m)|\mathbf{x}(m)). \tag{6.68}$$

Applying Bayes' rule yields an expression for the a posteriori probabilities in which only the modelled observation probabilities and state probabilities are contained

$$\begin{aligned} P(\mathcal{S}_i(m)|\mathbf{x}(m)) &= \frac{p(\mathbf{x}(m)|\mathcal{S}_i(m))\; P(\mathcal{S}_i(m))}{p(\mathbf{x}(m))} \\ &= \frac{p(\mathbf{x}(m)|\mathcal{S}_i(m))\; P(\mathcal{S}_i(m))}{\sum_{j=1}^{N_S} p(\mathbf{x}(m)|\mathcal{S}_j(m))\; P(\mathcal{S}_j(m))}. \end{aligned} \tag{6.69}$$

First-order HMM (AK1) For a first-order hidden Markov model, the a posteriori PMF $P(\mathcal{S}_i(m)|\mathbf{X}(m_k))$ is expressed in terms of the joint PDF $p(\mathcal{S}_i(m), \mathbf{X}(m_k))$ and the PDF

$p(\mathbf{X}(m_k))$ of the observation sequence

$$P(\mathcal{S}_i(m)|\mathbf{X}(m_k)) = \frac{p(\mathcal{S}_i(m), \mathbf{X}(m_k))}{\sum_{j=1}^{N_S} p(\mathcal{S}_j(m), \mathbf{X}(m_k))}. \tag{6.70}$$

The joint probability density function $p(\mathcal{S}_i(m), \mathbf{X}(m_k))$ can be determined as the product of the observation probability density of the current feature vector $\mathbf{x}(m)$ and of two components $\alpha_i(\cdot)$ and $\beta_i(\cdot)$ that comprise the contributions of the observed feature vectors from preceding and subsequent signal frames

$$p(\mathcal{S}_i(m), \mathbf{X}(m_k)) = \alpha_i(m)\, \beta_i(m)\, p(\mathbf{x}(m)|\mathcal{S}_i(m)). \tag{6.71}$$

The two quantities $\alpha_i(m)$ and $\beta_i(m)$ are calculated via forward and backward recursion, respectively, thereby utilizing the complete available observation sequence.

The quantity $\alpha_i(m)$ can be interpreted as the a priori probability $p(\mathcal{S}_i(m), \mathbf{X}(m-1))$ of the ith state of the HMM, considering all *past* observed feature vectors. The successive calculation of $\alpha_i(m)$ is based on the recursive equation

$$\alpha_i(m+1) = \sum_{j=1}^{N_S} \alpha_j(m)\, p(\mathbf{x}(m)|\mathcal{S}_j(m))\, P(\mathcal{S}_i(m+1)|\mathcal{S}_j(m))$$

$$\alpha_i(1) = P(\mathcal{S}_i). \tag{6.72}$$

Because there exists no predecessor for the very first frame of the input speech, the recursion has to be initialized with the non-conditional state probabilities $P(\mathcal{S}_i)$. Owing to the recursive definition of $\alpha_i(m)$ in Eqn. 6.72, it is not necessary to store all past frames but it suffices to pass the a priori knowledge $\alpha_i(m+1)$ from one frame to the other.

If future observations shall also be taken into account for the a posteriori probability, that is, if $m_k > m$, the terms $\beta_i(m)$ can likewise be calculated recursively

$$\beta_i(m-1) = \sum_{j=1}^{N_S} \beta_j(m)\, p(\mathbf{x}(m)|\mathcal{S}_j(m))\, P(\mathcal{S}_j(m)|\mathcal{S}_i(m-1)). \tag{6.73}$$

The initialization of the recursion has to be performed for the most recently observed signal frame with the index m_k by $\beta_i(m_k) = 1$. If no future observations shall be considered by the estimation, the quantities $\beta_i(m)$ also have to be set to a value of $\beta_i(m) = 1$ in the calculation of the a posteriori probabilities (Eqn. 6.71).

6.9.2.2 Maximum Likelihood Classification

A widely used classification method is the *maximum likelihood* (ML) approach. This estimation rule does not take the a priori knowledge on the state sequence into account.

That codebook entry of $\mathcal{C}_{\mathbf{y}}$ is selected which corresponds to that state of the HMM for which the observation density of the currently observed feature vector is maximized

$$\tilde{\mathbf{y}}_{\text{ML}}(m) = E\left\{\mathbf{y}|\mathcal{S}_{i_{\text{ML}}(m)}\right\} = \hat{\mathbf{y}}_{i_{\text{ML}}(m)}$$

with

$$i_{\text{ML}}(m) = \arg\max_{i=1}^{N_S} p(\mathbf{x}(m)|\mathcal{S}_i(m)). \tag{6.74}$$

Consequently, the range of possible output values of the ML estimator is limited to the entries of the codebook. Note that the a posteriori probabilities from the previous section, that is, the state and transition probabilities of the HMM, are not utilized.

There are, in fact, certain parallels between the ML classification rule and the codebook mapping approach from Sec. 6.6: as with the primary codebook in Sec. 6.6, with Eqn. 6.74, fixed regions of the b-dimensional feature space are assigned to fixed pre-trained spectral envelope representatives[8]. The major difference is that the use of GMMs for $p(\mathbf{x}|\mathcal{S}_i)$ allows for a much more flexible definition of those regions, and the HMM-based training is more directly targeted on maximizing the estimation performance.

6.9.2.3 Maximum A Posteriori Classification

The goal of the *maximum a posteriori* (MAP) rule is to maximize the a posteriori *probability mass function* (PMF). Accordingly, that entry of the codebook $\mathcal{C}_{\mathbf{y}}$ is selected which is assigned to the state of the HMM for which the a posteriori PMF $P(\mathcal{S}_i|\mathbf{X}(m_k))$ is maximum

$$\tilde{\mathbf{y}}_{\text{MAP}}(m) = E\left\{\mathbf{y}|\mathcal{S}_{i_{\text{MAP}}(m)}\right\} = \hat{\mathbf{y}}_{i_{\text{MAP}}(m)}$$

with

$$i_{\text{MAP}}(m) = \arg\max_{i=1}^{N_S} P(\mathcal{S}_i(m)|\mathbf{X}(m_k)). \tag{6.75}$$

Because the normative factor in the denominator of the fraction in Eqn. 6.70 is identical for all states of the HMM, its value is irrelevant for the classification such that

$$i_{\text{MAP}}(m) = \arg\max_{i=1}^{N_S} p(\mathcal{S}_i(m), \mathbf{X}(m_k)). \tag{6.76}$$

In contrast to the ML approach, a priori knowledge about the state sequence of the HMM is utilized by the MAP method. The range of results of the estimation is, however, still limited to the contents of the codebook $\mathcal{C}_{\mathbf{y}}$. The MAP rule minimizes the number of mis-classifications of the HMM state.

[8] Actually, codebook mapping with Euclidian distance criterion (e.g. $d(\mathbf{x}, \hat{\mathbf{x}}) = \|\mathbf{x} - \hat{\mathbf{x}}\|^r$ with $r > 0$) can be interpreted as a special case of Eqn. 6.74 if the covariance matrices in $\tilde{p}(\mathbf{x}|\mathcal{S}_i)$ are fixed to $\mathbf{V}_{il} = \sigma^2\mathbf{I}$ in the EM algorithm.

6.9.2.4 Minimum Mean Square Error Estimation

Now, the range of results of the estimation shall no longer be limited to the contents of the codebook $\mathcal{C}_{\mathbf{y}}$, but all values in the d-dimensional Euclidian space $\mathbb{R}^d$ are allowed. The aim of the *minimum mean square error* (MMSE) optimization is to minimize the error criterion

$$\begin{aligned}\mathcal{D}_{\text{MSE}}(\mathbf{y}(m), \tilde{\mathbf{y}}(m)|\mathbf{X}(m_k)) &= E\left\{\|\mathbf{y}(m) - \tilde{\mathbf{y}}(m)\|^2 \big| \mathbf{X}(m_k)\right\} \\ &= \int_{\mathbb{R}^d} p(\mathbf{y}(m)|\mathbf{X}(m_k))\, \|\mathbf{y}(m) - \tilde{\mathbf{y}}(m)\|^2 \,\mathrm{d}\mathbf{y}(m). \qquad (6.77)\end{aligned}$$

The integral in Eqn. 6.77 has to be solved for the complete d-dimensional parameter space. The solution is the conditional expectation

$$\begin{aligned}\tilde{\mathbf{y}}_{\text{MMSE}}(m) &= E\left\{\mathbf{y}(m)\big|\mathbf{X}(m_k)\right\} \\ &= \int_{\mathbb{R}^d} \mathbf{y}(m)\, p(\mathbf{y}(m)|\mathbf{X}(m_k))\,\mathrm{d}\mathbf{y}(m). \qquad (6.78)\end{aligned}$$

Because we do not have a model of the conditional PDF $p(\mathbf{y}(m)|\mathbf{X}(m_k))$ in closed form, this quantity has to be expressed indirectly via the states of the HMM

$$\begin{aligned}p(\mathbf{y}(m)|\mathbf{X}(m_k)) &= \sum_{i=1}^{N_{\mathcal{S}}} p(\mathbf{y}(m), \mathcal{S}_i(m)|\mathbf{X}(m_k)) \\ &= \sum_{i=1}^{N_{\mathcal{S}}} p(\mathbf{y}(m)|\mathcal{S}_i(m), \mathbf{x}(m))\, P(\mathcal{S}_i(m)|\mathbf{X}(m_k)). \qquad (6.79)\end{aligned}$$

The second line in Eqn. 6.79 results from the model assumption that the vectors $\mathbf{x}(m)$ and $\mathbf{y}(m)$ exclusively depend on the state $\mathcal{S}(m)$ of the source in the mth signal frame. Inserting Eqn. 6.79 into Eqn. 6.78 yields the general state-based rule

$$\tilde{\mathbf{y}}_{\text{MMSE}}(m) = \sum_{i=1}^{N_{\mathcal{S}}} P(\mathcal{S}_i(m)|\mathbf{X}(m_k)) \int_{\mathbb{R}^d} \mathbf{y}(m)\, p(\mathbf{y}(m)|\mathcal{S}_i(m), \mathbf{x}(m))\,\mathrm{d}\mathbf{y}(m). \qquad (6.80)$$

Depending on the available statistical model of the emission probability (compare Sec. 6.9), that is, whether only the state-specific expectation of $\mathbf{y}$ or the a priori knowledge on the joint PDF $p(\mathbf{x}, \mathbf{y}|\mathcal{S}_i)$ is at hand, two variants of MMSE estimators can be formulated. In addition, the AK0 and AK1 assumptions can be used. We will, however, not distinguish between these in the following.

Variant I: 'soft classification' For the first variant of MMSE estimation, the emission probability shall be modelled without taking the observed feature vectors $\mathbf{x}$ into account.

By this, the conditioning on the feature vector $\mathbf{x}(m)$ within the integral on the right-hand side of Eqn. 6.80 is neglected, and the integral reflects the expectation of the coefficient vector $\mathbf{y}$ on the condition that the source is in the state $\mathcal{S}_i$

$$\int_{\mathbb{R}^d} \mathbf{y}(m)\, p(\mathbf{y}(m)|\mathcal{S}_i(m))\, \mathrm{d}\mathbf{y}(m) = E\{\mathbf{y}(m)|\mathcal{S}_i(m)\} = \hat{\mathbf{y}}_i. \tag{6.81}$$

The integral can be replaced by the corresponding entry $\hat{\mathbf{y}}_i$ of the pre-trained codebook. Substituting the a posteriori probability defined in Eqn. 6.67, we derive the MMSE classification rule

$$\tilde{\mathbf{y}}_{\text{MMSE}^{\text{I}}}(m) = \sum_{i=1}^{N_\mathcal{S}} \hat{\mathbf{y}}_i\, P(\mathcal{S}_i(m)|\mathbf{X}(m_k)). \tag{6.82}$$

Hence, the estimated coefficient set $\tilde{\mathbf{y}}_{\text{MMSE}^{\text{I}}}$ is calculated by the sum of the individual codebook entries that are weighted by the respective a posteriori probabilities of the corresponding states of the HMM. Accordingly, the described MMSE estimator can be interpreted as a *soft classification*. Note that with AK0 (with $\tilde{P}(\mathcal{S}_i) = 1/N_\mathcal{S}$) and fixed diagonal covariance matrices $\mathbf{V}_{il} = \sigma^2\,\mathbf{I}$, the MMSE rule (Eqn. 6.82) results in an MMSE-optimized codebook mapping approach with interpolation, compare Sec. 6.6.

Variant II: 'cascaded estimation' The second variant of the MMSE estimation rule shall take the state-specific joint PDF $p(\mathbf{x}(m), \mathbf{y}(m)|\mathcal{S}_i(m))$ into consideration (Jax and Vary [132]). Then, the integral on the right-hand side of Eqn. 6.80 reflects the conditional expectation $E\big\{\mathbf{y}(m)\big|\mathcal{S}_i(m), \mathbf{x}(m)\big\}$. This conditional expectation can be calculated from the parameters of a Gaussian mixture model of the joint PDF $p(\mathbf{x}(m), \mathbf{y}(m)|\mathcal{S}_i(m))$ as described in Eqn. 6.59. Inserting the conditional expectation into Eqn. 6.80 leads to the second MMSE estimation rule

$$\tilde{\mathbf{y}}_{\text{MMSE}^{\text{II}}}(m) = \sum_{i=1}^{N_\mathcal{S}} E\big\{\mathbf{y}(m)|\mathcal{S}_i(m), \mathbf{x}(m)\big\}\, P(\mathcal{S}_i(m)|\mathbf{X}(m_k)). \tag{6.83}$$

This estimation rule can be interpreted as a cascaded estimation: first the state-dependent expectation of $\mathbf{y}$ is calculated for each state, followed by an individual weighting with the respective a posteriori probabilities.

Compared to the first variant of the MMSE estimation from Eqn. 6.82, the second variant (Eqn. 6.83) should always provide better performance because additional information is exploited from the observed features $\mathbf{x}$. This advantage does not come for free, however, since the calculation of the expectation operation for GMMs with full covariance matrices implies a higher computational complexity.

It can easily be seen that the GMM-based algorithm from Sec. 6.8 is a special case of Eqn. 6.83, if only a single state $N_\mathcal{S} = 1$ is employed in the HMM. In this case, the sum in Eqn. 6.83 degenerates to the conditional expectation $\tilde{\mathbf{y}}(m) = E\{\mathbf{y}(m)|\mathcal{S}_1(m), \mathbf{x}(m)\}$ because $P(\mathcal{S}_1(m)|\mathbf{X}(m)) = 1$. This conditional expectation is identical to the estimation rule (Eqn. 6.59) in Sec. 6.8.

There are also certain parallels of special cases of variant II of the MMSE rule to the linear mapping and piecewise-linear mapping methods from Sec. 6.7: if there is only a single state in the HMM and one Gaussian in the mixture model of the emission probability, that is, if $N_{\mathcal{S}} = 1$ and $L = 1$, the MMSE rule (Eqn. 6.83) also leads to estimation of $\tilde{\mathbf{y}}$ by linear transformation, though with consideration of the mean vectors of $\mathbf{x}$ and $\mathbf{y}$ (cf. Eqn. 6.59). With one Gaussian in the emission PDF models $\tilde{p}(\mathbf{x}, \mathbf{y}|\mathcal{S}_i)$ and more than one state in the HMM, that is, if $L = 1$ and $N_{\mathcal{S}} > 1$, the MMSE rule of Eqn. 6.83 resembles a piecewise-linear mapping approach with soft decision.

6.10 DISCUSSION

In the past, bandwidth extension algorithms for speech have reached a stable baseline quality: the artificial wideband output of a BWE system is in general preferred to narrowband telephone speech. Nevertheless, the quality of the enhanced speech is far from reaching the quality of the original wideband speech. It would be desirable to further improve the subjective speech quality of BWE systems.

With respect to the performance of wideband spectral envelope estimation, comparison of the theoretical performance bound (e.g. Fig. 6.19) with the best actually achieved estimation results (e.g. Fig. 6.20) yields a performance gap of about 3.2 and 2.3 dB for high-frequency (3.4–7 kHz) and low-frequency (50–300 Hz) BWE of telephone speech, respectively (Jax [128]). It is unclear, unfortunately, whether this gap can be closed by more sophisticated estimation schemes, because the theoretical bound, in general, is not tight. Some authors have come to the conclusion that improving the objective performance, as, for example, measured in terms of log spectral distortion, of (memoryless) BWE for speech may be very intricate (Epps [65], Nilsson *et al.* [186]).

To date, BWE for speech has mostly been developed for clean input speech. The vast majority of the published approaches do not consider any adverse conditions such as additive background noise or distortion of the narrowband input signal. To improve the acceptance in practice in the wide range of possible applications, the robustness of BWE for speech schemes has to be increased. Important issues in this respect are robustness against additive background noises, and against input signals that differ from the model assumptions, like music and so on. In such circumstances, at least the bandwidth extension system should switch to a secure fallback mode, for example, similar to one of the generic BWE algorithms as described in Chapters 3 and 5.

7

Noise Abatement

7.1 A SPECIAL KIND OF NOISE REDUCTION

Pop-music reproduction or sound reinforcement in discos or at concerts at a very high SPL is highly appreciated by the so-called 'target group' audience. For the neighbouring community however, this can be very annoying, especially when these music sessions take place during the night. A poor sound insulation (between adjacent houses, but also for buildings at some distance) creates an inadmissible sound emission level in, for example, bedrooms. Noise reduction methods of a constructional nature are, in most cases, very expensive, and also take a considerable amount of time to be realized. This often causes a temporary closing of the venue that creates the problem.

Another kind of option is to make use of 'active' noise reduction by signal-processing means. In Aarts *et al.* [7], two options are discussed, which were applied in a real-life situation of 'noise pollution'. The first method uses 'anti-sound', which will not be discussed here; the second method is based on low-frequency psychoacoustic BWE technology, as discussed previously in Chapter 2.

7.2 THE NOISE POLLUTION PROBLEM – CASE STUDY

As mentioned above, sound insulation between the community homes and sound-producing locations, such as clubs or entertainment venues, can be very poor. These places may produce SPLs that can go far beyond 100 dB(A) in the evening hours. At these levels, local authorities can force these sound producers to stop the music, to set a penalty in case of transgression of the (local) laws, or even force them to close the premise. Sometimes, the annoying sound follows paths that are unpredictable, and venue owners are not willing, or not able, to make investments in constructional investigations and their solutions.

As a case example, we present a situation that existed at a club in The Netherlands, where high SPLs were produced in the dwellings opposite of the club. The irritation of the surrounding population was reinforced by the interrupted character of the disturbance (the rhythmic 'thumping' that penetrated into the houses) and the time of day (late evening). The spectrum of the 'thumping' was roughly 50–100 Hz. Beyond these frequencies, no

Audio Bandwidth Extension E. Larsen and R. M. Aarts
 ISBN 0-470-85864-8

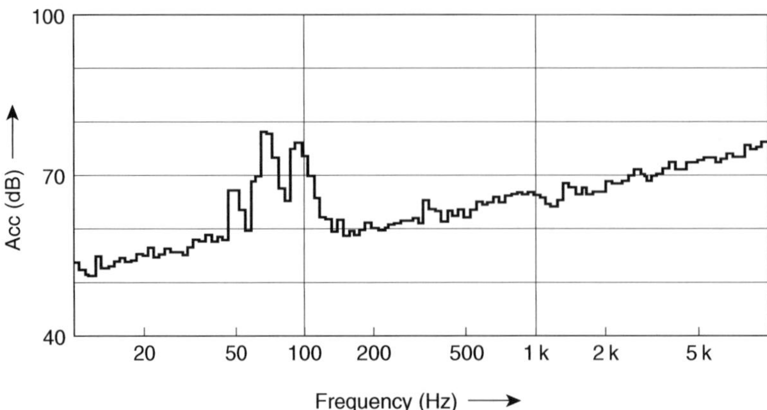

Figure 7.1 Structural excitation measurement (dB rel. 1 μm/sec^2) in a room of a housing located near the club

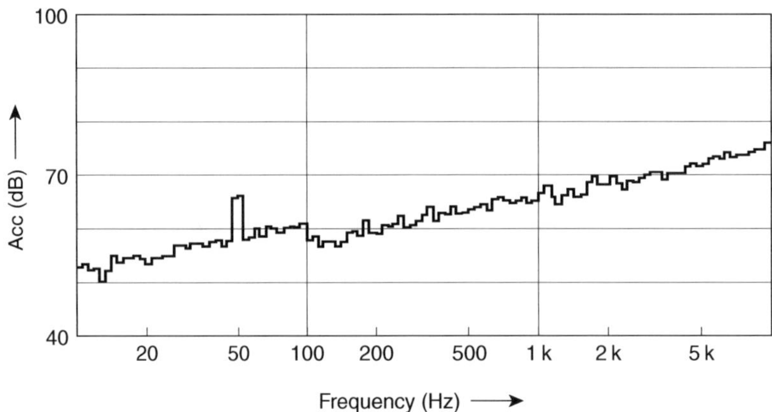

Figure 7.2 Structural background spectrum measurement (dB rel. 1 μm/sec^2); same room as in Fig. 7.1. At the time of this measurement, there was no music production in the club (the excitation at 50 Hz was continuously present, and caused by a factory in the neighbourhood – there had never been complaints about this signal)

transmission had been measured, see Fig. 7.1. Note that this is a spectrum of structural vibration, not the spectrum as measured in air.

At the time of measurement, the SPL in the club was 112 dB. Figure 7.2 shows the background signal in the same room, when there was no disturbance from the music produced in the club. The annoyance was not caused by air transmission, but by vibrations that travelled through the soil, or other underground structures, into the houses. Of course, these vibrations were transformed to airborne sound by vibrations of the house construction. The transmission paths were not known and would have been very difficult to detect, which is why it was chosen to opt for an electro-acoustical/signal-processing approach and not for constructional measures.

A possible solution was to cut out the disturbing part of the frequency spectrum. However, this was judged to be intolerable from a musical point of view. The alternate method proposed was based on the principle of the 'missing fundamental' (see Sec. 1.4.5), whereby the perception of pitch is not disturbed when low-number harmonics are eliminated. The actual method closely follows the concept of low-frequency psychoacoustic BWE as discussed in Chapter 2. In the next section, we shall discuss the implementation and result of applying this method to the described situation.

7.3 THE APPLICATION LOW-FREQUENCY PSYCHOACOUSTIC BANDWIDTH EXTENSION TO NOISE POLLUTION

The concept of a noise-abatement low-frequency psychoacoustic BWE processor is based on the principles of pitch perception (Sec. 1.4.5). In summary, the perceived pitch of a signal consisting of a fundamental at f_0 and higher harmonics will not change when the fundamental at f_0 is completely removed. The remaining harmonics will still mediate the same strong pitch percept.

The algorithm follows the structure shown in Fig. 2.4 and the discussion presented there. The main application area that was considered there was for enhanced low-frequency reproduction on small loudspeakers. Thus, the goal was to emphasize all frequencies below a certain value, for example, 100 Hz. Here, the goal is to emphasize signals with pitches in the range 50–100 Hz by emphasizing their harmonics. Therefore, in the lower branch of Fig. 2.4, the input signal is processed by a band-stop filter of 50–100 Hz, see

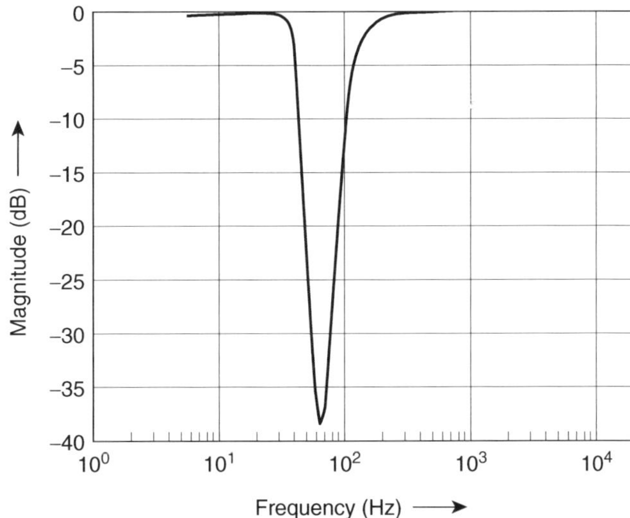

Figure 7.3 Band-stop filter for the low-frequency psychoacoustic BWE noise-abatement system. This filter eliminates the disturbing frequency components. This is compensated by a complementary filter that feeds into a non-linear device that generates a harmonics signal. This harmonics signal emphasizes the same low pitch of the bass sounds, but does not cause any disturbance in the neighbouring community

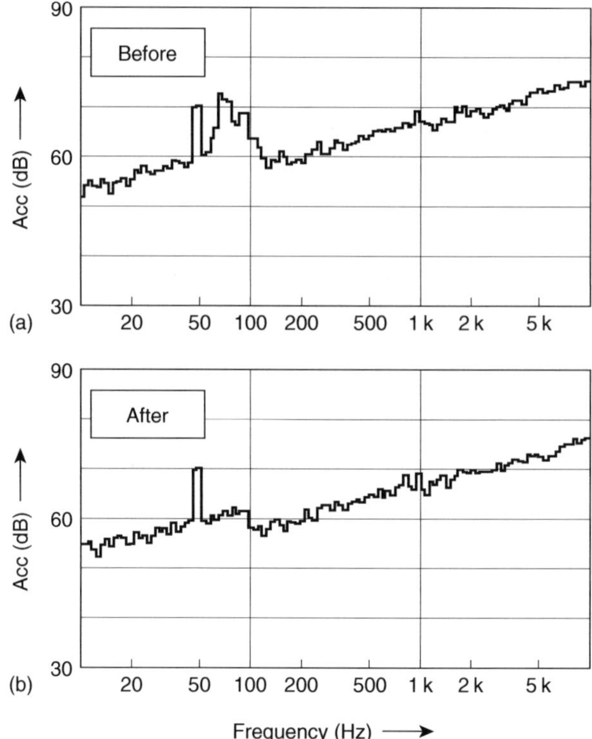

Figure 7.4 Structural excitation spectrum measurement (dB rel. 1 μm/sec^2) during music production in the club. The upper panel shows 'before', that is, without low-frequency psychoacoustic BWE noise-abatement processing. The lower panel shows 'after', that is, with low-frequency psychoacoustic BWE noise-abatement processing. Note that the peak around 100 Hz had been completely removed, while the perceptual difference between the two situations was judged to be small

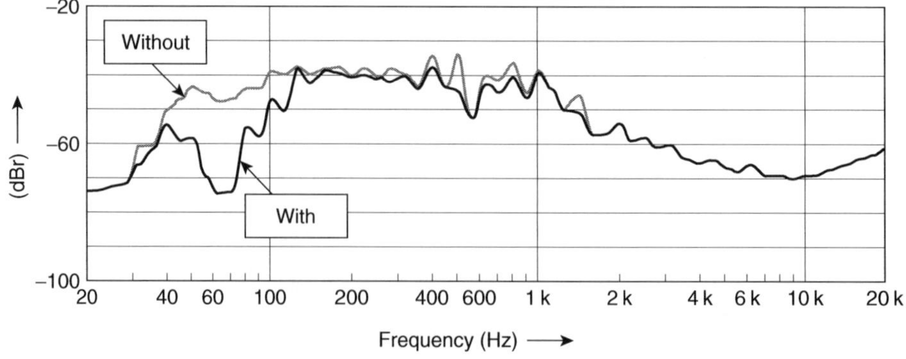

Figure 7.5 Short-term spectrum of representative music fragment measured (in air) in the club, with and without low-frequency psychoacoustic BWE processing

Fig. 7.3. This step eliminates the disturbing frequency components. To compensate for this, the upper branch generates harmonics of the bass input fundamental frequencies. The first filter complements the band-stop filter of the lower branch, thus it is a band-pass filter of 50–100 Hz. The NLD generates higher harmonics, and FIL2 shapes the thus-generated spectrum. An appropriate gain is applied to the harmonics signal after which it is added back to the band-stop-filtered input signal. Using this approach, it was possible to stay within the limits of the law in The Netherlands, even when the measured SPL in the club reached 105 dB(A). An additional advantage was that frequencies below 50 Hz did not cause annoyance and were thus not processed by dB. This is important as these very low frequency components add 'feel' or 'impact' to the music sensation, which is appreciated by the audience.

This procedure had completely removed the annoyance in the neighbouring houses. Although annoyance is a subjective variable, it is made plausible by comparing measurements shown in Fig. 7.4, which shows that the originally disturbing components between 50–100 Hz are removed after the dB processing. The quality of music as processed by the low-frequency psychoacoustic BWE noise-abatement system was judged to be sufficiently high such that this solution was adopted. The difference with the unprocessed sound could be heard, with some effort, when doing an informal A/B test (rapidly switching the processing on and off), but the club visitors would not normally notice any difference, in particular because the processing would always be turned on. Figure 7.5 shows an illustrative example of the difference between the short-term spectra of the same music signal with and without low-frequency psychoacoustic BWE processing.

In conclusion, low-frequency psychoacoustic BWE can be successfully applied to treat noise pollution problems as caused by entertainment venues or concerts, and so on, if the disturbing frequencies are relatively narrowband. The perceptual impairment is small, and the main advantages are low cost, rapid implementation, and robustness. The alternative method of constructional modifications can be very high cost, requires a much longer implementation period, and success is sometimes difficult to guarantee. For constructional measures, the advantage is obviously that the sound within the venue is not modified in any way, but the disadvantages would usually be more important than this minor advantage. Another signal-processing approach (discussed in the original treatment by Aarts *et al.* [7]) is using 'anti-sound', which only modifies the standing wave pattern inside the enclosure (but not the power spectrum of the sound) and thus attempts to minimize the SPL in those areas where acoustic energy is thought to propagate out of the enclosure, was found to be much more delicate and time consuming than the low-frequency psychoacoustic BWE approach.

8

Bandwidth Extension Patent Overview

Here we present a chronological overview of BWE-related patents, from 1943 to 2004. This overview has resulted from searches by the authors, and is not assumed to cover *all* BWE patents. Specifically, the overview is limited to US patents; also, the kind of patents included cover those areas that are more or less closely related to the material covered in this book (including both low- and high-frequency BWE methods). As a resource, it is hoped that this patent list will complement the list of cited references (these patents are therefore not separately listed in the bibliography).

Each item in the list presents the key data for the patent (title and US patent number, inventor, assignee, and date[1]), and is believed to be accurate but not guaranteed to be so. The abstract is in most cases directly copied from the published abstract; deleted passages are marked as (...). Full text for US patents can be obtained from the US Patent and Trademark Office (USPTO), which also maintains a website with a searchable database (http://www.uspto.gov).

PSEUDO-EXTENSION OF FREQUENCY BANDS

Title: Pseudo-extension of frequency bands (2,315,248).

Inventor: Louis A. de Rosa.

Date: March 30, 1943

This invention deals with the pseudo-extension of frequency bands and particularly with improvements in the method and means wherein an audio signal, at some point or at some time in its transmission either directly or indirectly to the ear, is modified so that, while all the composite frequencies present in the original audio signal are not present in the signal ultimately transmitted to the ear, the auditory perception is of a sound that has substantially all the sonant characteristics of the original audio signal.

[1] Note that before June 8, 1995, patent protection expired 17 years after the patent was granted. After June 8, 1995, patent protection expires 20 years after the filing date.

Audio Bandwidth Extension E. Larsen and R. M. Aarts
 ISBN 0-470-85864-8

APPARATUS FOR IMPROVING SOUNDS OF MUSIC AND SPEECH

Title: Circuit for simulating string bass sound (2,866,849).
Inventor: Charles D. Lindridge.
Assignee: one-fourth to L.C. Krazinski.
Date: December 30, 1958.

This invention relates to sound-reproducing systems, sound-reinforcing systems, music and speech, and specifically to apparatus for improving sounds of music and speech in which there is a deficiency of high frequencies received from the sound source. The improvement is made by producing harmonics of frequencies higher than 3 kc in the sounds and producing sounds at frequencies higher than 6 kc at a loudspeaker.

An object of this invention is to compensate for loss in sound at high frequencies due to the use of a narrowband of frequencies in a transmission system or due in part to the greater directivity of high frequencies in air relative to that at low frequencies, and in part to the greater absorption of sound in air at normal room temperature and humidity at high frequencies than at low frequencies.

(. . .)

AUDIO TRANSMISSION NETWORK

Title: Audio transmission network (2,379,714).
Inventor: R.L. Hollingsworth.
Assignee: Radio Corporation of America, Del.
Date: July 3, 1945

It is an object of this invention to improve the intelligibility or overall quality of a band-limited audio signal, such as that received from radio transmissions.

CIRCUIT FOR USE IN MUSICAL INSTRUMENTS

Title: Circuit for use in musical instruments (3,006,228).
Inventor: J. P. White.
Date: October 31, 1961.

This invention relates to certain novel circuit arrangements that may be used to produce a pleasing musical effect (. . .) to produce tones rich in overtones or quality.

ARTIFICIAL RECONSTRUCTION OF SPEECH

Title: Artificial reconstruction of speech (3,127,476).
Inventor: Edward E. David.
Assignee: Bell Telephone Laboratories, Incorporated, New York, N.Y.
Date: March 31, 1964.

This invention relates to the reconstruction of artificial speech from narrowband transmitted signals, and has for its principal object the improvement of quality of such artificial speech.

TONE GENERATION SYSTEM

Title: Tone generation system (3,213,180).

Inventor: Jack C. Cookerly and George R. Hall.

Date: October 19, 1965.

The object of the invention is 'to provide a novel tone generation system in which output tones are derived from the normal sound of the instrument, but in which such output tones may have an entirely different quality so that the known instrument may be used to generate tones sounding completely different from those characterizing the instrument', but are 'nevertheless characterized by the manner in which the initiating natural tone of the instrument is played by the musician'.

ELECTRICAL WOODWIND MUSICAL INSTRUMENT

Title: Electrical woodwind musical instrument having electronically produced sounds for accompaniment (3,429,976).

Inventor: Daniel J. Tomcik.

Assignee: Electro-Voice Incorporated, Buchanan, Mich.

Date: February 25, 1969.

A monophonic wind-type instrument is disclosed, employing a piezoelectric pickup communicating with the air column of the instrument. The piezoelectric pickup is utilized to generate electrical signals that are amplified, filtered as to tone, and reproduced by a loudspeaker. The electrical signals are also shaped and utilized to drive a pulse generator to produce multiples and sub-multiples of the frequency of the tone produced by the woodwind instrument. The output of the divider is tone filtered to produce a voice independent of the wind instrument. A gate circuit is provided between the trigger circuit and divider to delay in actuation of the divider following initial production of a tone by the instrument in order to avoid spurious mechanically excited electrical outputs.

MUSICAL INSTRUMENT ELECTRONIC TONE PROCESSING SYSTEM

Title: Musical instrument electronic tone processing system (3,535,969).

Inventor: David A. Bunger.

Assignee: D.H. Baldwin Company, Cincinnati, Ohio.

Date: October 27, 1970.

An audio tone from a musical instrument is applied to a phase splitter to provide two oppositely phased tone signals, each of which is passed to a respective field effect transistor transmission gate. In addition, the audio tone is converted to a square wave having the same frequency as the fundamental frequency of the tone. The square wave is passed to a frequency divider, which provides a pair of gating signals for the respective transmission gates, the gating signals being oppositely phased and at half the fundamental frequency of the tone. The oppositely phased tone signals are alternately passed by the gates and combined in a tone colour filter circuit, which imparts specified musical tone qualities to the combined signal. In addition, the original tone may be gated via a further field effect transistor gate by a signal having a frequency that is one-quarter of the tone fundamental frequency, the gated signal being passed to an appropriate tone colour filter. The original tone is also passed directly to a tone colour filter. All of the filtered signals are then amplified and passed to a loudspeaker system to provide an acoustic signal of substantially greater tonal complexity than the original tone and which is controlled in frequency and amplitude by the frequency and amplitude respectively in the input tone.

OCTAVE JUMPER FOR MUSICAL INSTRUMENTS

Title: Octave jumper for musical instruments (3,651,242).

Inventor: Chauncey R. Evans.

Assignee: Columbia Broadcasting System Inc. N.Y.

Date: March 21, 1972.

A bass or other guitar has a transducer for each of its strings, and each transducer is connected to an octave-jumping circuit that lowers the musical tone produced by the individual string, all without loss of either harmonics or amplitude variations. The waveform of the fundamental frequency of each musical tone is squared, divided by two and then amplitude modulated to follow the amplitude envelope of the original tone. The modulated square wave contains only odd harmonics of the lowered frequency fundamental. The missing even harmonics are restored by combining with the modulated square wave the original tone containing all of its harmonics.

SPEECH QUALITY IMPROVING SYSTEM

Title: Speech quality improving system utilizing the generation of higher harmonic components (3,828,133).

Inventors: Hikoichi Ishigami *et al.*

Assignee: Kokusai Denshin Denwa Kabushiki Kaisha, Tokyo-To, Japan.

Date: August 6, 1974.

A speech quality–improving system for a band-limited voice signal, comprising a branching circuit for dividing the band-limited voice signal into two branched signals, each having the same waveform as the band-limited voice signal, a higher harmonic signal generator for generating higher harmonic components of one of the two branched signals, and a combining circuit for combining the other of the two branched signals with the generated higher harmonic components to provide a combined voice signal having an improved speech quality realized by increasing the higher harmonic components of the bandlimited voice signal. The higher harmonic generator comprises a cascade combination of an instantaneous compressor and a level range expander having reciprocal power characteristics of the compression ratio of the instantaneous compressor.

AUDIO SIGNAL PROCESSOR

Title: Audio signal processor (4,144,581).
Inventor: Andrew Prudente.
Date: March 13, 1979.

Random audio signals are converted into sine waves of corresponding frequency and duration, and harmonics are derived from the sine waves. The harmonics are controllably attenuated and selectively inverted and combined to form an output signal. The conversion of the random signals is effected by squaring the same to drive a Schmitt trigger that feeds into a levelled integrator that leads to a diode function generator. The harmonics are generated with the use of four-quadrant multipliers.

CIRCUIT FOR SIMULATING STRING BASS SOUND

Title: Circuit for simulating string bass sound (4,175,465).
Inventor: George F. Schmoll.
Assignee: CBS Inc., New York, N.Y.
Date: November 27, 1979.

In a circuit for simulating the sound produced when a stringed instrument, such as a bass viol, is plucked, square wave signals of different frequencies from a tone generator are combined to produce a synthesized saw-tooth wave form, which is applied to a low-pass filter to remove the extremely high order harmonics, and then applied to an amplifier the gain of which is controlled in accordance with an envelope signal having a fast attack and a relatively slow decay. The resulting amplified signal is applied to an off-centre-biased amplifier that alters the harmonic content of the output signal as a function of decay time such that when the signal is acoustically reproduced it closely simulates the sound produced when a bass viol string is plucked.

DETECTION AND MONITORING DEVICE

Title: Detection and monitoring device (4,182,930).
Inventor: David E. Blackmer.
Assignee: dbx Inc., Newton, MA.
Date: January 8, 1980.

An improved audio signal processing system synthesizes from an audio signal, an enhanced audio signal by sensing signal energy of the audio signal within a preselected frequency portion of the audio signal, dividing the sensed signal energy into a plurality of discrete bands according to the frequency thereof and generating, responsively to the signal energy in each of the bands, a like plurality of second signals each of which includes frequency components that are sub-harmonics of the frequencies of the corresponding frequency band. The second signals are combined so as to provide a combined signal and the latter is added to the audio signal to provide the enhanced audio signal.

FREQUENCY CONVERSION SYSTEM

Title: Frequency conversion system of tone signal produced by electrically picking up mechanical vibrations of musical instruments (4,233,874).

Inventor: Rokurota Mantani.

Assignee: Nippon Gakki Seizo Kabushiki Kaisha, Hamamatsu, Japan.

Date: November 18, 1980.

An octave conversion system of the fundamental frequency of an audible tone signal produced by electrically picking up mechanical vibration of a musical instrument in which the audible tone signal and an audible modulation signal having a frequency in a preselected relation to the fundamental frequency of the tone signal are applied to a multiplier, which is preferably constituted by a voltage-controlled amplifier. When the modulation signal has a frequency half that of the tone signal, the tone signal is one-octave down-converted, while, when the modulation frequency is equal to the tone signal frequency the tone signal is one-octave upconverted. With this frequency conversion system, the fundamental wave component of the octave-converted tone signal has the same envelope as that of the original tone signal. This frequency conversion system is advantageous in attaining small-size versions of electric musical instruments and extension of inherent compasses of electric musical instruments.

AUDIO PROCESSING SYSTEM FOR RESTORING BASS FREQUENCIES

Title: Audio processing system for restoring bass frequencies (4,698,842).

Inventor: Gregory C. Mackie *et al.*

Assignee: Electronic Engineering and Manufacturing, Inc., Lynnwood, Wash.

Date: October 6, 1987.

An audio processing system for injecting left and right channel audio signals with a signal having a fundamental frequency component that is half the frequency of the highest amplitude low-frequency component in the left and right channel audio signals. The left and right channel audio signals are combined to form a monaural signal that is low-pass filtered and applied to a demodulator circuit. The demodulator circuit generates a control signal having a frequency that is half the frequency of the highest amplitude frequency component in the signal at the output of the band-pass filter. The control signal varies the phase of the signal at the output of the band-pass filter according to the polarity of the control signal. The resulting signal is selectively added to the left and right input signals. In order to prevent the audio processing circuit from producing annoying artefact when the audio signals are vocally generated, a voice detector determines that the input signals are from a vocal source and then disables the audio processing circuit. The voice detector operates by comparing the monaural (left plus right) signal to a differential signal (left minus right). Vocal source material has a relatively higher monaural signal, while a musical source has a relatively higher differential signal.

SIGNAL SYNTHESIZER

Title: Signal synthesizer (4,700,390).

Inventor: Kenji Machida.

Date: October 13, 1987.

To enhance low- and high-frequency components in a sound signal, low-frequency components are used to generate new yet lower frequencies (sub-harmonics), and high-frequency components are used to generate new yet higher frequencies (harmonics), the new frequencies added to the original signal thereby increasing the original signal bandwidth.

METHOD TO ELECTRONICALLY CLARIFY SOUND OR PICTURE INFORMATION

Title: Method to electronically clarify sound or picture information and an arrangement to carry out the method (4,731,852).

Inventor: Liljeryd, Lars G.

Date: March 15, 1988

A method for electronically clarifying sound or picture information and an arrangement for carrying out the method. It was previously known to generate harmonics and sub-harmonics of a useful signal within an audio or video frequency band and to add these to the useful signal in order to improve the perceptibility. Undesirable intermodulation products are generated, however, particularly the difference intermodulation products and the non-linear amplitude ratio between generated harmonic components related to the input signal. The suggested method eliminates substantially all of these undesirable intermodulation products completely and provides a linear amplitude ratio by forming two orthogonal components from the useful signal, compressing one or both of these components and multiplying the result to form the harmonics that are thereafter mixed with the useful signal.

LOW-PITCHED SOUND CREATOR

Title: Low-pitched sound creator (4,790,014).
Inventors: Koji Watanabe *et al.*
Assignee: Matsushita Electric Industrial Co., Ltd., Osaka, Japan.
Date: December 6, 1988.

An analog sound signal is outputted. Low-frequency components are selected from the outputted analog sound signal so that a low-pitched sound signal is derived from the analog sound signal. A key of the low-pitched sound signal is lowered so that a very low pitched sound signal is derived from the low-pitched sound signal. The analog sound signal and the very low pitched sound signal may be converted into corresponding sounds respectively.

SUB-HARMONIC TONE GENERATOR FOR BOWED MUSICAL INSTRUMENTS

Title: Sub-harmonic tone generator for bowed musical instruments (4,856,401).
Inventor: Richard E. D. McClish.
Date: August 15, 1989.

A device to produce sub-harmonic tone signals in response to a tone signal from a transducer having preferably maximum sensitivity in the plane of bowing of a bowed musical instrument by passing selected cycles of the transducer signal through signal gates that are controlled jointly by sub-harmonic control signals at sub-multiples of the fundamental frequency of the transducer signal and by a signal indicative of the detection of a fundamental frequency. Each sub-harmonic tone signal thus produced has a tone colour, which approximates that of the corresponding bowed musical instrument of the same frequency range and which is independent of the direction of bowing.

LOW FREQUENCY AUDIO DOUBLING AND MIXING CIRCUIT

Title: Low frequency audio doubling and mixing circuit (EP0546619).
Inventor: Wayne Schott.
Assignee: US Philips Corporation, NY
Priority Date: December 9, 1991.

A circuit for doubling and mixing low-frequency audio signals includes an input for receiving an audio signal having a substantially wide frequency range, a circuit coupled to said input means for separating signal components in a low-frequency band of the audio signal from the wide frequency range thereof, a frequency doubler coupled to the separating circuit for doubling the frequencies of the signal components in the low-frequency band, and a mixer for mixing the frequency-doubled signal components with the input audio signal, whereby the signal components in the low-frequency band now also appear one octave higher.

MUSIC TONE PITCH SHIFT APPARATUS

Title: Music tone pitch shift apparatus (5,131,042).
Inventor: Mikio Oda.
Assignee: Matsushita Electric Industrial Co., Ltd., Osaka, Japan.
Date: July 14, 1992.

A music tone pitch shift apparatus that converts an original audio signal into digital data by way of pulse code modulation (PCM), shifting the pitch, and converting the pitch-shifted digital data into an analog signal. The PCM digital data is stored in a ring memory at a given sampling speed, and is read out of the memory by a pair of identical read circuits at a common read-addressing speed corresponding to the desired pitch. One of the read circuits starts reading from the opposite address location to the other on the ring memory. Since the read-addressing speed is set faster than the write-addressing speed when increasing the pitch, and vice versa, overtaking or lapping between the addresses could occur. In switching the read circuits from a now-outputting side to a switching-to side alternately, the read address on the switching-to side circuit is stopped increasing at an address location where a zero-amplitude data has been read, until a zero-amplitude data in phase with that which the switching-to side circuit has read is read by the now-outputting side circuit and the switching is made, immediately before the overtaking or lapping occurs on the now-outputting side circuit. Thus, a smooth connection of the pitch-shifted audio signals can be made without including such amplitude-modulated components as in the cross fade method, and therefore, a high-quality music tone pitch shift operation can be realized.

STRING INSTRUMENT SOUND ENHANCING METHOD AND APPARATUS

Title: String instrument sound enhancing method and apparatus (5,218,160).
Inventor: Matthias Grob-Da Veiga.
Date: June 8, 1993.

A sound-enhancing apparatus for use with a string instrument has separate tone pickups for picking up the tones of individual strings and circuits for, determining the fundamental tones of these tones, multiplying the frequencies of these fundamental tones by small integers, and/or dividing them by small integers and consequently producing harmonic overtones and/or undertones. The thus-produced harmonic undertones and/or overtones are selected and amplified according to fixed and/or adjustable criteria and finally admixed with the original sound. The electronic apparatus for performing the process can operate in analog or digital manner.

DIGITAL RECONSTRUCTING OF HARMONICS TO EXTEND BAND OF FREQUENCY RESPONSE

Title: Digital reconstructing of harmonics to extend band of frequency response (5,267,095).
Inventors: T. Hasegawa *et al.*
Assignee: Pioneer Electronic Corporation, Tokyo, Japan.
Date: November 30, 1993.

A PCM digital audio signal playback apparatus is provided for extracting from the digital audio signal read out from a recording medium an original signal component ranging lower than $\frac{1}{2}$ of its sampling frequency f_s, producing a harmonic from the original signal component, extracting a harmonic component ranging higher than $\frac{f_s}{2}$ from the harmonic, and adding the harmonic component to the original signal component. Accordingly, a high-frequency-carrying signal, for example, an impulse, is processed without causing ringings in the waveform response.

TRANSIENT DISCRIMINATE HARMONICS GENERATOR

Title: Transient discriminate harmonics generator (5,424,488).
Inventor: Donn Werrbach.
Assignee: Aphex Systems, Ltd., Sun Valley, Ca.
Date: June 13, 1995.

A transient discriminate harmonics generator that receives an audio input signal and produces an output signal containing harmonics of the input signal. The output signal is amplitude shaped as a function of the input signal's time and amplitude envelope. The present invention, the transient discriminate harmonics generator generally comprises a control circuit for determining a control parameter, and a harmonics-generating circuit regulated by the control circuit for producing an output signal containing harmonics of an input signal, where the transient discriminate harmonics generator first generates a relatively high level of harmonics at an initial occurrence of the input signal, then incrementally reduces the level of harmonies generated during a time period determined by the control parameter following the initial occurrence of the input signal, and finally produces a relatively low level of harmonics after the end of the time period.

SPEECH BANDWIDTH EXTENSION METHOD AND APPARATUS

Title: Speech bandwidth extension method and apparatus (5,455,888).
Inventor: Vasu Iyengar *et al.*
Assignee: Northern Telecom Ltd., Montreal, Canada.
Date: October 3, 1995.

A speech bandwidth extension method and apparatus analyses narrowband speech sampled at 8 kHz using LPC (*authora comment*: linear predictive coding) analysis to determine its spectral shape and inverse filtering to extract its excitation signal. The excitation signal is interpolated to a sampling rate of 16 kHz and analysed for pitch control and power level. A white noise–generated wideband signal is then filtered to provide a synthesized wideband excitation signal. The narrowband shape is determined and compared with templates in respective vector quantizer codebooks, to select respective high-band shape and gain. The synthesized wideband excitation signal is then filtered to provide a high-band signal which is, in turn, added to the narrowband signal, interpolated to the 16-kHz sample rate, to produce an artificial wideband signal. The apparatus may be implemented on a digital signal processor chip.

MUSICAL TONE GENERATING APPARATUS EMPLOYING MICRORESONATOR ARRAY

Title: Musical tone generating apparatus employing microresonator array (5,569,871).
Inventors: James A. Wheaton *et al.*
Assignee: Yamaha Corporation, Japan
Date: October 29, 1996

A musical tone–generating apparatus employs an array of microresonant structures to generate the harmonic component signals of a musical tone to be generated. The microresonant structures produce high-frequency signals that are down-converted to audio-frequency range by mixing them with a high-frequency reference signal. The desired tone colour is achieved by modifying the relative amplitudes of the harmonic component signals to produce a desired tone colour. A large number of microresonators are preferably integrated on a single integrated circuit substrate to provide a variable tone–generating system in a relatively compact environment.

HARMONIC TONE GENERATOR FOR LOW LEVEL INPUT AUDIO SIGNALS

Title: Harmonic tone generator for low level input audio signals and small amplitude input audio signals (5,578,948).
Inventor: Soichi Toyama.
Assignee: Pioneer Electronic Corporation, Tokyo, Japan.
Date: November 26, 1996.

A harmonic tone generator produces a harmonics signal even for input audio signals of small amplitude. Conversion of a digitized audio signal in accordance with a predetermined non-linear function is also performed for an audio signal of small amplitude. According to the second aspect of the invention, a level difference between the digital audio signal level in the present sampling time and the audio signal level in the preceding sampling time is detected and the detected level difference is converted to an output value in accordance with a predetermined non-linear function by a non-linear converting circuit. The converted output value is accumulated. According to the third aspect of the invention, the detected level difference is converted to a function conversion output in accordance with a predetermined function by a non-linear converting circuit. A gain of an amplifier to amplify the audio signal in the present sampling time is changed in accordance with the function conversion output.

LOW FREQUENCY AUDIO CONVERSION CIRCUIT

Title: Low frequency audio conversion circuit (5,668,885).
Inventor: Mikio Oda.
Assignee: Matsushita Electric Industrial Co., Ltd., Osaka, Japan.
Date: September 16, 1997.

A low-frequency audio conversion circuit for converting the frequency of low-frequency audio components. An input audio signal includes a low-frequency audio component lower than the frequency a speaker can reproduce. The low-frequency audio component is filtered and extracted by a low-pass filter and full-wave rectified to generate even-numbered harmonics of the low-frequency audio component. Secondary harmonics are extracted from the even-numbered harmonics and added to the input audio signal after being amplified to an appropriate level. When a speaker whose low-frequency sound reproduction characteristics are poor is used, and a low-frequency component lower than the frequency the speaker can reproduce is supplied, the low-frequency audio component is reproduced as secondary harmonics, which fall within the frequency range of the speaker. Thus, the low-frequency audio component is compensated, and a powerful sound is reproduced at a low cost without degrading the sound.

DIGITAL SIGNAL PROCESSOR FOR ADDING HARMONIC CONTENT

Title: Digital signal processor for adding harmonic content to digital audio signal (5,748,747).
Inventor: Dana C. Massie.
Assignee: Creative Technology, Ltd., Singapore.
Date: May 5, 1998 (The term of this patent shall not extend beyond the expiration date of 5,524,074, which was patented on June 4, 1996).

A digital audio signal processor for adding harmonic content to an input audio signal through a non-linear transfer function with discontinuities. The discontinuities are generated by bit shifting each input value by an amount that is dependent on the sign and magnitude of the input value. The amount by which the input value is shifted is roughly inversely related to the magnitude of the logarithm of the input value. The transfer function is fractal and so provides increased harmonic content for all signal amplitudes.

AUDIO CIRCUIT

Title: Audio circuit (5,771,296).
Inventor: Toyoaki Unemura.
Assignee: Matsushita Electric Industrial Co., Ltd., Osaka, Japan.
Date: June 23, 1998.

An audio circuit for use in a television receiver and the like compensates for the capacity shortage of a speaker box or low-frequency characteristic of the speaker to reproduce vivid and voluminous low-frequency sound. L and R signals of the audio signal are mixed, and then an arbitrary low-frequency band component is extracted therefrom by a filter having an arbitrary frequency characteristic, and extracted component is bisected by a distribution means, and only low-frequency band component is added to the original L and R signals to reproduce the audio signal that is voluminous in a low-frequency band. With a low-frequency band that is difficult to be reproduced by a speaker, the harmonic is stressed by full-wave rectification means to stress low-frequency sound feeling, and when a switching circuit is provided, low-frequency sound stressing by an amplifier and low-frequency sound harmonic stressing by full-wave rectification means can be easily switched.

METHOD AND DEVICE FOR PROCESSING SIGNALS

Title: Method and device for processing signals (5,828,755).
Inventor: E.E. Feremans and F. De Smet.
Date: October 27, 1998.

A method is set forth for processing signals, in particular for treating audio signals, characterized in that it mainly consists in the supply of an input signal to be treated; in the isolation of a number of signals from the input signal, which are mainly situated in a predetermined part of the sound range; in the additional generation of higher harmonics based on the isolated signals; and in the formation of an output signal by combining the signal that contains the generated higher harmonics with at least a part of the above-mentioned input signal, this input signal is either treated or not treated before being combined.

AUDIO SIGNAL PROCESSING CIRCUIT FOR CHANGING THE PITCH OF RECORDED SPEECH

Title: Audio signal processing circuit for changing the pitch of recorded speech (5,848,392)
Inventor: Katsuyuki Shudo.
Assignee: Victor Company of Japan, Ltd., Yokohama, Japan.
Date: December 8, 1998.

A memory has storage segments at different addresses respectively. A write address signal represents an address that is periodically updated at a first frequency. Samples of an audio signal are sequentially written into storage segments of the memory at addresses represented by the write address signal, respectively. A read address signal represents an address that is periodically updated at a second frequency lower than the first frequency. Samples of the audio signal are sequentially read out from storage segments of the memory at addresses represented by the read address signal, respectively. After the address represented by the write address signal overtakes the address represented by the read address signal and until the address represented by the read address signal reaches the address represented by the write address signal that occurs when the address represented by the write address signal overtakes the address represented by the read address signal, inhibition is given of writing of samples of the audio signal into storage segments of the memory at addresses different from the address represented by the write address signal that occurs when the address represented by the write address signal overtakes the address represented by the read address signal.

METHOD AND SYSTEM FOR ENHANCING QUALITY OF SOUND SIGNAL

Title: Method and system for enhancing quality of sound signal (5,930,373)
Inventor: Meir Shashoua and Daniel Glotter.
Assignee: K.S. Waves Ltd., Tel Aviv, Israel.
Date: July 27, 1999.

An apparatus for conveying to a listener a pseudo low-frequency psycho-acoustic sensation (Pseudo-LFPS) of a sound signal, including: frequency unit capable of deriving from the sound signal high-frequency signal and low-frequency signal (LF signal) that extends over a low-frequency range of interest. Harmonics generator coupled to the frequency generator and being capable of generating, for each fundamental frequency within the low-frequency range of interest, a residue harmonic signal having a sequence of harmonics. The sequence of harmonics, generated with respect to each fundamental frequency contains a first group of harmonics that includes at least three consecutive harmonics from among a primary set of harmonics of the fundamental frequency. Loudness generator coupled to the harmonics generator and being capable of bringing the loudness of the residue harmonics signal to match the loudness of the low-frequency signal. Summation unit capable of summing the residue harmonic signal and the high-frequency signal so as to obtain psycho-acoustic alternative signal.

ULTRA BASS

Title: Ultra bass (6,134,330).
Inventor: Gerrit F.M. De Poortere *et al.*
Assignee: US Philips Corporation, NY
Date: October 17, 2000

To improve the perceived audio signal, it is known to use a harmonics generator to create the illusion that the perceived audio includes lower-frequency signal parts than really available. In addition to improving the perceived so-called ultra bass signals (for example 20–70 Hz), the signals in the frequency band between the ultra bass signal and the normal audio signal are also improved.

IMPROVING THREE DIMENSIONAL AUDIO POSITIONING

Title: Method for introducing harmonics into an audio stream for improving three dimensional audio positioning (6,215,879)
Inventor: M.J. Dempsey
Date: April 10, 2001

Method for introducing harmonics into an audio stream for improving three-dimensional audio positioning. The method adds high-frequency harmonics into sampled sound signals to replace high-frequency sound components eliminated before sampling. By adding high-frequency harmonics into the sampled sound signals, a 'richer sound' will be produced. The resulting sampled sound signals will have a frequency spectrum containing a larger number of frequencies. Thus, the ear will have more cues to better position the sampled sound signals.

SYSTEM AND METHOD FOR IMPROVING CLARITY OF AUDIO SYSTEMS

Title: System and method for improving clarity of audio systems (6,335,973)
Inventor: Eliot M. Case
Assignee: Qwest Communications International Inc.
Date: January 1, 2002

A system and method for improving the clarity of an audio signal selects frequencies of the audio signal for processing and adds even harmonic distortion to the selected frequencies, preferably, of at least the second order. The system and method are particularly suited for hearing aid, voice messaging, and telephony applications. In addition, the system and method may be applied to other very low bandwidth signals, such as data-compressed audio signals, computer voice files, computer audio files, and numerous other technologies that have an audio response less than normal human perception. The technique also applies to the use of perceptually coded audio.

PSEUDO-EXTENSION OF FREQUENCY BANDS

Title: Pseudo-extension of frequency bands (6,424,939).
Inventor: Jürgen Herre *et al.*
Assignee: Fraunhofer-Gesellschaft
Date: July 23, 2002

A method for coding or decoding an audio signal combines the advantages of TNS processing and noise substitution. A time-discrete audio signal is initially transformed to the frequency domain in order to obtain spectral values of the temporal audio signal. Subsequently, a prediction of the spectral values in relation to frequency is carried out in order to obtain spectral residual values. Within the spectral residual values, areas are detected encompassing spectral residual values with noise properties. The spectral residual values in the noise areas are noise-substituted, whereupon information concerning the noise areas and noise substitution is incorporated into side information pertaining to a coded audio signal. Thus, considerable bit savings in case of transient signals can be achieved.

AUDIO SYSTEM

Title: Audio system (6,678,380).
Inventor: R.M. Aarts
Assignee: US Philips Corporation, NY
Date: January 13, 2004.

An audio system includes a circuit for processing an audio signal, this circuit having an input for receiving the audio signal and an output for supplying an output signal. The circuit further includes a harmonics generator coupled to the input for generating harmonics of the audio signal, and an adding circuit coupled to the input as well as to the harmonics generator for supplying a sum of the audio signal and the generated harmonics to the output. The harmonics generator is embodied so as to limit the amplitude of the generated harmonics.

SOURCE CODING ENHANCEMENT USING SPECTRAL-BAND REPLICATION

Title: Source coding enhancement using spectral-band replication (6,680,972).
Inventor: L.G. Liljeryd *et al.*
Assignee: Coding Technologies Sweden AB
Date: January 20, 2004

The present invention proposes a new method and apparatus for the enhancement of source-coding systems. The invention employs bandwidth reduction prior to or in the encoder, followed by spectral-band replication at the decoder. This is accomplished by the use of new transposition methods, in combination with spectral envelope adjustments. Reduced bit rate at a given perceptual quality or an improved perceptual quality at a given bit rate is offered. The invention is preferably integrated in a hardware or software codec, but can also be implemented as a separate processor in combination with a codec. The invention offers substantial improvements practically independent of codec type and technological progress.

SOUND AND VISION SYSTEM

This is a European patent application, included for reference (see Sec. I.2.1).

Title: Sound and vision system (European patent application EP02708577)
Inventor: R.M. Aarts and M.T. Johnson
Assignee: Royal Philips Electronics N.V.
Date: Filed March 20, 2002

A sound and vision system comprising a display device and an acoustic transducer means, such as a loudspeaker or a microphone. The display device includes display cells having opposite electrodes and includes a conductive means connected to the electrodes in order to address said display cells. The acoustic transducer means is formed by a display cell and the conductive means is electrically coupled to the acoustic transducer means in order to convey signals, as a result of which the acoustic transducer means is an integral part of the display device.

Appendix A

Multidimensional Scaling

A.1 INTRODUCTION

A problem encountered in many disciplines is how to measure and interpret the relationships between objects (Aarts [4]). A second problem is the general lack of a mathematical relationship between the perceived response and the actual physical measure. Sometimes relationships are rather vague. How much does the character of one person resemble that of another? Or in the case of this book, to what extent are various processing methods alike? How do we measure and what scale do we need? In the following text, we discuss some scales and techniques and give some examples.

A short but authoritative introduction to multidimensional scaling(MDS) is Kruskal's book [151]. A comprehensive survey of the development of MDS is that of Caroll & Arabie [46], which cites 334 references, mostly published during the 1970s. More recent are the surveys by Young [303] and (on general scaling) by Gescheider [88]. The latter is more about sensory and cognitive factors that affect psychophysical behavior than about measurement and computational aspects. A review intended for a wide, generally scientific audience, concerning the models and applications of MDS and cluster analysis, has been provided by Shepard [246].

A.2 SCALING

The purpose of scaling is to quantify qualitative data. Scaling procedures attempt to do this by using rules that assign numbers to qualities of things or events. Multidimensional scaling is an extension of univariate scaling (App. A.7); it is simply a useful mathematical tool that enables us to represent the similarities of objects spatially as in a map.

MDS models may be either metric or non-metric, depending on the scale of measurement used in collecting the data. For metric scaling, the collected data should be measured using an interval or ratio scale. In the former case, the unit of the 'yardstick', used for measuring the phenomenon, as well as the zero point (offset) are unknown. For a ratio scale, the zero point is known but there is an unknown scaling factor. For non-metric MDS, only the ranking order of the data values is used; the data are, or are used at, an

Audio Bandwidth Extension E. Larsen and R. M. Aarts
 ISBN 0-470-85864-8

ordinal level. However, it is sometimes possible to recover metric distances obtained by non-metric MDS, as will be shown in an example later on.

In order to obtain a spatial map from an MDS computer program, we only need to apply a set of numbers as the input. To all (or most) combinations of pairs out of a group of objects, a number is assigned, which expresses the similarity between the objects of that pair. Such numbers are sometimes referred to as proximities. MDS procedures will then represent objects judged similar to one another as points close to each other in the resulting spatial map. Objects judged to be dissimilar are represented as points distant from each other.

MDS programs that use direct similarity measures as input have the advantage of being low in experimental contamination. They do not require a prior knowledge of the attributes of the stimuli to be scaled.

A.3 EXAMPLE

An obvious procedure for obtaining similarity data is to ask people directly to judge the 'psychological distance' of the stimulus objects. Another way is the method of triadic comparisons (Levelt *et al.* [159]). This has the advantage that it simplifies the subject's task, because only the ranking order of three presented stimuli is asked for. However, there can be some drawbacks, as pointed out by Roskam [229]. A practical problem arises when for a complete experiment the number of triads (all possible combinations of three stimuli out of the set of all stimuli) is considered too large. It can be reduced by using an incomplete balanced block design (BIBD).

As an example of MDS using triadic comparisons, consider the following. Suppose someone with a good topographical knowledge of The Netherlands is asked to give the nearest city and the most distant city out of three given cities. The same question is asked for three other cities, and so on, until each distance from one city to another (each out of a total list of 14 cities) is considered. A matrix M can be constructed so that for the three cities (i,j,k) the two closest together, for example, (i,j) contribute 0 points to the matrix element M(i,j), the next closest pair, for example, (j,k), adds 1 point, and the remaining pair adds two points to M(i,k). The (dissimilarity) matrix obtained in this way resembles an ordinary distance table. If the phrases most distant and nearest in the question are interchanged, one obtains a similarity (data) matrix.

Instead of relying on a topographer, we used an ordinary distance table as input for the program. The program we applied was KYST-2a, pronounced 'kissed', formed from the names Kruskal, Young, Shepard, and Torgerson. It gives the coordinates of the cities in one or more dimensions. The analysis was carried out for both the metric case (with linear regression) and the non-metric case. In the latter case, the actual number of miles was not used. However, the ranking order of the calculated interpoint distances should be, as far as possible, the same as the ranking order of the interpoint distances in the given distance matrix. The results of both the metric and the non-metric cases were practically the same. Only the results of the latter case will be discussed in the following.

All calculations were carried out in the Euclidian space (Minkowski's parameter = 2). A measure of the goodness of fit between both rankings is called stress, which can to some extent be compared with a least-squares sum in an ordinary fitting procedure. The stress value in this particular case is 0.249 for one dimension, 0.028 for two dimensions, and 0.013 for three dimensions. It appears that a two-dimensional fit is a good one. The decrease of stress in three dimensions is rather weak, while the deterioration due to a low dimensionality is obvious. The results of the calculations are plotted in Fig. A.1; the solid points are the real locations, whereas the small circles represent the calculated places. As the figure shows, in this particular case it is possible to derive metric data from non-metric input data. The orientation of the map is arbitrary; there is no real North–South axis. For convenience only, the contour of the Netherlands and a compass needle are drawn.

A second example is from Ekman's [63] similarity judgement among 14 colours varying in hue. Subjects made ratings of qualitative similarity for each pair of combinations of colours ranging in wavelength from 434 to 674 nm. Shepard [245] applied a non-metric MDS procedure to the similarity ratings and extracted the underlying structure depicted by Fig. A.2. The underlying structure recovered from Ekman's similarity data was simply

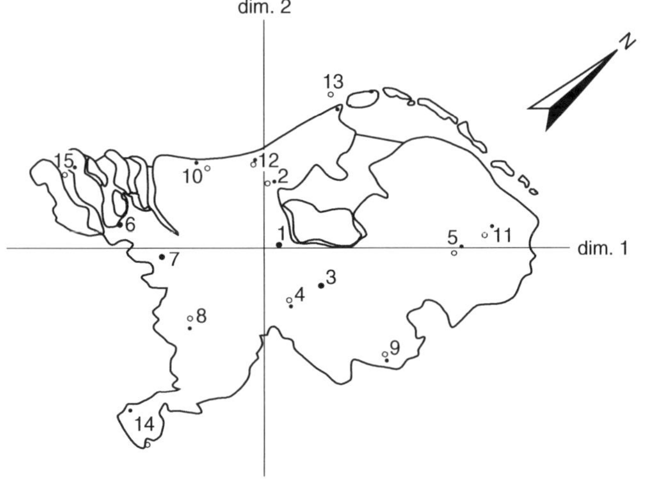

Figure A.1 Configuration with real (solid) and calculated (circles) locations

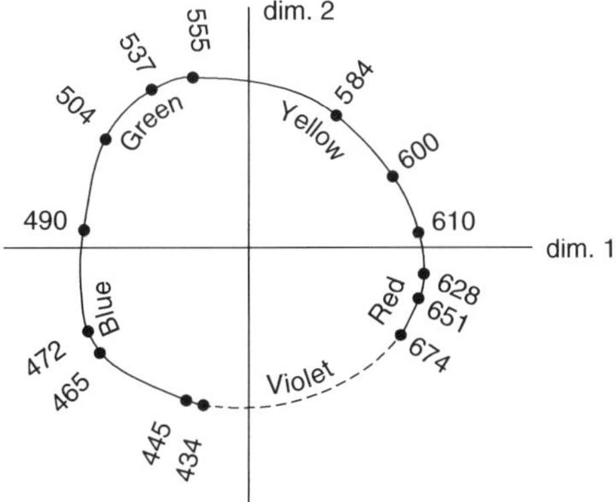

Figure A.2 Colour circle (From Shepard's [245] analysis of Ekman's data)

the conventional colour circle, with the colours arranged along the smooth contour in order of increasing wavelength.

A.4 PROCEDURE

The input data, δ_{ij} or *proximities*, are numbers that indicate how similar or how different two objects are, or are perceived to be. The distance between the points i and j in the configuration, which reflects the 'hidden structure' is denoted by d_{ij}. The basic concept takes the form that

$$f(\delta_{ij}) = d_{ij}, \tag{A.1}$$

where f is of some specified type. The discrepancy between $f(\delta_{ij})$ and d_{ij} is then

$$f(\delta_{ij}) - d_{ij}. \tag{A.2}$$

An objective function that is called stress is

$$S_1 = \sqrt{\frac{\sum_i \sum_j [f(\delta_{ij}) - d_{ij}]^2}{\sum_i \sum_j {d_{ij}}^2}} \tag{A.3}$$

The values $f(\delta_{ij})$ are often called fitted distances and denoted by $\hat{d_{ij}}$; sometimes, they are also called 'disparities'. When the *only* restriction for f is that it has to be monotonous, then the procedure is of the non-metric type.

A.5 PRECAUTIONS CONCERNING THE SOLUTION

The interpretation and generation of the configuration map should both be monitored carefully, as undesirable results can occur, which render any use of the configuration inadvisable (Kruskal *et al.* [152]). *Before* attempting to interpret the configuration, the user should *always* check for these possibilities, beginning with the inspection of $\delta - d$, the Shepard diagram (i.e. the scatter plot of recovered distances versus data values). Some anomalous $\delta - d$ configurations are discussed below. The relation between stress and significance is studied by Wagenaar and Padmos [292], and is discussed in the next section.

Jaggedness of the fitted function: The function relating distances to data values will always be somewhat jagged. However, this function should in fact approximate a smooth and continuous curve. Since the user is assuming an underlying continuous function, a configuration associated with a step function is undesirable. There are, however, two possible remedies for step-functions. One is the possibility of a local minimum, hence different initial values have to be tried. The second remedy is to specify a stronger form of regression.

Clumping of stimulus points: This is when several distinct objects occur at the same position in the Shepard diagram. The phenomenon is associated with undesirable behaviour of the fitted function (i.e. the d values) in the region of the smallest dissimilarities. The antidote for undesirable clumping is to declare a different form of regression, perhaps with a preliminary transformation of the data as well.

Degeneracy: A degenerate solution is an extreme case of both clumping (clustering) and jaggedness. Usually the stress is in this case zero, or nearly so. Hence, a very small stress value can indicate an utterly useless solution instead of an exceptionally good one. This result often occurs when, for one or more subsets of the stimuli, the dissimilarities within that subset are smaller than the dissimilarities between stimuli in that subset and the remaining stimuli (Shepard [245]). Possible solutions are (1) separate the clusters, (2) scale them separately and combine them in a later run (3) use the FIX option, or (4) a form of regression stronger than monotone can be specified.

A.6 SIGNIFICANCE OF STRESS

For the representation of an arbitrary dissimilarity matrix by distances between n points in m dimensions, a probability $p(s)$ exists that a stress value $\leq s$ will be obtained by chance. For the determination of the probability distributions $p(s)$, the dissimilarity matrices contained the numbers from 1 to $0.5n(n-1)$ and were attributed randomly to the cells. In this way, 100 'random scatters' were produced and analyzed by the MDS technique in various dimensions. The results are in Table A.1. The empty cells in Table A.1 correspond to conditions where more than 5% of the scatters have a stress smaller than 0.5%; it is advisable never to use MDS in these conditions.

A.7 UNIVARIATE SCALING

The purpose of scaling is to quantify the qualitative relationships between objects by scaling data. Scaling procedures attempt to do this by using rules that assign numbers to qualities of things or events. Here we discuss univariate scaling, in contrast to multidimensional scaling. Univariate scaling is usually based on the law of comparative judgement

Table A.1 The maximum stress in percentages, which can be accepted at a significance level of $\alpha = 0.05$ for n points in m dimensions, from Table III of Wagenaar and Padmos [292]

	m = 1	2	3	4	5
n = 7	20	7	–	–	–
8	27.5	10	1.5	–	–
9	30.5	13	5.5	1	–
10	34	15	7	3	–
11	35	18	9.5	4.5	1
12	39.5	20.5	10	6.5	3.5

(Torgerson [276], Thurstone [270]). It is a set of equations relating the proportion of times any stimulus i is judged greater, or has higher appreciation for a given attribute, than any other stimulus j. The set of equations is derived from the postulates presented in Torgerson [276]. In brief, these postulates are:

1. Each stimulus when presented to an observer gives rise to a discriminal process, which has some value on the psychological continuum of interest (e.g. in the context of this book, appreciation for a particular processing method as judged by listening to a processed signal).
2. Because of momentary fluctuations in the organism, a given stimulus does not always excite the same discriminal process. This can be considered as noise in the process. It is postulated that the values of the discriminal process are such that the frequency distribution is normal on the psychological continuum.
3. The mean and standard deviation of the distribution associated with a stimulus are taken as its scale value and discriminal dispersion respectively.

Consider the theoretical distributions S_j and S_k of the discriminal process for any two stimuli j and k respectively, as shown in the upper panel of Fig. A.3. Let $\overline{S}_j$ and $\overline{S}_k$ correspond to the scale values of the two stimuli and σ_j and σ_k to their discriminal dispersion caused by noise.

Now we assume that the standard deviations of the distributions are all equal and constant (as in Fig. A.3), and that the correlation between the pairs of discriminal processes is constant; this is called 'Condition C' by Torgerson [276]. Since the distribution of the difference of the normal distributions is also normal, we get

$$\overline{S_k} - \overline{S_j} = cx_{jk}, \tag{A.4}$$

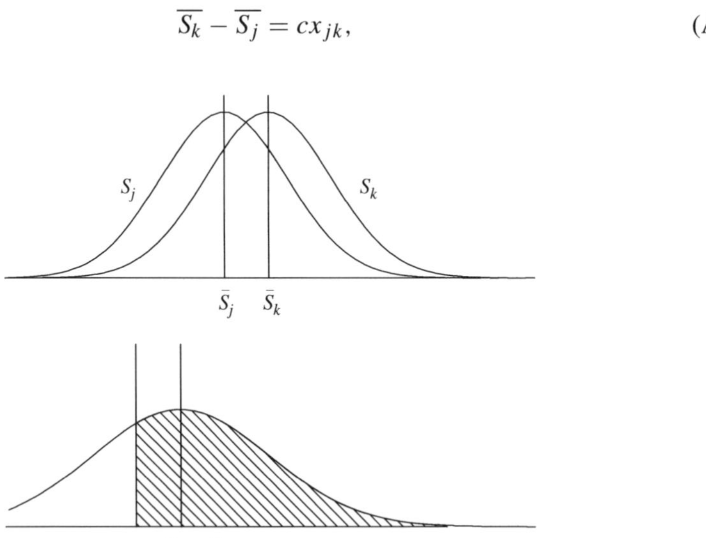

Figure A.3 The upper panel shows two Gaussian distributions corresponding to stimuli j and k, having different mean values ($\overline{S}_j$ and $\overline{S}_k$). The probability that k is judged to be larger than j is given by the shaded area in the lower panel

where c is a constant and x_{jk} is the transformed (see Eqn. A.7) proportion of times stimulus k is more highly appreciated (or judged greater) than stimulus j. Equation A.4 is also known as Thurstone's case V. The distribution of the discriminal differences is plotted in the lower panel of Fig. A.3. Equation A.4 is a set of $n(n-1)$ equations with $n+1$ unknowns, n scale values and c. This can be solved with a least-squares method. Setting $c = 1$, and the origin of the scale to the mean of the estimated scale values, that is,

$$1/n \sum_{j=1}^{n} s_j = 0, \tag{A.5}$$

we get

$$s_k = 1/n \sum_{j=1}^{n} x_{jk}. \tag{A.6}$$

Thus, the least-square solution of the scale values can be obtained simply by averaging the columns of matrix $\mathbf{X}$; however, the elements x_{jk} of $\mathbf{X}$ are not directly available. With paired comparisons we measure the proportion p_{kj} that stimulus k was judged greater than stimulus j. This proportion can be considered as a probability that stimulus k is in fact greater than stimulus j. This probability is equal to the shaded area in Fig A.3, or

$$x_{jk} = \mathrm{erf}(p_{jk}), \tag{A.7}$$

where erf is the error function (Abramowitz and Stegun [12, 7, 26.2]), which can easily be approximated (Abramowitz and Stegun [12, 26.2.23]). A problem may arise if $p_{jk} \approx \pm 1$ since $|x_{jk}|$ can be very large. In this case, one could simply replace x_{jk} by a large value.

It may be noted that this type of transformation is also known as Gaussian transform, where instead of the symbol x, z is used, known as the z scores. Instead of using Eqn. A.7, other models are used, for example, the Bradley–Terry model, see David [56]. All forms of the law of comparative judgement assume that each stimulus has been compared with other stimuli a large number of times. The direct method of obtaining the values of p_{jk} is known as the method of paired comparisons, see, for example, David [56].

References

[1] 3GPP TR 26.976. AMR-WB speech codec performance characterisation, December 2001.

[2] 3GPP TS 26.171. AMR wideband speech codec; general description, March 2001.

[3] 3GPP TS 26.190. AMR wideband speech codec; transcoding functions, December 2001.

[4] R.M. Aarts. *On the Design and Psychophysical Assessment of Loudspeaker Systems*. PhD thesis, Delft University of Technology, 1995.

[5] R.M. Aarts. Low-complexity tracking and estimation of frequency and amplitude of sinusoids. *Digital Signal Processing*, **14**(4):372–378, July 2004.

[6] R.M. Aarts and R.T. Dekkers. A real-time speech-music discriminator. *J. Audio Eng. Soc.*, **47**(9):720–725, 1999.

[7] R.M. Aarts, H. Greten, and P. Swarte. A special form of noise reduction. In *Proceedings of 21st AES Conference*, St. Petersburg, Russia. Audio Engineering Society, 2002.

[8] R.M. Aarts, R. Irwan, and A.J.E.M. Janssen. Efficient tracking of the cross-correlation coefficient. *IEEE Trans. Speech Audio Process.*, **10**(6):391–402, 2002.

[9] R.M. Aarts and A.J.E.M. Janssen. Approximation of the struve function H1 occurring in impedance calculations. *J. Acoust. Soc. Am.*, **113**(5):2635–2637, 2003.

[10] R.M. Aarts, E. Larsen, and O. Ouweltjes. A unified approach to low- and high-frequency bandwidth extension. In *115th AES Convention*, New York. Audio Engineering Society, 2003.

[11] M. Abe and Y. Yoshida. More natural sounding voice quality over the telephone! *NTT Rev.*, **7**(3):104–109, 1995.

[12] M. Abramowitz and I.A. Stegun. *Handbook of Mathematical Functions*. Dover, 1972.

[13] A.J. Abrantes, J.S. Marques, and I.M. Trancoso. Hybrid sinusoidal modeling of speech without voicing decision. In *Proceedings of EUROSPEECH*, volume 1, pages 231–234, Genova, Italy, September 1991.

[14] R.M. Adelson. Frequency estimation from few measurements. *Digital Signal Process.*, **7**:47–54, 1997.

[15] R.M. Adelson. Rapid power-line frequency monitoring. *Digital Signal Process.*, **12**:1–11, 2002.

Audio Bandwidth Extension E. Larsen and R. M. Aarts

ISBN 0-470-85864-8

[16] A.M. Abdelatty Ali and J. Van der Spiegel. Acoustic-phonetic features for the automatic classification of fricatives. *J. Acoust. Soc. Am.*, **109**(5):2217–2235, 2001.

[17] A.M. Abdelatty Ali, J. Van der Spiegel, and P. Mueller. An acoustic-phonetic feature-based system for the automatic recognition of fricative consonants. In *Proceedings of ICASSP*, volume 2, pages 961–964, Seattle, WA, May 1998.

[18] J.B. Allen. Short-term spectral analysis, synthesis, modification by discrete Fourier transform. *IEEE Trans. Speech Audio Process.*, **25**:235–238, 1977.

[19] J.B. Allen and S.T. Neely. Modeling the relation between the intensity just-noticeable difference and loudness for pure tones and wideband noise. *J. Acoust. Soc. Am.*, **102**(6):3628–3646, 1997.

[20] American Standards Association. *Standard Acoustical Terminology (S1.1-1994 (ASA 111-1994))*. ASA, New York, 1994.

[21] G.B. Arfken and H.J. Weber. *Mathematical Methods for Physicists*. John Wiley & Sons, 4th edition, 1995.

[22] B.S. Atal and L.R. Rabiner. A pattern recognition approach to voiced-unvoiced-silence classification with applications to speech recognition. *IEEE Trans. Acoust. Speech Signal Process.*, **ASSP-24**(3):201–212, 1976.

[23] C. Avendano, H. Hermansky, and E.A. Wan. Beyond nyquist: towards the recovery of broad-bandwidth speech from narrow-bandwidth speech. In *Proceedings of EUROSPEECH*, volume 1, pages 165–168, Madrid, Spain, September 1995.

[24] L.R. Bahl, P.F. Brown, P.V. de Souza, and R.L. Mercer. Maximum mutual information estimation of hidden markov model parameters for speech recognition. In *Proceedings of ICASSP*, pages 49–52, Tokyo, Japan, April 1986.

[25] I. Barrodale and C. Phillips. An improved algorithm for discrete Chebyshev linear approximation. In *Proceedings of 4th Manitoba Conference on Numerical Mathematics*, pages 177–190. University of Manitoba, Canada, 1974.

[26] D.W. Batteau. The role of the pinna in human localization. *Proc. R. Soc. London B*, **168**:158–180, 1967.

[27] A.G. Bell. Telegraphy (Filed Morning of 14 February 1876), US Patent 174,465, Issued March 7, 1876.

[28] L.L. Beranek. *Acoustics*. McGraw-Hill, New York, 1954. (Reprinted by ASA 1986).

[29] A.J. Berkhout. Least-squares inverse filtering and wavelet deconvolution. *Geophysics*, **42**(7):1369–1383, 1977.

[30] A.J. Berkhout, D. de Vries, and M.M. Boone. A new method to acquire impulse responses in concert halls. *J. Acoust. Soc. Am.*, **68**(1):179–183, 1980.

[31] B. Bessette, R. Lefebvre, M. Jelínek, J. Rotola-Pukkila, H. Mikkola, and K. Järvinen. The adaptive multirate wideband speech codec (AMR-WB). *IEEE Trans. Speech Audio Process.*, **10**(8):620–636, 2002.

[32] F.A. Bilsen. *On the Interaction of a Sound with its Repititions*. PhD thesis, Delft University of Technology, 1968.

[33] F.A. Bilsen and R.J. Ritsma. Some parameters influencing the perceptibility of pitch. *J. Acoust. Soc. Am.*, **47**(2)(Part 2):469–475, 1970.

[34] J. Blauert. *Spatial Hearing. The Psychophysics of Human Sound Localization*. MIT Press, 2nd edition, 1984.

[35] P. Boersma and D. Weenink. http://www.fon.hum.uva.nl/praat/, University of Amsterdam, The Netherlands, Retrieved July 2003.

[36] J. Borwick, editor. *Loudspeaker and Headphone Handbook*. Butterworths, London, 1988.

[37] M. Bosi and R.E. Goldberg. *Introduction to Digital Audio Coding and Standards*. Kluwer Academic Publishers, 2003.

[38] A.S. Bregman. *Auditory Scene Analysis*. MIT Press, 1990.

[39] F.H. Brittain. The loudness of continuous spectrum noise and its application to loudness measurements. *J. Acoust. Soc. Am.*, **11**:113–117, 1939.

[40] A.W. Bronkhorst. Localization of real and virtual sound sources. *J. Acoust. Soc. Am.*, **98**:2542–2553, 1995.

[41] O. Brosze, K.O. Schmidt, and A. Schmoldt. Der Gewinn an Verständlichkeit beim "Fernsehsprechen". *Nachrichtentech. Z. (NTZ)*, **15**(7):349–352, 1962 (in German).

[42] D. Byrne, H. Dillon, A.S. Khanh Tran *et al.*. An international comparison of long-term average speech spectra. *J. Acoust. Soc. Am.*, **96**(4):2108–2120, 1994.

[43] J.P. Campbell and T.E. Tremain. Voiced/unvoiced classification of speech with applications to the U.S. government LPC-10E algorithm. In *Proceedings of ICASSP*, pages 473–476, Tokyo, Japan, April 1986.

[44] H. Carl. *Untersuchung Verschiedener Methoden der Sprachkodierung und eine Anwendung zur Bandbreitenvergrößerung von Schmalband-Sprachsignalen*. PhD thesis, Ruhr-Universität Bochum, Bochum, Germany, 1994 (in German).

[45] H. Carl and U. Heute. Bandwidth enhancement of narrow-band speech signals. In *Proceedings of EUSIPCO*, volume 2, pages 1178–1181, Edinburgh, Scotland, September 1994.

[46] J.D. Carroll and P. Arabie. Multidimensional scaling. *Annu. Rev. Psychol.*, **31**:607–649, 1980.

[47] C.-F. Chan and W.-K. Hui. Wideband re-synthesis of narrowband CELP coded speech using multiband excitation model. In *Proceedings of ICSLP*, volume 1, pages 322–325, Philadelphia, PA, October 1996.

[48] C.-F. Chan and W.-K. Hui. Quality enhancement of narrowband CELP-coded speech via wideband harmonic re-synthesis. In *Proceedings of ICASSP*, volume 2, pages 1187–1190, Munich, Germany, April 1997.

[49] Y.M. Cheng, D. O'Shaughnessy, and P. Mermelstein. Statistical recovery of wideband speech from narrowband speech. In *Proceedings of ICSLP*, pages 1577–1580, Edmonton, Canada, 1992.

[50] S. Chennoukh, A. Gerrits, G. Miet, and R. Sluijter. Speech enhancement via frequency bandwidth extension using line spectral frequencies. In *Proceedings of ICASSP*, volume 1, pages 665–668, Salt Lake City, UT, May 2001.

[51] D. Clark. High-resolution subjective testing using a double-blind comparator. *J. Audio Eng. Soc.*, **30**(5):330–338, 1982.

[52] T.M. Cover and J.A. Thomas. *Elements of Information Theory*. Wiley Series in Telecommunications, 1991.

[53] R.E. Crochiere and L.R. Rabiner. *Multirate Signal Processing*. Prentice Hall, Englewood Cliffs, NJ, 1983.

[54] M. G. Croll. Sound-quality improvement of broadcast telephone calls. Technical Report 1972/26, The British Broadcasting Corporation (BBC), 1972.

[55] C. Cuttriss and J. Redding. Telephone (Filed 28 November 1877), US Patent 242 816, Issued June 14, 1881.

[56] H.A. David. *The Method of Paired Comparisons*. Ch. Griffin, 2nd edition, London, 1988.

[57] S.B. Davis and P. Mermelstein. Comparison of parametric representations for monosyllabic word recognition in continuously spoken sentences. *IEEE Trans. Acoust. Speech Signal Process.*, **ASSP-28**(4):357–366, 1980.

[58] E. de Boer. *On the 'Residue' in Hearing*. PhD thesis, University of Amsterdam, 1956.

[59] A.P. Dempster, N.M. Laird, and D.B. Rubin. Maximum likelihood from incomplete data via the EM algorithm. *J. R. Stat. Soc. Ser. B*, **39**(1):1–38, 1977.

[60] M. Dietrich. Performance and implementation of a robust ADPCM algorithm for wideband speech coding with 64 kbit/s. In *Proceedings of International Zürich Seminar on Digital Communications*, Zürich, Switzerland, March 1984.

[61] M. Dietz, L. Liljeryd, K. Kjorling, and O. Kunz. Spectral band replication, a novel approach in audio coding. In *Proceedings of AES 112th Convention*, Paper 5553, Munich, Germany. Audio Engineering Society, 2002.

[62] A. Ehret, M. Dietz, and K. Kjorling. State-of-the-art audio coding for broadcasting and mobile applications. In *Proceedings of AES 114th Convention*, Paper 5834, Amsterdam, The Netherlands. Audio Engineering Society, 2003.

[63] G. Ekman. Dimensions of color vision. *J. Psychol.*, **38**:467–474, 1954.

[64] N. Enbom and W.B. Kleijn. Bandwidth expansion of speech based on vector quantization of the mel frequency cepstral coefficients. In *IEEE Speech Coding Workshop*, pages 171–173, Porvoo, Finland, September 1999.

[65] J. Epps. *Wideband Extension of Narrowband Speech for Enhancement and Coding*. PhD thesis, School of Electrical Engineering and Telecommunications, The University of New South Wales, 2000.

[66] J. Epps and W.H. Holmes. A new technique for wideband enhancement of coded narrowband speech. In *IEEE Speech Coding Workshop*, pages 174–176, Porvoo, Finland, September 1999.

[67] C. Erdmann, P. Vary, K. Fischer, W. Xu, M. Marke, T. Fingscheidt, I. Varga, M. Kaindl, C. Quinquis, B. Koevesi, and D. Massaloux. A candidate proposal for a 3GPP adaptive multi-rate wideband speech codec. In *Proceedings of ICASSP*, volume 2, pages 757–760, Salt Lake City, UT, May 2001.

[68] ETSI Rec. GSM 03.50. Digital cellular telecommunication system (phase 2+); transmission planning aspects of the speech service in the GSM public land mobile (PLMN) system. Version 8.1.1, 2000.

[69] ETSI Rec. GSM 06.10. GSM full rate speech transcoding. Version 3.2.0, February 1992.

[70] L.D. Fielder and E.M. Benjamin. Subwoofer performance for accurate reproduction of music. *J. Audio Eng. Soc.*, **36**(6):443–456, 1988.

[71] J.L. Flanagan. *Speech Analysis, Synthesis and Perception*. Springer-Verlag, 2nd edition, Berlin, Heidelberg, New York, 1972.

[72] H. Fletcher. Auditory patterns. *Rev. Mod. Phys.*, **12**:47–65, 1940.

[73] H. Fletcher and W.A. Munson. Loudness, its definition, measurement and calculation. *J. Acoust. Soc. Am.*, **5**:82–108, 1933.

[74] W. Flügge. *Viscoelasticity*. Blaisdell Publishing Company, 1967.

[75] F.J.M. Frankort. *Vibration and Sound Radiation of Loudspeaker Cones*. PhD thesis, Delft University of Technology, 1975.

[76] N.R. French and J.C. Steinberg. Factors governing the intelligibility of speech sounds. *J. Acoust. Soc. Am.*, **19**:90–119, 1947.

[77] J.A. Fuemmeler and R.C. Hardie. Techniques for the regeneration of wideband speech from narrowband speech. In *IEEE Workshop on Nonlinear Signal and Image Proceedings*, Baltimore, MD, June 2001.

[78] J.A. Fuemmeler, R.C. Hardie, and W.R. Gardner. Techniques for the regeneration of wideband speech from narrowband speech. *EURASIP J. Appl. Signal Process.*, **2001**(4):266–274, 2001.

[79] K. Fukunaga. *Introduction to Statistical Pattern Recognition*. Morgan Kaufmann, Academic Press, 2nd edition, San Francisco, San Diego, 1990.

[80] S. Furui. *Digital Speech Processing, Synthesis and Recognition*. Marcel Dekker, 1989.

[81] K.R. Gabriel. The biplot graphical display of matrices with application to principal component analysis. *Biometrika*, **58**:453–467, 1971.

[82] A. Gabrielsson and B. Lindström. Perceived sound quality of high-fidelity loudspeakers. *J. Audio Eng. Soc.*, **33**(1/2):33, 1985.

[83] W.S. Gan, S.M. Kuo, and C.W. Toh. Virtual bass for home entertainment, multimedia PC, game station and portable audio systems. *IEEE Trans. Cons. Electron.*, **47**(4):787–793, 2001.

[84] M.R. Gander. Fifty years of loudspeaker developments as viewed through the perspective of the audio engineering society. *J. Audio Eng. Soc.*, **46**(1/2):43–58, 1998.

[85] E. Geddes and L. Lee. *Audio Transducers*. 2002.

[86] C.D. Geisler. *From Sound to Synapse: Physiology of the Mammalian Ear*. Oxford University Press, 1998.

[87] A. Gersho and R.M. Gray. *Vector Quantization and Signal Compression*. Kluwer Academic Publishers, Boston, Dordrecht, London, 1992.

[88] G.A. Gescheider. Psychological scaling. *Annu. Rev. Psychol.*, **39**:169–200, 1988.

[89] B.R. Glasberg and B.C.J. Moore. Derivation of auditory filter shapes from notched-noise data. *Hear. Res.*, **47**:103–138, 1990.

[90] B.R. Glasberg and B.C.J. Moore. A model of loudness applicable to time-varying sounds. *J. Audio Eng. Soc.*, **50**(5):331–342, 2002.

[91] J.L. Goldstein. Auditory nonlinearity. *J. Acoust. Soc. Am.*, **41**(3):676–689, 1967.

[92] J.L. Goldstein. An optimum processor theory for the central formation of the pitch of complex tones. *J. Acoust. Soc. Am.*, **54**(6):1496–1516, 1973.

[93] G.H. Golub and C.F. van Loan. *Matrix Computations*. Johns Hopkins University Press, 1989.

[94] P.S. Gopalakrishnan, D. Kanevsky, A. Nádas, and D. Nahamoo. A generalization of the Baum algorithm to rational objective functions. In *Proceedings of ICASSP*, volume 1, pages 631–634, Glasgow, Scotland, May 1989.

[95] A.H. Gray and J.D. Markel. Distance measures for speech processing. *IEEE Trans. Acoust. Speech Signal Process.*, **24**(5):380–391, 1976.

[96] R.M. Gray, A. Buzo, A.H. Gray, and Y. Matsuyama. Distortion measures for speech processing. *IEEE Trans. Acoust. Speech Signal Process.*, **ASSP-28**(4):367–376, 1980.

[97] R.M. Gray and D.L. Neuhoff. Quantization. *IEEE Trans. Inf. Theory*, **44**(6):2325–2383, 1998.

[98] M. Greenspan. Piston radiator: some extensions of the theory. *J. Acoust. Soc. Am.*, **65**(3):608–621, 1979.

[99] R.A. Greiner and J. Eggers. The spectral amplitude distribution of selected compact discs. *J. Audio Eng. Soc.*, **37**(4):246–275, 1989.

[100] D.W. Griffin and J.S. Lim. Multiband excitation vocoder. *IEEE Trans. Acoust. Speech Signal Process.*, **36**(8):1223–1235, 1988.

[101] A. Gröschel, M. Schug, M. Beer, and F. Henn. Enhancing audio coding efficiency of MPEG layer-2 with spectral band replication for digitalradio (DAB) in a backwards compatible way. In *Proceedings of AES 114th Convention, Amsterdam*. Audio Engineering Society, 2003.

[102] N. Guttman and S. Pruzansky. Lower limits of pitch and musical pitch. *J. Speech Hear. Res.*, **5**(3):207–214, 1962.

[103] R. Hagen. Spectral quantization of cepstral coefficients. In *Proceedings of ICASSP*, volume 1, pages 509–512, Adelaide, Australia, April 1994.

[104] A. Härmä, M. Karjalainen, L. Savioja, V. Välimäki, U.K. Laine, and J. Huopaniemi. Frequency-warped signal processing for audio applications. *J. Audio Eng. Soc.*, **48**:1011–1031, 2000.

[105] W.M. Hartmann. The effect of amplitude envelope on the pitch of sine wave tones. *J. Acoust. Soc. Am.*, **63**:1105–1113, 1978.

[106] P. Hedelin and J. Skoglund. Vector quantization based on Gaussian mixture models. *IEEE Trans. Speech Audio Process.*, **8**(4):385–401, 2000.

[107] D.A. Heide and G.S. Kang. Speech enhancement for bandlimited speech. In *Proceedings of ICASSP*, volume 1, pages 393–396, Seattle, WA, May 1998.

[108] H. Helmholtz. *Die Lehre von den Tonempfindungen [The Theory of Tone Perception]*. Vieweg, 1954. English translations of 4th German edition of 1877.

[109] W. Hess. *Pitch Determination of Speech Signals*. Springer, Berlin, 1983.

[110] V. Hohmann. Frequency analysis and synthesis using a gammatone filterbank. *Acta Acoust.*, **88**:433–442, 2002.

[111] D. Homm, T. Ziegler, R. Weidner, and R. Bohm. Bandwidth extension of audio signals by spectral band replication. In *Proceedings of 1st IEEE Benelux Workshop on MPCA, Louvain, Belgium*. IEEE, 2002.

[112] D. Homm, T. Ziegler, R. Weidner, and R. Bohm. Implementation of a DRM audio cecoder (aacPlus) on ARM architecture. In *Proceedings of AES 114th Convention, Paper 5833, Amsterdam, The Netherlands*. Audio Engineering Society, 2003.

[113] A.J.M. Houtsma and J.L. Goldstein. The central origin of the pitch of complex tones: evidence from musical interval recognition. *J. Acoust. Soc. Am.*, **51**(2)(Part 2):520–529, 1972.

[114] F.V. Hunt. *Electroacoustics*. John Wiley & Sons, 1954.

[115] C. Huygens. En envoiant le problème d'Alhazen en France ... '. *Oevres Complètes*, Vol. 10. Société Hollandaises de Sciences, Haarlem, The Netherlands, 1905 (Originally published 1693).

[116] A. Illényi and P. Korpássy. Correlation between loudness and quality of stereophonic loudspeakers. *Acoustica*, **49**(4):334–336, 1981.

[117] International Standard ISO 226-1987(E), Acoustics–normal equal-loudness level contours, 1987.

[118] International Standard ISO 7029-1984(E), Acoustics–Threshold of hearing by air conduction as a function of age and sex for otologically normal persons, 1984.

[119] B. Iser and G. Schmidt. Neural networks versus codebooks in an application for bandwidth extension of speech signals. In *Proceedings of EUROSPEECH*, pages 565–568, Geneva, Switzerland, September 2003.

[120] F. Itakura. Line spectrum representation of linear predictor coefficients of speech signals. *J. Acoust. Soc. Am.*, **57**(Suppl. 1):35, 1975 (89th Meeting of the Acoustical Society of America).

[121] ITU-T Rec. G.132. Attenuation performance. In Blue Book, vol. Fascicle III.1 (General Characteristics of International Telephone Connections and Circuits), 1988.

[122] ITU-T Rec. G.151. General performance objectives applicable to all modern international circuits and national extension circuits. In Blue Book, vol. Fascicle III.1 (General Characteristics of International Telephone Connections and Circuits), 1988.

[123] ITU-T Rec. G.711. Pulse code modulation (PCM) of voice frequencies, 1972.

[124] ITU-T Rec. G.712. Performance characteristics of PCM channels between 4-wire interfaces at voice frequencies. In Blue Book, vol. Fascicle III.4 (General Aspects of Digital Transmission Systems; Terminal Equipments), 1988.

[125] ITU-T Rec. G.722. 7 khz audio coding within 64 kbit/s. In Blue Book, vol. Fascicle III.4 (General Aspects of Digital Transmission Systems; Terminal Equipments), 1988.

[126] ITU-T Rec. G.722.1. Coding at 24 and 32 kbit/s for hands-free operation in systems with low frame loss, September 1999.

[127] ITU-T Rec. G.722.2. Wideband coding of speech at around 16 kbit/s usign adaptive multi-rate wideband (amr-wb), July 2003.

[128] P. Jax. *Enhancement of Bandlimited Speech Signals: Algorithms and Theoretical Bounds*. PhD thesis, Aachen University (RWTH), Aachen, Germany, 2002.

[129] P. Jax and P. Vary. Wideband extension of telephone speech using a hidden Markov model. In *IEEE Speech Coding Workshop*, pages 133–135, Delavan, WI, September 2000.

[130] P. Jax and P. Vary. Enhancement of band-limited speech signals. In *Proceedings of Aachen Symposium on Signal Theory*, pages 331–336, Aachen, Germany, September 2001.

[131] P. Jax and P. Vary. An upper bound on the quality of artificial bandwidth extension of narrowband speech signals. In *Proceedings of ICASSP*, volume 1, pages 237–240, Orlando, FL, May 2002.

[132] P. Jax and P. Vary. Artificial bandwidth extension of speech signals using MMSE estimation based on a hidden Markov model. In *Proceedings of ICASSP*, volume 1, pages 680–683, Hong Kong SAR, China, April 2003.

[133] P. Jax and P. Vary. On artificial bandwidth extension of telephone speech. *Signal Process.*, **83**(8):1707–1719, 2003.

[134] F. Jay, editor. *IEEE Standard Dictionary of Electrical and Electronics Terms*. IEEE, 2nd edition, 1978.

[135] A.J. Jerri. The shannon sampling theorem – its various extensions and applications: a tutorial review. *Proc. IEEE*, **65**:1565–1596, 1977.

[136] J. D. Johnston. Transform coding of audio signals using perceptual noise criteria. *IEEE J. Select. Areas Commun.*, **6**(2):314–323, 1988.

[137] E. Joliveau, J. Smith, and J. Wolfe. Tuning of vocal tract resonances by sopranos. *Nature*, **427**:116, 2004.

[138] E.I. Jury. *Theory and Application of the z-Transform Method*. John Wiley & Sons, New York, 1964.

[139] M. Kahrs and K. Brandenburg. *Applications of Digital Signal Processing to Audio and Acoustics*. Kluwer Academic Publishers, 1998.

[140] A.J.M. Kaizer. *On the Design of Broadband Electrodynamical Loudspeakers and Multiway Loudspeaker Systems*. PhD thesis, Eindhoven University of Technology, 1986.

[141] B. Kedem. *Time Series Analysis by Higher Order Crossings*. IEEE Press, New York, 1994.

[142] L.E. Kinsler, A.R. Frey, A.B. Coppens, and J.V. Sanders. *Fundamentals of Acoustics*. John Wiley & Sons, 1982.

[143] W. Klippel. Dynamic measurement and interpretation of the nonlinear parameters of electrodynamic loudspeakers. *J. Audio Eng. Soc.*, **38**(12):944–955, 1990.

[144] W. Klippel. The nonlinear large-signal behavior of electrodynamic loudspeakers at low frequencies. *J. Audio Eng. Soc.*, **40**(6):483–496, 1992.

[145] M.H. Knudsen and J.G. Jensen. Low-freqency loudspeaker models that include suspension creep. *J. Audio Eng. Soc.*, **41**(1/2):3–18, 1993.

[146] A. Kohlrausch and A.J.M. Houtsma. Pitch related to spectral edges of broadband signals. *Philos. Trans. R. Soc. London B*, **336**:81–88, 1992.

[147] U. Kornagel. Spectral widening of the excitation signal for telephone-band speech enhancement. In *Proceedings of IWAENC*, pages 215–218, Darmstadt, Germany, September 2001.

[148] W. Krebber. *Sprachübertragungsqualität von Fernsprech-Handapparaten*. PhD thesis, RWTH Aachen, 1995 (in German).

[149] K.R. Krishnamachari, R.E. Yantorno, J.M. Lovekin, D.S. Benincasa, and S.J. Wendt. Use of local kurtosis measure for spotting usable speech segments in co-channel speech. In *Proceedings of ICASSP*, volume 1, pages 649–652, Salt Lake City, UT, May 2001.

[150] P. Kroon, E.F. Depettere, and R.J. Sluyter. Regular-pulse excitation – a novel approach to effective and efficient multipulse coding of speech. *IEEE Trans. Acoust. Speech. Signal Process.*, **34**(5):1054–1063, 1986.

[151] J.B. Kruskal and M. Wish. *Multidimensional Scaling, Quantitative Applications in the Social Sciences*. Sage Publishers, 10th edition, 1983.

[152] J.B. Kruskal, F.W. Young, and J.B. Seery. *How to Use the KYST2A, A Very Flexible Program to do Multidimensional Scaling and Unfolding*. Bell Telephone Laboratories, 1978.

[153] O. Kunz. SBR Explained: White Paper, December 2003, http://www.codingtechnologies.com.

[154] H.J. Landau and H.O. Pollack. Prolate spheroidal wave functions, Fourier analysis and uncertainty, II. *Bell Syst. Tech. J.*, **40**:65–84, 1961.

[155] H.J. Landau and H.O. Pollack. Prolate spheroidal wave functions, Fourier analysis and uncertainty, III. *Bell Syst. Tech. J.*, **41**:1295–1336, 1962.

[156] E. Larsen and R.M. Aarts. Reproducing low-pitched signals through small loudspeakers. *J. Audio Eng. Soc.*, **50**(3):147–164, 2002.

[157] E. Larsen, R.M. Aarts, and M. Danessis. Efficient high-frequency bandwidth extension of music and speech. In *proceedings of 112th AES Convention*, Munich, Germany. Audio Engineering Society, 2002.

[158] N. Le Goff, R.M. Aarts, and A.G. Kohlrausch. Thresholds for hearing mistuning of the fundamental component in a complex sound. In *Proceedings of the 18th International Congress on Acoustics (ICA2004)*, Paper Mo. P3.21, p. I-865, Kyoto, Japan, 2004.

[159] W.J.M. Levelt, J.P. van de Geer, and R. Plomp. Triadic comparisons of musical intervals. *Br. J. Math. Stat. Psychol.*, **19** (Part 2):163–179, 1966.

[160] B.G. Levi. Acoustic experiments shows why it's so hard to make out the heroine's words at the opera. *Phys. Today*, **57**(3):23–25, March 2004.

[161] J.C.R. Licklider. Auditory frequency analysis. In C. Cherry, editor, *Information Theory*. Academic Press, New York, NY, 1956.

[162] Y. Linde, A. Buzo, and R.M. Gray. An algorithm for vector quantizer design. *IEEE Trans. Commun.*, **28**(1):84–95, 1980.

[163] S.P. Lloyd. Least squares quantization in pcm. *IEEE Trans. Inf. Theory*, **IT-28**(2):129–137, 1982.

[164] E.M. Long and R.J. Wickersham. Method and apparatus for operating a loudspeaker below resonant frequency. US patent 4,481,662, 1984. Filing year 1982.

[165] X. Maitre. 7 kHz audio coding within 64 kbit/s. *IEEE J. Select. Areas Commun.*, **6**(2):283–298, 1988.

[166] J. Makhoul. Linear prediction: a tutorial review. *Proc. IEEE*, **63**:561–580, 1975.

[167] J. Makhoul and M. Berouti. High-frequency regeneration in speech coding systems. In *Proceedings of ICASSP*, pages 428–431, Washington, DC, April 1979.

[168] J.D. Markel and A. H. Gray. *Linear Prediction of Speech*. Springer-Verlag, Berlin, Heidelberg, New York, 1976.

[169] S.L. Marple. Computing the discrete-time "analytic" signal via FFT. *IEEE Trans. Inf. Theory*, **47**(9):2600–2603, 1999.

[170] R.J. McAulay and T.F. Quatieri. Sinusoidal coding. In W. Bastiaan Kleijn and K.K. Paliwal, editors, *Speech Coding and Synthesis*, chapter 4, pages 121–173. Elsevier, 1995.

[171] A. McCree, T. Unno, A. Anandakumar, A. Bernard, and E. Paksoy. An embedded adaptive multi-rate wideband speech coder. In *Proceedings of ICASSP*, volume 2, pages 761–764, Salt Lake City, UT, May 2001.

[172] N.W. McLachlan. *Loudspeakers*. Oxford at the Clarendon Press, 1934.

[173] J. Merhaut. *Theory of Electroacoustics*. McGraw-Hill, 1981.

[174] G. Miet, A. Gerrits, and J.C. Valière. Low-band extension of telephone-band speech. In *Proceedings of ICASSP*, volume 3, pages 1851–1854, Istanbul, Turkey, June 2000.

[175] T.K. Moon. The expectation-maximization algorithm. *IEEE Signal Process. Mag.*, **13**(6):47–60, 1996.

[176] B.C.J. Moore. Frequency difference limens for short-duration tones. *J. Acoust. Soc. Am.*, **54**:610–619, 1973.

[177] B.C.J. Moore. *Handbook of Perception and Cognition: Hearing*. Academic Press, 1995.

[178] B.C.J. Moore. *An Introduction to the Psychology of Hearing*. Academic Press, 5th edition, 2003.

[179] B.C.J. Moore, R.W. Peters, and B.R. Glasberg. Thresholds for the detection of inharmonicity in complex tones. *J. Acoust. Soc. Am.*, **77**:1861–1867, 1985.

[180] P.M. Morse and K.U. Ingard. *Theoretical Acoustics*. McGraw-Hill, 1968.

[181] E. Moulines and W. Verhelst. Time-domain and frequency-domain techniques for prosodic modification of speech. In W. Bastiaan Kleijn and K.K. Paliwal, editors, *Speech Coding and Synthesis*, chapter 15, pages 519–555. Elsevier, 1995.

[182] *NAG Library Manual*. Numerical Algorithm Group Ltd, Oxford, 1999. Mark 19, Chapter E02.

[183] Y. Nakatoh, M. Tsushima, and T. Norimatsu. Generation of broadband speech from narrowband speech using piecewise linear mapping. In *Proceedings of EUROSPEECH*, volume 3, pages 1643–1646, Rhodos, Greece, September 1997.

[184] S.T. Neely and J.B. Allen. Invertibility of a room impulse response. *J. Acoust. Soc. Am.*, **66**(1):165–169, 1979.

[185] M. Nilsson, S.V. Andersen, and W.B. Kleijn. On the mutual information between frequency bands in speech. In *Proceedings of ICASSP*, volume 3, pages 1327–1330, Istanbul, Turkey, June 2000.

[186] M. Nilsson, H. Gustafsson, S.V. Andersen, and W.B. Kleijn. Gaussian mixture model based mutual information estimation between frequency bands in speech. In *Proceedings of ICASSP*, volume 1, pages 525–528, Orlando, FL, May 2002.

[187] M. Nilsson and W.B. Kleijn. Avoiding over-estimation in bandwidth extension of telephony speech. In *Proceedings of ICASSP*, volume 2, pages 869–872, Salt Lake City, UT, May 2001.

[188] F. Nordén, T. Eriksson, and P. Hedelin. An information theoretic perspective on the speech spectrum process. In *IEEE Speech Coding Workshop*, pages 93–95, Delavan, WI, September 2000.

[189] G. Oetken and W. Schüßler. On the design of digital filters for interpolation. *Arch. Elektron. Übertragungstech. (AEÜ), Electron. Commun.*, **27**(11):471–476, 1973.

[190] G.S. Ohm. Über die definition des tones, nebst daran geknüpfter theorie der Sirene and ähnlicher tonbildender Vorrichtungen [On the definition of the tone and the related theory of the siren and similar tone-producing devices]. *Ann. Phys. Chem.*, **59**:513–565, 1843.

[191] J.P. Olive, A. Greenwood, and J. Coleman. *Acoustics of American English Speech*. Springer-Verlag, 1993.

[192] H.F. Olson. *Acoustical Engineering*. Van Nostrand, 1957.

[193] H.F. Olson. Analysis of the effects of nonlinear elements upon the performance of a back-enclosed, direct radiator loudspeaker mechanism. *J. Audio Eng. Soc.*, **10**(2):156–162, 1962.

[194] A.V. Oppenheim and R.W. Schafer. *Discrete-Time Signal Processing*. Prentice Hall, Englewood Cliffs, NJ, 1989.

[195] A.J. Oxenham and C.A. Shera. Estimates of human cochlear tuning at low levels using forward and simultaneous masking. *J. Assoc. Res. Otolaryngol.*, **4**(4):541–554, 2003.

[196] K.K. Paliwal. Interpolation properties of linear prediction parametric representations. In *Proceedings of EUROSPEECH*, volume 2, pages 1029–1032, Madrid, Spain, September 1995.

[197] K.K. Paliwal and W.B. Kleijn. Quantization of LPC parameters. In W. Bastiaan Kleijn and K.K. Paliwal, editors, *Speech Coding and Synthesis*, chapter 12, pages 433–466. Elsevier, 1995.

[198] W.J. Palm III. *Modeling, Analysis, and Control of Dynamic Systems*. John Wiley & Sons, 2000.

[199] A. Papoulis. *Probability, Random Variables, and Stochastic Processes*. McGraw-Hill, 3rd edition, New York, 1991.

[200] K.-Y. Park and H.S. Kim. Narrowband to wideband conversion of speech using GMM-based transformation. In *Proceedings of ICASSP*, volume 3, pages 1847–1850, Istanbul, Turkey, June 2000.

[201] P.J. Patrick. *Enhancement of Bandlimited Speech Signals*. PhD thesis, Loughborough University of Technology, 1983.

[202] R.D. Patterson. http://www.mrc-cbu.cam.ac.uk/cnbh/aimmanual/. AIM Manual, Retrieved September 2003.

[203] R.D. Patterson, M.H. Allerhand, and C. Giguère. Time-domain modeling of peripheral auditory processing: a modular architecture and a software platform. *J. Acoust. Soc. Am.*, **98**(4):1890–1894, 1998.

[204] R.D. Patterson, J. Nimmo-Smith, J. Holdsworth, and P. Rice. An efficient auditory filterbank based on the gammatone function. In *Paper Presented at a Meeting of the IOC Speech Group on Auditory Modelling at RSRE*, December 1987.

[205] E. Paulus and E. Zwicker. Programme zur automatischen Bestimmung der Lautheit aus Tertzpegeln oder Frequenzgruppenpegeln. *Acoustica*, **27**(5):253–266, 1972.

[206] J. Paulus. *Codierung Breitbandiger Sprachsignale Bei Niedriger Datenrate*. PhD thesis, RWTH Aachen, 1997 (in German).

[207] J.W. Paulus. Variable rate wideband speech coding using perceptually motivated thresholds. In *IEEE Speech Coding Workshop*, pages 35–36, Annapolis, MD, September 1995.

[208] A.D. Pierce. *Acoustics, An Introduction to Its Physical Principles and Applications*. ASA, 1989.

[209] R. Plomp. Pitch of complex tones. *J. Acoust. Soc. Am.*, **41**:1526–1533, 1967.

[210] R. Plomp and H.J.M. Steeneken. Effect of phase on the timbre of complex tones. *J. Acoust. Soc. Am.*, **46**:409–421, 1969.

[211] F.J. Pompei. The use of airborne ultrasonics for generating audible sound beams. *J. Audio Eng. Soc.*, **47**(9):726–731, 1999.

[212] D. Povey and P.C. Woodland. Improved discriminative training techniques for large vocabulary continuous speech recognition. In *Proceedings of ICASSP*, volume 1, pages 45–48, Salt Lake City, UT, May 2001.

[213] S.R. Powell and P.M. Chau. A technique for realizing linear phase IIR filters. *IEEE Trans. Signal Process.*, **39**(11):2425–2435, 1991.

[214] G. Quinn and E.J. Hannan. *The Estimation and Tracking of Frequencies*. Cambridge University Press, Cambridge, 2001.

[215] L.R. Rabiner. A tutorial on hidden Markov models and selected applications in speech recognition. *Proc. IEEE*, **77**(2):257–286, 1989.

[216] L.R. Rabiner and B.-H. Juang. *Fundamentals of Speech Recognition*. Prentice Hall International, 1993.

[217] L.R. Rabiner and R.W. Schafer. *Digital Processing of Speech Signals*. Prentice Hall, 1978.

[218] L.R. Rabiner and B. Gold. *Theory and Application of Digital Signal Processing*. Prentice Hall, Englewood Cliffs, NJ, 1975.

[219] J.W.S. Rayleigh. *The Theory of Sound*, Vol. 2. Dover, 1945.

[220] D.G. Raza and C.F. Chan. Quality enhancement of CELP coded speech by using an MFCC based Gaussian mixture model. In *Proceedings of EUROSPEECH*, pages 541–544, Geneva, Switzerland, September 2003.

[221] O. Read and W.L. Welch. *From Tin Foil to Stereo: Evolution of the Phonograph*. Howard Sams and Bobbs-Merill, New York, IN, 1959.

[222] D.A. Reynolds and R.C. Rose. Robust text-independent speaker identification using Gaussian mixture speaker models. *IEEE Trans. Speech Audio Process.*, **3**(1):72–83, 1995.

[223] C.W. Rice and E.W. Kellog. Notes on the development of a new type of hornless loud speaker. *Trans. Am. Inst. Electron. Eng.*, **44**:982–991, 1925.

[224] L.F. Richardson and J.S. Ross. Loudness and telephone current. *J. Gen. Psychol.*, 121–164, 1916.

[225] R.J. Ritsma. Existence region of the tonal residue. I. *J. Acoust. Soc. Am.*, **34**(9):1224–1229, 1962.

[226] R.J. Ritsma. Existence region of the tonal residue. II. *J. Acoust. Soc. Am.*, **35**(8):1241–1245, 1963.

[227] R.J. Ritsma. Frequencies dominant in the perception of the pitch of complex sounds. *J. Acoust. Soc. Am.*, **42**:191–198, 1967.

[228] R. Roberts and C. Mullis. *Digital Signal Processing*. Addison-Wesley, 1987.

[229] E.E. Roskam. The method of triads for nonmetric multidimensional scaling. *Ned. Tijdschr. Psychol. Grensgeb.*, **25**:404–417, 1970.

[230] R. Russell. http://home.earthlink.net/ rogerr7/ionovac.htm. About the Ionophone Loudspeaker, Retrieved April 2004.

[231] B. Scharf. In E.C. Carterette editor, *Handbook of Perception*, Vol. IV. Academic Press, New York, 1978.

[232] B. Scharf and A.J.M. Houtsma. In K.R. Boff, L. Kaufman, and J.P. Thomas, eds, *Handbook of Perception and Human Performance*, Vol. I, *Sensory Processes and Perception*, chapter 15, Audition III: Loudness, pitch, localization, aural distortion, pathology. John Wiley & Sons, 1986.

[233] L.L. Scharf. *Statistical Signal Processing. Detection, Estimation, and Time Series Analysis*. Addison-Wesley, Reading, MA, 1990.

[234] F. Schiel. Speech and speech-related resources at BAS. In *Proceedings of International Conference on Language Resources and Evaluation*, Granada, Spain, May 1998.

[235] R. Schlüter and W. Macherey. Comparison of discriminative training criteria. In *Proceedings of ICASSP*, volume 1, pages 493–496, Seattle, WA, May 1998.

[236] K.-O. Schmidt. Neubildung von unterdrückten Sprachfrequenzen durch ein nichtlinear verzerrendes Glied. *Telegraphen- Fernsprech-Tech.*, **22**(1):13–22, 1933 (in German).

[237] K.-O. Schmidt and O. Brosze. *Fernsprech-Übertragung*. Fachverlag Schiele & Schön, Berlin, 1967 (in German).

[238] J. Schnitzler. *Breitbandige Sprachcodierung: Zeitbereichs- und Frequenzbereichskonzepte*. PhD thesis, RWTH Aachen, 1999 (in German).

[239] J.F. Schouten. The perception of pitch. *Philips Tech. Rev.*, **5**(10):286, 1940.

[240] J.F. Schouten, R.J. Ritsma, and B. Lopes Cardozo. Pitch of the residue. *J. Acoust. Soc. Am.*, **34**(8(Part 2)):1418–1424, 1962.

[241] M. Schug, A. Groschel, M. Beer, and F. Henn. Enhancing audio coding efficiency of MPEG layer-2 with spectral band replication (SBR) for digital radio (EUREKA 147/DAB) in a backwards compatible way. In *Proceedings of AES 114th Convention*, Paper 5850, Amsterdam, The Netherlands. Audio Engineering Society, 2003.

[242] H.W. Schüßler. *Digitale Signalverarbeitung*, Band I. Springer-Verlag, 2nd edition, Berlin, 1988 (in German).

[243] A. Seebeck. Beobachtungenüber einige Bedingungen der Entstehung von Tönen [Observations on some conditions for the creation of tones]. *Ann. Phys. Chem.*, **53**:417–436, 1841.

[244] A. Sek and B.C.J. Moore. Frequency discrimination as a function of frequency, measured in several ways. *J. Acoust. Soc. Am.*, **97**:2479–2486, 1995.

[245] R.N. Shepard. The analysis of proximities: multidimensional scaling with an unknown distance function. *Psychometrica*, **27**:Part I 125–140, Part II 219–246, 1962.

[246] R.N. Shepard. Multidimensional scaling, tree-fitting and clustering. *Science*, **210**:390–398, 1980.

[247] C.A. Shera, J.J. Guinan Jr., and A.J. Oxenham. Revised estimates of human cochlear tuning from otoacoustic and behavioral measurements. *Proc. Natl. Acad. Sci.*, **99**(5):3318–3323, 2002.

[248] L.J. Sivian, H.K. Dunn, and S.D. White. Absolute amplitudes and spectra of certain musical instruments and orchestras. *J. Acoust. Soc. Am.*, **2**(3):330–371, 1931.

[249] D. Slepian. On bandwidth. *Proc. IEEE*, **64**:292–300, 1976.

[250] D. Slepian and H.O. Pollack. Prolate spheroidal wave functions, Fourier analysis and uncertainty, I. *Bell Syst. Tech. J.*, **40**:43–64, 1961.

[251] A.M. Small and R.G. Daniloff. Pitch of noise bands. *J. Acoust. Soc. Am.*, **41**:506–512, 1967.

[252] R.H. Small. Vented-box loudspeaker systems part I: small-signal analysis. *J. Audio Eng. Soc.*, **21**(5):363–372, 1973.

[253] G.F. Smoorenburg. Combination tones and their origin. *J. Acoust. Soc. Am.*, **52**(2(Part 2)):615–632, 1972.

[254] H.W. Sorenson and D.L. Alspach. Recursive Bayesian estimation using Gaussian sums. *Automatica*, **7**:465–479, 1971.

[255] Sound Quality Assessment Material (recordings for subjective tests). European Broadcasting Union, 1988, No. 422 204-2.

[256] S.S. Stevens. Procedure for calculating loudness: mark VI. *J. Acoust. Soc. Am.*, **33**(11):1577–1585, 1961.

[257] S.S. Stevens. Perceived level of noise by mark VII and Decibels (E). *J. Acoust. Soc. Am.*, **51**(2(Part 2)):575–601, 1972.

[258] I. Stylianou. *Harmonic Plus Noise Models for Speech, Combined with Statistical Methods, for Speech and Speaker Modification*. PhD thesis, Ecole Nationale Supérieure des Télécommunications, Paris, 1996.

[259] H. Suzuki and J. Tichy. Sound radiation from convex and concave domes in an infinite baffle. *J. Acoust. Soc. Am.*, **69**(1):41–49, 1981.

[260] H. Suzuki and J. Tichy. Sound radiation from an axis symmetric radiator in an infinite baffle. *J. Acoust. Soc. Jpn. (E)*, **3**(3):167–172, 1982.

[261] S.-E. Tan, W.-S. Gan, C.-W. Toh, and J. Yang. Application of virtual bass in audio cross-talk cancellation. *IEEE Electron. Lett.*, **36**(17):1500–1501, 2000.

[262] Y. Tanaka and N. Hatazoe. Reconstruction of wideband speech from telephone-band speech by multilayer neural networks. In *Spring Meeting of ASJ*, pages 255–256, 1995.

[263] Y. Tannaka and T. Koshikawa. Correlations between soundfield characteristics and subjective ratings on reproduced music quality. *J. Acoust. Soc. Am.*, **86**(2):603–620, 1989.

[264] Y. Tannaka, K. Muramori, M. Kohashi, and T. Koshikawa. Correlations between harmonic distortion, sound field characteristics and reproduced sound quality change in listening tests for loudspeakers. *J. Acoust. Soc. Jpn. (E)*, **11**(1):29–42, 1990.

[265] R. Taori, R. J. Sluijter, and A. J. Gerrits. Hi-BIN: an alternative approach to wideband speech coding. In *Proceedings of ICASSP*, volume 2, pages 1157–1160, Istanbul, Turkey, June 2000.

[266] E. Terhardt. Pitch, consonance, and harmony. *J. Acoust. Soc. Am.*, **55**:1061–1069, 1974.

[267] E. Terhardt. *Akustische Kommunikation: Grundlagen Mit Hörbeispielen*. Springer, Berlin, 1998 (in German).

[268] A.N. Thiele. Loudspeakers in vented boxes: Part I. *J. Audio Eng. Soc.*, **19**(5):382–392, 1971.

[269] E.A. Thompson. *The Soundscape of Modernity: Architectural Acoustics and the Culture of Listening in America, 1900-1933*. MIT Press, 1st edition, ISBN: 0 26 22 01 380, 2002.

[270] L.L. Thurstone. A law of comparative judgement. *Psychol. Rev.*, **34**:273–286, 1927.

[271] P. Tichavsky and A. Nehorai. Comparative study of four adaptive frequency trackers. *IEEE Trans. Signal Process.*, **45**(6):1473–1484, 1997.

[272] F.E. Toole. Listening tests–turning opinion into fact. *J. Audio Eng. Soc.*, **30**(6):431–445, 1982.

[273] F.E. Toole. Subjective measurements of loudspeaker sound quality and listener performance. *J. Audio Eng. Soc.*, **33**(1/2):2–32, 1985.

[274] F.E. Toole. Loudspeaker measurements and their relationship to listener preferences Part I. *J. Audio Eng. Soc.*, **34**(4):227–235, 1986.

[275] F.E. Toole. Loudspeaker measurements and their relationship to listener preferences Part II. *J. Audio Eng. Soc.*, **34**(5):323–348, 1986.

[276] W.S. Torgerson. *Theory and Methods of Scaling*. John Wiley & Sons, 1958.

[277] A. Uncini, F. Gobbi, and F. Piazza. Frequency recovery of narrow-band speech using adaptive spline neural networks. In *Proceedings of ICASSP*, volume 2, pages 997–1000, Phoenix, AZ, May 1999.

[278] P.P. Vaidyanathan. Homogeneous time-invariant systems. *IEEE Signal Process. Lett.*, **6**(4):76–77, 1999.

[279] J.-M. Valin and R. Lefebvre. Bandwidth extension of narrowband speech for low bit-rate wideband coding. In *IEEE Speech Coding Workshop*, pages 130–132, Delavan, WI, September 2000.

[280] V. Valtchev, J.J. Odell, P.C. Woodland, and S.J. Young. MMIE training of large vocabulary recognition systems. *Speech Commun.*, **22**(4):303–314, 1997.

[281] A.W.M. van den Enden and N.A.M. Verhoeckx. *Discrete-Time Signal Processing: An Introduction*. Prentice Hall, 1989.

[282] L.J. van der Pauw. The trapping of acoustical energy by a conical membrane and its implications for loudspeaker cones. *J. Acoust. Soc. Am.*, **68**(4):1163–1168, 1980.

[283] J. Vanderkooy. A model of loudspeaker impedance incorporating eddy currents in the pole structure. *J. Audio Eng. Soc.*, **37**:119–128, 1989.

[284] J. Vanderkooy, P.M. Boers, and R.M. Aarts. Direct-radiator loudspeaker systems with high *Bl*. *J. Audio Eng. Soc.*, **51**(7/8):625–634, 2003.

[285] P. Vary, K. Hellwig, R. Hofmann, R.J. Sluyter, C. Galand, and M. Rosso. Speech codec for the European mobile radio system. In *Proceedings of ICASSP*, volume 1, pages 227–230, New York, April 1988.

[286] P. Vary, U. Heute, and W. Hess. *Digitale Sprachsignalverarbeitung*. Teubner-Verlag, Stuttgart, 1998 (in German).

[287] S.V. Vaseghi. *Advanced Signal Processing and Digital Noise Reduction*. John Wiley & Sons, Teubner, 1996.

[288] W. Verhelst. Overlap-add methods for time-scaling of speech. *Speech Commun.*, **30**(4):207–221, 2000.

[289] W. Verhelst and M. Roelands. An overlap-add technique based on waveform similarity (WSOLA) for high quality time-scale modification of speech. In *Proceedings of ICASSP*, volume 2, pages 554–557, Minneapolis, MN, April 1993.

[290] G. von Békésy. *Experiments in Hearing*. Acousticl Society of America, 1989. Originally published in 1960; Reprinted by ASA in 1989.

[291] S. Voran. Listener ratings of speech passbands. In *IEEE Speech Coding Workshop*, pages 81–82, Pocono Manor, PA, September 1997.

[292] W.A. Wagenaar and P. Padmos. Quantitative interpretation of stress in Kruskal's multidimensional scaling technique. *Br. J. Math. Stat. Psychol.*, **24**:101–110, 1971.

[293] K. Walliser. Über ein Funktionsschema für die Bildung der Periodentonhöhe aus dem Schallreiz. *Cybernetik*, **6**:65–72, 1969.

[294] S. Wang. *Low Bit-Rate Vector Excitation Coding of Phonetically Classified Speech*. PhD thesis, University of California, Santa Barbara, CA, August 1991.

[295] R.L. Wegel and C.E. Lane. The auditory masking of one sound by another and its probable relation to the dynamics of the inner ear. *Phys. Rev.*, **23**:266–285, 1924.

[296] P.J. Westervelt and R.S. Larson. Laser-excited broadside array. *J. Acoust. Soc. Am.*, **54**(1):121–122, 1973.

[297] H. Yang, S. van Vuuren, and H. Hermansky. Relevancy of time-frequency features for phonetic classification measured by mutual information. In *Proceedings of ICASSP*, volume 1, pages 225–228, Phoenix, AZ, May 1999.

[298] Z.R. Yang and M. Zwolinski. Mutual information theory for adaptive mixture models. *IEEE Trans. Pattern Anal. Machine Intell.*, **23**(4):396–403, 2001.

[299] H. Yasukawa. Spectrum broadening of telephone band signals using multirate processing for speech quality enhancement. *IEICE Trans. Fundam. Electron. Commun. Comput. Sci.*, **E78-A**(8):996–998, 1995.

[300] M. Yoneyama and J-I. Fujimoto. The audio spotlight: an application of nonlinear interaction of sound waves to a new type of loudspeaker design. *J. Acoust. Soc. Am.*, **73**(5):1532–122, 1983.

[301] Y. Yoshida and M. Abe. An algorithm to reconstruct wideband speech from narrowband speech based on codebook mapping. In *Proceedings of ICSLP*, pages 1591–1594, Yokohama, Japan, 1994.

[302] W.A. Yost, A.N. Popper, and R.R. Fay. *Human psychophysics*. Springer-Verlag, 1993.

[303] F.W. Young. Scaling. *Ann. Rev. Psychol.*, **35**:55–81, 1984.

[304] J. Zera and D.M. Green. Detecting temporal onset and offset asynchrony in multicomponent complexes. *J. Acoust. Soc. Am.*, **12**:47–65, 1940.

[305] E. Zwicker. Ein Verfahren zur Berechnung der Lautstärke. *Acustica*, 10: 1960.

[306] E. Zwicker. 'Negative afterimage' in hearing. *J. Acoust. Soc. Am.*, **36**:2413–2415, 1964.

[307] E. Zwicker. Dependence of level and phase of the $(2f_1 - f_2)$ cancellation tone on frequency range, frequency difference, level of primaries, and subject. *J. Acoust. Soc. Am.*, **70**(5):1277–1288, 1981.

[308] E. Zwicker. *Psychoakustik*. Springer-Verlag, New York, 1982.

[309] E. Zwicker and H. Fastl. *Psychoacoustics. Facts and Models*. Springer, 2nd edition, Berlin, Heidelberg, New York, 1999.

[310] E. Zwicker, H. Fastl, and C. Dallmayr. Letter to the editors: BASIC-program for calculating the loudness of sounds from their 1/3-oct band spectra according to ISO 532 B. *Acoustica*, **55**(1):63–67, 1984.

[311] E. Zwicker and R. Feldtkeller. *Das Ohr als Nachrichtenempfänger*. S. Hirzel Verlag, Stuttgart, 1967.

Index

Audio Bandwidth Extension E. Larsen and R. M. Aarts
 ISBN 0-470-85864-8

Printed and bound by CPI Group (UK) Ltd, Croydon, CR0 4YY

Printed and bound by CPI Group (UK) Ltd, Croydon, CR0 4YY

17/04/2025

14658869-0001